On the Roads to Modernity

On the Roads to Modernity

CONSCIENCE, SCIENCE AND CIVILIZATIONS

Selected Writings by
BENJAMIN NELSON

Edited by Toby E. Huff

LEXINGTON BOOKS
Lanham • Boulder • New York • Toronto • Plymouth, UK

Published by Lexington Books
A wholly owned subsidiary of The Rowman & Littlefield Publishing Group, Inc.
4501 Forbes Boulevard, Suite 200, Lanham, Maryland 20706
www.lexingtonbooks.com

Estover Road, Plymouth PL6 7PY, United Kingdom

First published in 1981 by Rowman & Littlefield

British Library Cataloguing in Publication Information Available

Library of Congress Cataloging-in-Publication Data

Nelson, Benjamin N. 1911–1977
On the roads to modernity.

Includes bibliographies and index.
1. Civilization, Modern—Collected works.
2. Science and civilization—Collected works.
3. History—Periodization—Collected works.
I. Huff, Toby E., 1942– II. Title.
CB358.N44 901 79-21321
ISBN 978-0-7391-7278-0 (cloth : alk. paper)
ISBN 978-0-7391-7294-0 (pbk. : alk. paper)

∞™ The paper used in this publication meets the minimum requirements of American National Standard for Information Sciences—Permanence of Paper for Printed Library Materials, ANSI/NISO Z39.48-1992.

Printed in the United States of America

Contents

Illustrations

Preface

Benjamin Nelson—historian, sociologist and interpreter of diverse fields of learned endeavor—began his intellectual career as a medievalist trained in those exacting skills of sifting, comparing, and decoding documentary evidences of remote periods of time. In the process of tracing such esoteric matters as the sources and variations of "the legend of the divine surety," the official and private commercial dilemmas of "the Usurer and the Merchant Prince," Nelson was drawn into a domain which evidently must be termed, "systems of spiritual direction."

To those moderns who can scarcely fancy any connection between the down-to-earth pursuit of economic advantage and religious predicaments, it may come as a surprise that in the 12th and 13th centuries the restitution of usury—the return of "ill-gotten gains"—did in fact occur under the auspices of the Church and through its intermediaries. But there is a theoretical excitement behind these innocent facts. For this social orchestration of religion, economy, and morality represents a very special case of what Nelson called "the triangulated regulation of conscience, casuistry, and the cure of souls." Soon after it was fully recognized, that is, officially and publicly proclaimed, that everyone had a "conscience," an inner moral agency capable of applying reason to prospective moral dilemmas, "everyone in Christendom was made responsible under the Fourth Lateran Council of 1215 to confess at least once a year" (this volume, p. 224). And this fateful decision mandated a most peculiar sort of "social planning and engineering" which makes contemporary efforts along such lines appear feeble. Once everyone was supposed to be possessed of moral agency, and faith was not to be left to chance, "there was need for a highly elaborated moral theology in which all the cases of conscience were considered as they arose in the practical life of the times" (p. 224). Subsequently, these cases were worked out, even "alphabetically," so that spiritual directors guiding their charges could "do it by the book."

By the mid-1940s Nelson saw this societal (and later "civilizational") arrangement as a highly rationalized structuring of social life which might be taken as a paradigmatic "ideal type" which few societies and civilizations of the world have ever attained, but which, as an exercise in *comparative* sociological analysis, might be used in Weberian style as a heuristic model.

From this perspective, therefore, it might be said that a "prime clue" in the "understanding of successive historical eras" is to be found in the alternative ways in which societies and civilizations "of the several epochs scale and arrange the claims of religious convictions, moral sentiments, legal order, and psychic (spiritual) well-being."[1]

To reach such an insight is to invent an entire research program, one whose execution no single individual could be expected to carry out in a lifetime. But the idea of such a program, at least, as it applied to the West, did in fact occur to Nelson when he was writing his classic study, *The Idea of Usury: From Tribal Brotherhood to Universal Otherhood* (1949a; 2nd edition 1969). As Nelson later recounted, he asked himself the question:

> What if I were to consider all the developments from the Protestant Reformation forward into the twentieth century from the perspective of conscience, casuistry, and the cure of souls? What actually happened when this triangulated structure was broken up for a variety of very good reasons in the eyes of those who found every element and link in this structure a bar to pure religion and reform? What happened when great numbers would no longer tolerate casuistry, since it seemed to them in every particular to be a deviation from the moral principle which presented itself with absolute clarity in Scripture or by the Inner Light or through some other source? What took the place of casuistry? And then, how was the cure of souls taken care of when men came to have a horror of judges in the court of conscience? (1974g, pp. 257ff; cf. Nelson 1972a, p. 15)

As a result of posing the original question, the succeeding historical and sociological issues became the core of the agenda which directed Nelson's inquiries for virtually the rest of his life.

The weight of one aspect of this project was clearly on Nelson's mind when he wrote, slightly humorously, to his friend Morton White in 1948, "Give him [Nelson] another decade to live and he will release his *Conscience and Casuistry: The Itinerary of the Moral Sentiments in the West*" (April 27th).

To Nelson it was evident that the breakdown of the inherited structures of consciousness of the medieval period unleashed a "relentless surge forward of new religious emphases" which continued unabated after the 16th century. Moreover,

> the most powerful among the new religions which were released by the Reformation may be described as the religion of the transcendental self, the transmoral self beyond conscience. Its myriad expressions are elaborated in all the spheres of existence, experience, and expression since the sixteenth century, [and] its echoes will be found in all the masterworks of theology, philosophy, art, literature, and even science. Its rumblings are the sources of all the triumphs and tragedies of our collective and individual lives. (1974g, p. 258)

In the three decades remaining in Nelson's life after the first formulation of this research program,[2] Nelson plunged into virtually all of the domains listed in his itinerary. In the 1950s, he explored the heights and depths of Freudian theory and psychoanalysis. This resulted in his various papers (e.g., Nelson 1954, 1965c) and volumes on Freud and psychotherapy, especially the centenary volume, *Freud and the Twentieth Century* (Nelson 1957b), but also *Sigmund Freud on Creativity and the Unconscious* (Nelson 1958a, and 1957c).

As editor of multiple Harper Torchbook series (beginning in 1957), Nelson brought into print several dozen books on religion, literature, philosophy, the arts and sciences. These included works written by or on such figures as Adler, Bainton, Bentham, Burckhardt, Cassirer, Durkheim, Eliade, Feuerbach, Fingarette, Freud, Hegel, Kant, Kierkegaard, Kirk, Koyré, Lea, MacNeill, Mill, Panofsky, Parsons, Popkin, Santayana, Schliermacher, Tawney and many others.

In the late 1950s and early 1960s, Nelson simultaneously explored modes of psychotherapy (Nelson 1957a, 1960, 1962d,e), the arts, and literature (Nelson 1961c, 1963a,c, 1965f, as well as 1974f), and the problematics of sociological and psychoanalytic theory (Nelson 1962c, 1964a and 1965a [the latter two in this volume, chapters 2 and 3]).

Nelson's continuing interest in and exploration of Max Weber's work resulted in an invitation to join the centenary celebration of Weber's life and work in Heidelberg in 1964 (see Stammler 1971 [1965]). Yet, during this same period of time, Nelson delved into the history, philosophy, and sociology of science (Nelson 1962a, 1965b, 1967d, 1974b,e, 1975a,b; see chapters 6-10 of this volume). The originality and significance of these latter endeavors are reflected by the invitations to participate in two international symposia in commemoration of the 500th anniversary of the birth of Copernicus (held in Washington, D.C., December 1972 and April 1973; see Beer and Strand 1975, and Gingerich 1975).

To move with such ease and originality from canon law to psychoanalysis, from "philosophical paradoxes" to "psychological systems," to art and literature, to sociology, and thence to the sociology and history of science doubtless required the talents of a polymath. Yet the inspiration for taking up the challenge and the hope of success is to be found in the theoretical insights which tied it all together. Many of those insights are contained in this volume.

ON LIFE'S WAY: A BIOGRAPHICAL SKETCH

Benjamin Nelson was born in New York City on February 11, 1911. He was the eldest of three sons born to a Jewish family which had emigrated from Russia at the turn-of-the century. The senior Nelsons acquired a family store soon after their arrival in the United States. The proceeds of this enterprise were apparently invested in New York real estate which eventually made the family very wealthy. As a result, Mrs. Mary Nelson was freed to devote all of her energies to her family and her Yiddish poetry. And while Benjamin Nelson chose a career of scholarship, his two brothers entered the field of medicine.

After receiving an undergraduate degree in history from the City College of the City of New York (1931), Nelson went on to graduate study in history at Columbia. During his undergraduate years, Nelson apparently harbored aspirations for involvement in journalism. At the City College he worked on the college newspaper, and served from 1929–31 as a "college correspondent" and reporter on religious topics for the *New York Times*.

At Columbia Nelson wrote a Master's thesis on "Robert de Curzon's Campaign Against Usury" (1933). And then he began his large scale dissertation on "The Restitution of Usury in Late Medieval Ecclesiastical Law." A variety of intellectual circumstances converged which prevented this work from ever reaching the learned world. The dissertation that Nelson had in mind could be nothing less than a magnum opus. He was a great bibliophile, and intended his combing of the sources to be completely exhaustive. The generosity of his parents allowed him to engage a variety of "assistants" who served continually to check the sources, the footnotes, and to help in compiling the definitive bibliography on restitution and usury which was to accompany the final work. The family liberality also permitted Nelson to indulge his passion for books which began to reveal itself in the enormous library that he set out to accumulate. Eventually it was to reach 18,000 volumes and to include many rare books on medieval canon law, Roman law, moral theology, and religious history (now donated to Columbia's Lowe Library).

However, in the process of writing the dissertation, Nelson "had become aware of something else. I was working on another book: *The Idea of Usury: From Tribal Brotherhood to Universal Otherhood*" (Nelson 1972a, p. 14). Actually, it appears that Nelson and his friends had always referred to the dissertation as "the brother and the other." But as the dissertation began to take final form, Nelson's interests began to shift toward the elucidation of the larger story which concerned the development of Christian universalism and "the enlargement of the moral community" (1949a, p. 135). The story of the proscription of usury among brothers—but its allowance among strangers—such as the Deuteronomic passage enjoins, might be taken as an illustration of certain themes in Max Weber's inquiries. Weber, especially in his comparative studies of economic ethics, had pointed out the various sorts of invidious "dualisms" in different civilizational settings which had inhibited modern capitalist developments.

When *The Idea of Usury* was published in 1949, it elucidated the manner in which injunctions regarding usury, derived from Deuteronomy 23: 19–20, were refashioned by Christian thinkers so as to yield a "transvaluation of values." Calvin in particular struggled to realize a single universal economic ethic. In doing so, the original injunctions against usury were inverted by extending the privileges of "brotherhood" equally to all "others": "usury is permissible if it is not injurious to one's brother" (Nelson 1949a, p. 78). Consequently Nelson pointed out that "no one has noticed that Calvin, self-consciously and hesitantly, charted the path to Universal Otherhood, where all become 'brothers' in being equally 'others' "(p. 73).

The exhaustiveness of the research which had characterized the dissertation on restitution was evident here in the second book, and it prompted a reviewer to write that *The Idea of Usury*

> is based on what looks like an exhaustive analysis of hundreds of texts and yet the author never gets lost in their detail. He has read not only all the primary sources—at any rate it would be difficult to imagine any that he had not read—but has also modestly gone through a great number of dissertations on these primary sources. (George Boas in *The Journal of Philosophy* 47 [1950], p. 452)

Some readers of *The Idea of Usury* fell under the belief that this book was the final result of Nelson's research on restitution. But it was not: the dissertation still remains unpublished and unavailable to historians because *The Idea of Usury* was printed in its stead in fulfillment of doctoral requirements.

Despite Nelson's great erudition in medieval history (for which he received a Ph.D. in 1944, and a Guggenheim Fellowship in 1945), his first full academic appointment was as an Assistant Professor of Social Science in the new program at the University of Chicago (serving there from 1945–48). If the situation there was not quite ideal, Nelson surely found it stimulating. In November 1945, he wrote to a friend,

> In Chicago, nothing but the fundamental questions are raised. Indeed, they are raised with such insistence and frequency, that occasional flights into solitude become imperative, if only for the pasturing of the soul. . . . The campus bristles with opinionated men who are ready at a moment's notice to fly through the interstellar spaces.

In December of the same year, Nelson married Eleanor Rackow (later Eleanor Rackow Widmer), a talented woman who became a novelist. The next move was to the University of Minnesota (1948) where Nelson joined the Social Science Program in the College of Science, Literature and the Arts. Subsequently he was to become co-chairman of this program and the European Heritage sequence. A committee decision was made to create a social science text which would adequately serve the needs of the students and faculty in the newly integrated program which Nelson and his colleagues had created. The result of this effort was a three volume work called *Personality, Work, Community*. Its first edition reached completion by the Fall of 1950. Nelson had been responsible for the third "book," *Community*. In 1953 the whole endeavor was published in one volume by Lippincott under the title, *An Introduction to Social Science: Personality, Work, Community*. The work proved to be quite successful, and it went through numerous editions, the third appearing in 1961.

The decade of the fifties was one of transition and considerable difficulty for Nelson. Although he continued to be on the staff at Minnesota until 1956, from 1952 on he lived in New York. The large measure of success of the new social science program at Minnesota elicited the rather familiar intellectual tensions of university life which surface whenever intrepid souls cross disciplinary boundaries. When it was proposed that this successful "program" be elevated to departmental status, feelings began to run high. These tensions and those engendered by the ending of his first marriage seem to have pushed Nelson increasingly into the study of Freud and psychoanalysis. But his reading of Freud's *Civilization and Its Discontents* at Chicago had induced him to draft in 1951 the first version of his well-known paper, "The Future of Illusions."[3] The paper was finally published in *Psychoanalysis* in 1954.

Moreover, as Nelson reported in the addenda to the second edition of *The Idea of Usury*, he had in fact become increasingly interested in the complex issues which he called "The Moralities of Thought and the Logics of Action" (1951). This move brought Nelson's intensifying concern for "systems of spiritual direction" even closer to the surface. In his symposium remarks (November 1950), which gave rise to this paper, Nelson identified with the "continuing social effort to evolve a comprehensive fabric of attitudes and methods in virtue of which moral decision, whether personal or social, might become increasingly permeated by rational deliberation and uncoerced consensus" (p. 1). He then expressed the view that there is "a dire necessity" to elaborate "a multidimensional logic of moral decision." Such a matrix of decision, in Nelson's view, could be worked out only through efforts which took into consideration both the frames of reference of "the cases of conscience" and the newer ones emerging in psychiatry and even psychosomatic medicine (pp. 3–5). It is therefore significant in the light of Nelson's life-long efforts to expand the theoretical and empirical boundaries of the social sciences—both sociology and psychoanalysis—that in this 1951 paper Nelson stated his view that efforts designed to give each of the social sciences a complete theoretical autonomy served to limit our understanding of social and cultural process:

> However worthy the original and current demands of these disciplines for autonomy, it has become evident that the resultant atomization of the sciences (and, one might add, the humanities) is now greatly retarding men's capacities both to comprehend their predicaments, and their opportunities to promote the life of reason in society. (p. 2)

During 1952–53 Nelson served as a visiting professor in the History of Contemporary Civilization program at Columbia University. At about this same period of time he met (and later married) Marie L. Coleman, a psychoanalyst of distinction, and his occasional collaborator. From 1954 until his death Nelson served actively as Advisory Editor to *The Psychoanalytic Review,* whose chief editor was his wife Marie. In this capacity Nelson published numerous essays of his own and conducted an "Adventure of Ideas" section wherein he published and commented on noteworthy papers, and sought to link psychoanalytic concepts to broader dimensions of historical and sociocultural process. Here, as with the Harper Torchbook Series, Nelson exercised his outstanding editorial talents to attract noted authors to Special Issues of the *Review.*

In 1956 Dr. Nelson became chairman of the Department of Sociology and co-ordinator of the Social Science Program at Hofstra. Beginning in 1957, his association with Harper and Row as Editor of its noted Torchbook Series, Nelson quickly brought twenty or more forgotten classics back into print. In the meantime, in the late fifties and early sixties, Nelson and his wife led a study group in the development of paradigmatic psychotherapy, theory and technique (Nelson 1957a). Their efforts to broaden Freudian theory and therapy drew upon "J. M. Baldwin's idea of 'semblant behavior,' C. H. Cooley's concept of 'primary group' and 'looking-glass self'," as well as George Herbert Mead's notions of "role-playing" and "role-taking" in the formation of self (see Nelson 1957a, pp. 29ff, and 1968e).

From 1959 to 1966, Professor Nelson served as chairman of the Department of Sociology at the State University of New York at Stony Brook (first located at Oyster Bay). In 1966, Nelson was called to the Graduate Faculty of the New School for Social Research where he held the title, Professor of History and Sociology. In addition to serving in the Sociology Department, Nelson had been invited to design and chair a program called Master of Arts in Liberal Studies sponsored by the Ford Foundation. Professor Nelson remained at the New School until his sudden death due to a heart attack in September 1977 on board a train in West Germany. He was traveling from Tübingen to a conference in Freiburg where he was scheduled to give a paper on "Tradition and Innovation in Law and Society: Comparative Civilizational Perspectives."

From his earliest days at the City College, Nelson was a vital part of that insatiable group of intellects and talkers who have since been recalled by Alfred Kazin and others. This group of intimates and strangers included such figures as Paul Goodman, Meyer Schapiro, Saul Bellow, Dwight MacDonald, Delmore Schwartz, Harold Rosenberg, and a variety of others who took up posts on *The Partisan Review, The New Republic* and other vehicles of literary expression.

While Nelson cherished the bonds of friendship, he did not wish to be identified with any narrow cause, any single ethnic style, nor any parochial outlook. He considered himself a sort of ambassador at large to all those individuals and groups who expressed a desire to enter into free inquiry about the human condition wherever it existed. He was an emissary from the learned world of the past as well as the present—the world of Nicholas of Cusa, Abelard and Bernard, as well as the contemporary world of less steady commitments—to all those who wished to explore the dimensions of our collective lives. It would not be an exaggeration to say that Nelson felt himself to be "in the service of a higher cause."

His friendship with Paul Goodman, which began in boyhood, was above all a source of excitement, of intellectual and social challenge. And the friendship reveals both Nelson's openness to every variety of intellectual discourse and his devoted loyalty to friends. Their meeting as undergraduates at the City College began a friendship which endured to the end despite all impediments. In his memorial tribute to Goodman, Nelson recalled "Our Walks and Talks" during which all realms of discourse were opened to the passions of the intellect. "What did we talk about? We talked about Aeschylus and Euclid, Saul of Tarsus and St. Paul, Maimonides and St. Thomas, Goethe and Coleridge, Sweetness and Light" (October 22, 1972).

The depth of Nelson's intellectual resources revealed themselves at every turn. They stood out in oral and written exchanges with philosophers of the stamp of Sir Karl Popper, N. R. Hanson, Imré Lakatos, and Stephen Toulmin. There seemed to be no field in which he was not compelled to do intellectual battle. The same confidence and insightfulness reveals itself in his review of Bellow's *Herzog* (Nelson 1965f), in "Sartre, Genet, Freud" (Nelson 1963c), "*The Balcony* and Parisian Existentialism" (1963a), the sources of "probabilism" in science and moral theology (this volume, chapter 7), his introduction to Kierkegaard's writings (1962b), and in his review of R. de Roover's *Business, Banking, and Economic Thought in Late Medieval and Early Modern Europe* (Nelson 1977c).

Benjamin Nelson was possessed of a soaring intellect which could not be contained within the confines of narrow specialization. And he approached the world as if men everywhere shared this same consuming passion to explore and understand all the realms of "the spirit." In a notice to the reader posted on the new edition of Feuerbach's *The Essence of Christianity* (1957 [1841]), Nelson stated his faith: "Every generation needs access to the entire spectrum of the human spirit: that is its inalienable right and unavoidable responsibility." His life in every respect was committed to that responsibility.

During his lifetime Nelson was active in many professional associations including the American Sociological Association, the Society for the Scientific Study of Religion, the American Association for the Advancement of Science (especially section L on the History and Philosophy of Science), the International Sociological Association (Research Committee for the Sociology of Science), as well as the National Psychological Association for Psychoanalysis, of which he was an honorary member. Nelson was instrumental in the founding of the Society for the Scientific Study of Religion and served as its vice-president from 1976 until his death. He had in fact just been re-elected to that post at the time of his death. In 1971 Nelson and several colleagues revived the International Society for the Comparative Study of Civilizations (U.S.), and Nelson served as president of that society from 1971 to 1977 when he stepped down. Nelson's unending quest for wider horizons and new perspectives kept him perpetually active in these and several other collegial endeavors. But to a great many of his associates he seemed to be an institution in itself.

ON THE ROADS TO MODERNITY

The essays assembled in this volume seek to give the reader access to the wider theoretical and empirical perspectives which Benjamin Nelson brought to the study of sociocultural process.

The introductory chapter of Part I was written for a German edition of a collection of Nelson's papers which included all but chapter 9 of Part III, plus chapters 5 and 12 of this volume. The Introduction was written just three months before Nelson's death.

The essays in Part II represent the various stages and dimensions of Nelson's thought which eventuated in his insistence on the need to broaden our horizons to civilizational and intercivilizational dimensions. Yet this move to wider horizons did not lead Nelson to abandon "an interest in studying the fabrics and textures of institutions and interactions from the so-called *micro*-sociological and *micro*-historical points of view" (p. 3 this volume). In part this was due to Nelson's enduring interest in the problems of "conscience" and "spiritual direction" which are *construed* as "micro-level" problems.

But there is an even greater source of theoretical unity in Nelson's thought that readers may not detect on first reading. From at least the early 1960s onward, Nelson was intent on rethinking and recasting the theoretical perspectives of Durkheim *and* Weber. Aside from Talcott Parsons's *The Structure of Social Action* (1937), sociologists have been inclined to think that Durkheim and Weber represent opposed theoretical stances. This is in line with the "positivist" reading of Durkheim and his notion of "social facts." Weber, in contrast, is said to stress the "meaning" of social action.

Nelson, however, saw that Durkheim's discussion of "social facts" in *The Rules of Sociological Method* (1938) served to establish a "cultural" frame of reference for explaining social action. After all, Durkheim had said that social facts "are ways of acting, thinking, and feeling" (p. 2). To say that they are also "external" to the individual and "coercive" is only to say that all social patterns of existence tend to become normative (cf. Nelson below, p. 72).

In the paper reprinted below as chapter 2, Nelson sought to map out the varieties of "cultural cues" which tend to become established in the form of "directive systems." The *percipienda,* for example, refer to all "the ways in which we are instructed or compelled or persuaded" to perceive any and all aspects of our surroundings. And this cultural dimension of existence might be seen as another side of the thesis of the "theory-laden" character of perception that has been given such attention by philosophers of science since the work of N. R. Hanson (1958). And thus it would seem that Durkheim's "social facts" refer to these "symbolic" and "meaningful" dimensions of social interaction.

On the other side of the equation, interpreters of Weber have not generally perceived that Weber rejected simple statistical "explanations" of social action on the grounds that "meanings" and "motives" are the "causes" of social action (see Weber 1975, pp. 126ff, 154, 186ff, and 1977, p. 107 and passim). Weber's thinking took this turn both because his notion of "motive" did not have the depth-psychological connotation of today and because of his training in law. Hence "purposes," "intentions," and "reasons" may be cited as related aspects of the "causes" of an actor's (or actors') behavior. If actors were not guided in their actions by diverse reasons, then the social world would have to be judged to be a hopelessly chaotic domain.

Moreover, to sociologists and anthropologists presumably it is obvious that the sundry reasons which might be the stimulants of action, and the "rationality" of the social setting, are products of *particular* cultural assumptions and traditions. Nelson expressed this point of view by saying that civilizations are reared "on different geometries" (p. 92 below).

It may therefore be said that the study of these symbolic structures of meaning and their multiple determinants and unfoldings in the societal and civilizational settings of the world is not remote from the "causal" explanation of social action. In recent years a number of writers, including philosophers of the social sciences, have helped to explore and elucidate these connections between "rational" and "causal" explanations of human action (see among others, Hart and Honoré 1959; Davidson 1968; and White 1965).

In Part III Nelson's papers on the comparative and historical sociology of science provide a sustained set of discussions of the religious, philosophical and cultural backgrounds of modern science. They illustrate the reasons why a *comparative* and *historical* sociology of science must also open its vision to *civilizational* perspectives and intercivilizational encounters.

By introducing the phrase, civilizational frames of reference, Nelson sought to accentuate both the trans-historical (i.e., intertemporal) and supra-societal dimensions of thought and interaction. This move draws upon a similar turn to be found in the work of Durkheim and Mauss. In 1913 they were compelled to stress the fact that

> social phenomena that are not strictly attached to a determinate social organism do exist; they extend into areas that reach beyond the national territory or they develop over periods of time that exceed the history of a single society. They have a life which is in some ways supra-national. (Durkheim and Mauss 1971, p. 810)

Moreover, not all social phenomena have the same potential for expansion, the same "coefficient of expansion," beyond local boundaries. By focusing attention on aspects of what Durkheim and Mauss called "symbolic frontiers," "all sorts of problems" come into view regarding the relative propensity of diverse sociocultural phenomena to become internationalized and *universalized.*

As we have seen, from his earliest graduate school days, Nelson had been interested in the problems of the expansion of the "moral community" and the impediments in the path "from tribal brotherhood to universal otherhood." However, in the middle 1960s, Nelson's original fascination with problems of *conscience* and "moral regulation" bore surprising fruit. In an essay on Weber and *The Protestant Ethic,* Nelson took the opportunity to explain the differences in stress that he and Weber gave to some critical issues. He began by citing a passage from the New Postscript to *The Idea of Usury:*

> Given Weber's overriding interest in uncovering the "spiritual roots of the vocational asceticism of Occidental rationalism," and given his commitment to the conduct of "thought experiments" by the "ideal-type" method, Weber was correct, *from his point of view,* to stress the decisive importance of the notion of *innerweltliche Askese.* (1969a, p. 239)

Then Nelson continued:

> In this spirit he again and again minimized the relevance to his concern of the changes in the doctrines of usury which had been a point of departure of my own researches. As the years passed, my emphasis was bound to become different from Weber's. My growing concerns with the structures of consciousness, including the principal legitimations and casuistries of action, spurred me to press forward to explore the *rationales* proposed by the contending groups locked in struggles over a whole series of theological, social, economic, and other issues. I sought to understand how the norms governing the settlement of disputed questions in all spheres came to be decided; how the bases for political obligation were reared; how conflicts at the level of self, community, civilization were moderated. The more intensively I searched the documentary remains for answers to questions of this nature, the more I appreciated the central position traditionally given to *conscience* by older historians of the last century. The issue of usury implicated the entire fabric of *opinions and actions* rooted in *conscience* and all the forms of the imbeddedness of *conscience,* including casuistry and the cure of souls.
>
> Suddenly, indeed, the historical emphases ascribed to the notion of *con-*

> *science* gained fresh meaning. It was not that Luther or Calvin invented this fateful idea—a systematic exploration of that concept had already been started by others in the Middle Ages—but that these leaders of the Reformation gave everything a new axis and a new center by the scope of their attacks on every aspect of the culture of conscience—its structures, its *decision-matrices*, and its very *rationales*.
>
> My own point of departure came to rest on the following perspectives: To understand the *makings* of early modern cultures we need to understand the *makings* of early modern minds and, therefore, need to have a proper sense of the change in the central paradigms as well as the restructuring of axial institutions in the society. If I may use the language of my current research, the makings of early modern cultures are most clearly in evidence when we study them from the special point of view of the revolutions in the rationale-systems in the sphere of conscience in its dual bearings in the spheres of moral action and intellectual opinion. (1973b, p. 93)

Thus the "roads to modernity" involved a two-sided attack on the inherited logics of *act* and *opinion*. The one centered on the logics of *action* (namely moral regulation) which was seemingly won quickly with the Reformation when the rule of the Forum of Conscience was abolished (in Protestant lands) in favor of a self-regulation which was said to be "beyond conscience," and perfectly guided by the "Inner Light" which is said to shine in the heart of every man (chapter 3, pp. 46 and 49-51).

On the other hand, the revolution in the logics of *opinion* and intellectual decision—namely in the domain of science—constituted the other half of this revolution. It concerns the "fundamental canons governing the decision-making structures" (p. 70 below), and came to a head over the question of whether or not men in general should be intellectually free to advocate as true propositions about nature which contradict the official position of the Church. Hence it is not surprising that Church officials in the early seventeenth century feared that Galileo might be in league with Luther's movement which attacked the authority of the Church. Nelson's impressive explorations of these questions are reported in chapters 7, 8, and 9 of this volume.

In chapter 10, "Sciences and Civilizations, 'East' and 'West' . . .," Nelson carried out a "thought experiment" and "test case" in which a variety of his assumptions about the decisive social and cultural foundations of science in the West are explored in the "East"—that is, in the context of China's scientific development. A parallel comparative analysis using the case of Arabic science was high on Nelson's agenda. The present editor had undertaken to carry out this project in collaboration with Nelson, but his sudden death frustrated the endeavor. It may simply be said that Nelson's perspectives added a great deal to the elucidation of the Arabic/Islamic case. And this case too seems to reveal blockages to breakthroughs in the inherited structures of thought, action, and social organization similar to those discussed by Nelson in his analysis of the Chinese case.

Moreover, in comparison to the latter, where Joseph Needham tells us that the Chinese lacked several very critical formal scientific notions, such as the zero and the Greek conception of geometrical proof (see below, pp. 90f and 171f), the Arab world of the 12th and 13th centuries had acquired both Euclid's geometry and the Hindu (transformed into) Arabic numerals. In short, possession of many technically important scientific conceptions lacking in the West did not form the basis for a breakthrough to "modern" science as might have been expected in Arabic/Islamic scientific thought.

In the last section, the essays explore and clarify some additional aspects of Nelson's understanding of the need for a comparative historical sociology with civilizational perspectives. In chapter 11 Nelson strongly accents the need for drawing sociological attention to all the *existences, experiences,* and *expressions* of men in their cultural, historical, and civilizational multiplicities. This paper was written prior to the "Civilizational Complexes" statement and reveals Nelson thinking through the need to resort to multiple levels of sociological, psychological, and depth-historical analysis. His stress there on the *"contemporaneity of the non-contemporaneous and non-contemporaneity of the contemporaneous"* (p. 210) serves well to suggest the critical need to reorient sociological inquiry away from what Nelson called "the specious present." His use of a discussion of the "grotesque" to achieve his results here reveals his startling ability to see virtually every aspect of daily encounters as variations on themes of *Eros, Logos, Nomos,* and *Polis* which had been such intense matters of discussion in the 12th and 13th centuries (chapter 12).

Finally, chapter 13 sets forth Nelson's only concerted effort to review the last phase of his research after he had completed the paper on Needham and Weber on China, and to reply to critical objections. The Introduction to this volume builds on this piece, but Nelson seems not to have attempted to go beyond it. Chapter 13 therefore offers the reader Nelson's own views on some of the connections between the various papers collected in this volume. And it suggests the multiple advantages that a "civilization-analytic" offers over the many alternatives which abound in contemporary sociology.

Lastly, the "Epilogue" provides a unique orchestration of Nelson's insights into the depth-historical foundations of Western civilization, and highlights critical issues regarding religious commitment, science, technology and social harmony which are sure to be at issue in the decades and century ahead.

May, 1979 T.E.H.

NOTES

1. These citations are taken from an unpublished and undated draft bearing the title, "Conscience, Casuistry, and the Cure of Souls: Eight Centuries of Interaction, ca. 1150–ca. 1950." From exterior markings it appears to be of Chicago or Minnesota vintage (i.e., 1945–51). A copy of it was given to Professor James Luther Adams (along with a draft of a paper on "casuistry" [Nelson 1963b], before 1956 when Adams moved from Chicago to the Harvard Divinity School.
2. The paper listed in note 1 sketches out six distinct historical phases each of which "exhibit significant tensions in the patternings of the aforementioned elements." The paper thus anticipates the reporting of Nelson's research on this topic in 1965 and chapter 3 of this volume. Further evidence of this thrust of Nelson's work is found in the topic of his Guggenheim Fellowship (1945–48): "Studies in the Relations Between Conscience and Casuistry in the Law and Moral Philosophy of the Later Middle Ages and Early Modern Eras."
3. This first version of "The Future of Illusions" was printed in, *Conflict in the Social Order* (University of Minnesota, lithograph, pp. 30–47, 1951), volume one of the centenary proceedings commemorating the founding of the University of Minnesota. I am grateful to Professor John M. Bell for making a copy of this available to me.

Acknowledgments

It gives me pleasure to acknowledge the many forms of aid, assistance and counsel which the many friends and former colleagues of Benjamin Nelson gave me in preparation for this volume. Special thanks go to John Mundy of Columbia University, Morton White at The Institute for Advanced Study, as well as to Eleanor Rackow Widmer, Roland Robertson, Jerry Gittleman, Chuck Lawrence, Steve Kalberg, and David McCloskey.

To Mrs. Marie Nelson, for encouragement and exceeding generosity I owe very special thanks. Above all I wish to express my gratitude to the Institute for Advanced Study in Princeton, New Jersey, for its most generous support of my work during 1978-79. Without the Institute's support (and the proficiency of its typists, above all, Mrs. Catherine Rhubart), the present work would have remained but a promise for some years to come.

Finally I express thanks and appreciation to my wife Cynthia for continued understanding and her selfless devotion to the task of completing this volume.

Introduction to the Reprinted Edition

Toby E. Huff

The centenary anniversary of Benjamin Nelson's birth fell on February 11, 2011. During the three decades since this book was first published, and the thirty-four years since Nelson's death, the subject of civilizational analysis has become far more widespread. As is typical of academic and other endeavors, there is no central script. Both independent scholars and those associated with the *Society for the Comparative Study of Civilizations* that Nelson helped found, have followed their own lights. Beyond that, historic events such as the 9/11 attack on the Pentagon and the World Trade Center, the launching of the so-called "War on Terror" of the Bush Administration, along with the rise of China and India as economic powers, suggest that *civilizational* entities must be considered central to world affairs.

Nevertheless, Nelson's influence on this increasingly important area of study can be seen in a variety of publications, many of which are listed below in the references to this section. As was highlighted in the Preface, Nelson was trained in medieval history but had much broader interests in religion, science, psychoanalysis, philosophy as well as economic history. He maintained an interest in the study of *sociocultural process* at all levels. Yet he eschewed any attempt at "system building" as was attempted by Talcott Parson and the "world systems" theorists (cf. Wallerstein, 1974; Hall and Chase-Dunn 2006; and Sanderson 1995). Rather than attempting to review the range of Nelson's interests and their possible influence, in this Introduction I shall be content mainly to highlight the continuing influence of Nelson's thought on civilizational analysis and some of the key issues that seem alive today. For it is true that both the inner logic of Nelson's earlier inquiries and the drift of historical events led him to a preliminary articulation of civilizational frames of reference. The end result he hoped would be a *new science* of civilizational analysis.

In an important sense, Nelson's pathway to his civilizational orientation is a self-exemplifying hypothesis in Robert Merton's sense (Merton 1965). It is self-exemplifying in that other scholars, some of remarkably different disciplinary training, and apparently independently (see Huntington 1993), arrived at the same hypothesis. That hypothesis simply states that civilizational identities, those based on historical experiences of centuries mediated through language, custom, law and religion that transcend village and state, have powerful galvanizing effects that continue to shape international cooperation and conflict.

Indeed, Nelson believed that Max Weber himself had arrived at "comparative civilizational" analysis in the last essay from his hand, the Introduction to Weber's *Collected Essay in the Sociology of Religion* (1920) but which many unsuspecting readers only know as the Introduction to Weber's *Protestant Ethic and the Spirit of Capitalism* (Weber 1958), placed there by Talcott Parsons. Nelson wrote, Weber's Introduction "confirms his right to be ranked as a pioneer in the *comparative historical differential sociology of sociocultural process and civilizational complexes*" (Nelson 1974, p. 271).

More recently Syed Arjomand and Edward Tiryakian have argued that the concept of civilization and the prospects of civilizational analysis, now framed in the post-9/11 world, constitute a more critical mode of "sociological analysis than the nation-state or the world system" (Arjomand and Tiryakian 2004, p. 11).

However loosely or precisely civilizations in this sense are delimited, the decisive indicator is that the complex delineated is composed of *two or more peoples, societies or territories* that share higher level symbolic systems, those typically labeled religious, legal, or philosophic. Samuel Huntington suggested seven or possibly eight civilizational entities operative in our modern world. These included Western civilization, Confucian, Japanese, Islamic, Hindu, Slavic-Orthodox, Latin American, and possibly an African civilization. Scholars may disagree about these divisions and identities, but the point is to *study them* from this point of view. It is debatable that Japan represents a separate civilization. Until World War II, its religious foundations, law, and language had a great deal in common with China and Korea. Since then its legal system has been radically transformed with a new constitutional order based on Western principles. From that point of view it is a hybrid outlier, neither encompassing other societies, nor fitting easily into either the East Asian or Western molds.

Likewise, it's doubtful that Latin America represents a separate civilization as it has always been part of the Christian West, relied on the Civil Law tradition, and so on. Yet its economic policies until recently have been notably unsuccessful.

On the other hand, if other scholars want to study small and isolated societies and call them "civilizations," nothing in Nelson's writings prevents that. That undertaking is simply another level of analysis.

Nelson was keenly interested in all the "experiences and expressions" of peoples throughout history and around the world in the spirit of Goethe: "let nothing human be foreign to me."

Further discussions of such issues can be found in the essays appearing in the *Comparative Civilizations Review* as well as the *Journal of World History* and the books published by members of those organizations. At the same time it is true that a number of scholars, including this writer, have taken up aspects of Nelson's framework or fashioned their own and plunged ahead with civilizational analysis (among others, see Arjomand and Tiryakian 2004; Arnason 2001, 2003; Dushesne, 2011; Eisenstadt 1986; Huff 1993/2003, 2010; Melko and Leighton 1987; Nielsen 2004; Sanderson, 1995; Synder 1999).

ON THE ROADS TO CIVILIZATIONAL ANALYSIS

In the context of this reprint of Benjamin Nelson's essays, however, we should recall that his intellectual moorings, as noted, were located in deeper and broader *historical* encounters, both within the Western world, and between West and East, including encounters with China, India, and the Muslim world. In that context, we should highlight the fact that a central preoccupation of all Nelson's thought was how and why groups of people evolved into ever larger communities of discourse, reciprocity, and shared moral commitments. This process of *universalization*, that of expanding the boundaries of discourse and legitimized participation, was for him a central preoccupation. For civilizational phenomena per se are just that: cultural emanations that extend beyond the local tribe, ethnic group, community or society. How cohesive and long-lasting these identities and complexes become, and how they are transformed, is the subject of study (see Huff 2003, pp. 127).

Moreover, one of the critical insights for which Nelson has been recognized is the idea that *intercivilizational encounters* are strategic sites for understanding both the boundaries between civilizational entities and for revealing the decisive religious, legal, and philosophical principles that distinguish these contrasting entities (cf. Arjomand and Tiryakian, 2002, p. 4; Arnason 2003, p.140; Nielsen, 2004). I shall say more about that later.

In the meantime, it should be noted that because Nelson entered this line of inquiry via religious studies located in the Middle Ages, he concluded that the governing forensic structures of communities and ethnic groupings were actually religious structures, or "structures of consciousness." Consequently, he became fascinated by how these were constructed and how, at least in the Western world, the architects of these directive structures often moved to expand the boundaries of inclusion. As they did so, the very structures of consciousness were themselves transformed into new types (see below, pp. 93–97).

Furthermore, since he plunged into a comparative analysis of medieval Christian culture and Judaic culture in the course of pursuing his doctoral subject, "the idea of usury," he had a ready-made comparative case study of two great world religions, and much more. For it was also true that interest payments were forbidden by Islamic law.

The belief at the center of this analysis was that taking interest on money loaned was forbidden by God as revealed in a famous passage in the book of Deuteronomy. Eventually the prohibition was taken up by the Medieval Church fathers. That theological prohibition was seen by later economic historians as a significant inhibitor of capitalist development and therefore a major charge against Catholic Christianity. Nelson set about investigating just how the Church dealt with these critical issues, the forum of conscience, the process of restituting ill-gotten gains, and so on (Nelson 1949).

What is more, Max Weber's great study of *The Protestant Ethic and the Spirit of Capitalism* published in 1904–1905 (Weber 1958), asserted that some kind of breakthrough in Protestant theology resulted in an "inner worldly asceticism" that enjoined all believers to undertake diligent hard work in *this world*, not just in the cloister. This injunction according to Weber had a fructifying effect on capitalist development that seemed to be especially pronounced in Protestant communities (for new research on this see Becker and Woessman 2009). Although Nelson was unaware of Weber's thesis when he first launched his study in the early 1940s, once he found Weber, it transformed his outlook as well as reassured him that Weber himself had noticed the evolution that Nelson meticulously charted. That is, Weber commented on the contrast in business ethics that applied to brothers within religious communities and those that prevailed outside. The Reformation and the theological work of John Calvin that Nelson elucidated revealed the ways in which this dualistic ethical situation was undermined by the theological teachings of Calvin. In doing so Calvin and the Reformation paved the way toward "universal morality conducive to systematic capitalism" (Nelson 1949, p. xv). Calvin had done this by declaring that only extravagant usury was forbidden among co-religionists, that so long as the lender kept in mind the principle of charity while treating the "other" as a "brother," lending money on interest was not offensive to God. If the brother could not afford the interest on borrowed money, then the lender was enjoined to give the borrower the money outright. By this means, Nelson saw a broad historical shift whereby the *other* became equally a *brother* and the moral community was expanded.

From a broad economic point of view, the prohibition against usury was lifted, businessmen were freed to pursue gain, and entrepreneurs in Protestant lands were encouraged to be diligent, frugal, and relentlessly industrious. From Weber's and Nelson's point of view this was a major intellectual reorientation with economic consequences.

STRUCTURES OF CONSCIOUSNESS

Studying that huge transition was one pathway to thinking about comparative, international, and then civilizational processes and encounters. Yet Nelson remained fascinated by changes in structures of consciousness, first in the medieval period, and then with the transformation of the Western world that could be traced to the broad impact of Protestantism, both in Protestant and non-Protestant lands. For the new Protestant dispensation was said by the theologian Paul Tillich to have taken us to "the transmoral self beyond conscience" (below p. 53). As a description of the present condition this does not seem far off the mark. The liberation of conscience from the "forum of conscience" maintained by the Catholic Church was undermined and destroyed in all Protestant lands and in its place true believers were said to be guided by the "inner light" that God had shone in their hearts.

In a word, much of Nelson's attention was riveted to the transformations of conscience and consciousness and the resulting new structures of consciousness that he came back to repeatedly. Even his seminal essay on "Civilizational Complexes" was originally titled, "Structures of Consciousness from the Perspective of the Comparative Historical Sociology of Sociocultural Process." Likewise, when Nelson wrote his long essay on science in China and Joseph Needham's monumental work (Needham 1954–2004), Nelson attempted to incorporate the language of "structures of consciousness" but with little advantage (see Chapter 10 below, esp. pp. 164ff, and 191ff). Nelson was surely right that in the great encounter between China and the West contrasting "symbolic universes [and] symbolic technologies" were at issue when the Jesuits attempted to bring Western natural philosophy to China in the early 17th century. But whether knowledge of indigenous Chinese institutions and less orchestrated aspects of Chinese culture would prove more insightful for our understanding than reference to structures of consciousness is an open question.

It is questionable, for example, as to whether the path focusing on structures of consciousness was the best one for grasping the even larger world-transformation that now goes under the heading of *globalization* (among others, see Robertson and White 2007; Lechner and Boli

2000; Held and McGrew 2000; Ritzer 2007; Turner 2010). Surely the latter does involve a number of paradigm shifts in the terms of reference needed by indigenous peoples around the world for understanding and resolving conflicts of interest in the new global order. I mention this because there were in Nelson's writings other strands, fully compatible with this universalizing trend so dear to him, and that would more easily render visible those indispensable structures, fashioned in the West, that underlie virtually all of the present globalizing forces: that is, the development of international standards of *law, legal process and legal autonomy* that govern all international economic, political, electronic and other transactions. But first, let us turn to Nelson's analysis of the twelfth and thirteenth century "axial shift" that in his view was located in at least a three way set of civilizational encounters that transformed a broad range of institutions and structures of consciousness.

THE AXIAL SHIFT AND THE STRUCTURES OF CONSCIOUSNESS

When Nelson returned, nearly three decades later to the European Middle Ages in which he had studied the changing fate of the idea of usury, he became more fully aware of the existence of both cooperative and abrasive encounters between Christian Europe and other civilizational entities. These included Judaism, the ancient Greek heritage, expanding Islam, Byzantium, and marginally, China, thanks to Marco Polo's travels and reports. He came to think that major shifts were occurring in science, natural theology, philosophy, mathematics, and even commercial thought. Although the study of Arabic-Islamic science was then in its infancy, enough was known in some circles to believe that in optics, mathematics, medicine, and astronomy the Muslim world held the lead. With the arrival of the "New Aristotle" and his "natural books," this new scientific agenda was quickly placed at the center of the curriculum in the universities across Europe (cf. Grant 1996; and Huff 2003, Chapter 5; Weisheipl 1964). In doing so, it institutionalized an extraordinary level of curiosity in the natural world (on which more below).

Having already studied the fate of the forum of conscience (the confessional and its establishment in 1258), Nelson was inclined to think that here, despite the demise of the restitution of usury, was indeed a breakthrough in the structures of consciousness, a switch from *sacro-magical* structures, to *faith structures*, and then to *rationalized* structures (see below, pp. 93–97). If one were in that context inclined to ask how the modern scientific revolution could have emerged out of this, then one would as Nelson did, address it as another aspect of this transformation of the structures of consciousness in which the individual believer was enabled to claim true (scientific) propositions against the objections of the church—a conflict that did indeed explode around Galileo when he claimed to know the real constitution of the world that "is unique, true, real, and could not possibly be otherwise" (as cited in Huff 2010, p.13). Furthermore, Galileo claimed that the center of the universe was the sun, not the earth. This insistent claim repeated by Galileo in many contexts led to his eventual censorship and house arrest in 1633. Insofar as greater Europe and the progress of the scientific revolution itself is concerned, there are many reasons for believing that the emphasis on this debacle has been overstressed (Huff 2010, chapters 7–12).

Further stimulation toward thinking about these problems came with the first two installments of Joseph Needham's *Science and Civilization in China* (ca. 1954 and 1959). As Nelson began to digest this extraordinary work, he foresaw "Needham's Challenge" which was the attempt to understand *sociologically* why modern science arose only in the West and not in China, or for that matter, in the Muslim world. This was not on the radar of any sociologists in the late 1960s. Robert Merton's students were looking at the "reward system in science" (Cole and Cole 1967; Hagstrom 1965), not comparative historical sociology of science.[1] In Nelson's view, some of Needham's insights were more powerful than those of Weber, though in the end (after a decade of thinking about the problem), Nelson gave Max Weber the nod (see below, pp. 191–95).

In short, Benjamin Nelson's approach to this problem context was both broad and deep: the great struggle for the emergence of modern science in Europe was rooted in religious and philosophical contexts going back to the Middle Ages, especially to Aristotle. At the same time, Nelson suggested, the rise of early modern science was a great *civilizational* breakthrough not only for the West, but as Needham put it, it was a breakthrough for *ecumenical* science, a "universally valid world science" (as cited below, p. 168). Still, the conundrum was that Arabic-Islamic science up until the thirteenth or fourteenth century was more advanced than European science, while Needham at least thought Chinese science was also more advanced—which seems doubtful to this writer (see Huff 2010, especially chapter 4).

This was the context for Nelson's new insights. Consequently Nelson attempted to frame that blockage both in the Islamic world and in China as issues in the structures of consciousness because, unlike Western Europe, scholars in those civilizations had not been freed to speak their minds regarding natural phenomena, and hence could not get to the Copernican-Galilean-Newtonian worldview. Indeed, the flowering of modern science in Europe in the 17th century—in astronomy, optics, microscopy, pneumatics and hydraulics and electrical studies—has no counterpart in China, India, or the Ottoman empire of the time (Huff 2010).

There is much to be said for this approach and readers will surely benefit from reading Nelson's chapters assembled here (especially chapters 3, 8, and 10). There is no question but that the Western world did achieve a great breakthrough somewhere between the twelfth and thirteenth centuries and the time of Newton when the *role of the scientist* was firmly ensconced in the university and other institutional structures. This was something that no other part of the world was able to accomplish during this period. Indeed, the non-Western world has lagged extraordinarily behind the West in science and technology all the way to the end of the 20th century. Only Japan rivaled the United States in some aspects of technological innovation in the last quarter of the 20th century. It remains for unfolding history to tell us whether this is about to change in the 21st century.

On the other hand, Europeans achieved an enormous breakthrough with regard to the structures of law, legality, and self-governance in the "axial shift." And this breakthrough, too often neglected,[2] also created indispensable foundations for economic development. Nelson was indeed aware of these deep alterations undergirding the great axial transformation of the West. As he put it, the march toward greater *universalization* of sociocultural process in the West was powered, in addition to Christian universalism, by two rich cultural reservoirs on which so much was later to be built. As he put it, "another very powerful source of universalism was Greek philosophy and logic [as well as] Roman law. Wherever these had influence, movements to universality achieved great importance" (below, p.192).

REFRAMING NELSONIAN ISSUES

Here then, is an outline of that great movement: the legal transformation that provides the continuity between the fusion of Greek philosophy and Roman law (as well as Christian theology) and the sweeping globalization process that has now gripped the world.

Throughout his career Nelson maintained a delicate balance between his interests in strictly religious contexts, structures of consciousness, and advances in science, as well as law. His tracing of the trajectory of Christian universalism extended from the Reformation shift "from tribal brotherhood to universalism otherhood" all the way to Weber's further charting of the rise of "impersonal service" so daringly worked out in *The Protestant Ethic and the Spirit of Capitalism* (Weber 1958, pp.108–9). In Nelson's day the demise of this ethic was often announced, prematurely Nelson thought. For he wrote in the late 1960s,

this ethic is being exported to every corner of the earth along with the newest energy packages, automation, the computer, the bank, the department store, the central office and the Board room (below, p.79)

Moreover, Max Weber always stressed the importance of *political and legal stability* based on the rule of law to support and stabilize the efficient working out of the Protestant (or any other work) ethos (see Weber 1954 and1961). Here the fact is, and Weber was well aware of it, that Europeans throughout history and especially in the Middle Ages were constantly forging *new legal structures* that created *non-denominational* rules and procedures that applied to ever greater sections of the European community. Non-westerners and some polemical writers often forget that Western Europe was anything but a unified entity in the early modern period. The Reformation itself spawned both the Thirty Years War and the Hundred Years War. In the end, the resolution of those conflicts entailed major theological rethinking as well as the fashioning of new rules such as the *Laws of War and Peace* accomplished by Hugo Grotius (in 1625). Likewise, the English Act of Toleration (1689) was a step toward building harmony between various Anglican, Protestant, and Dissenting groups.

But much earlier than that, efforts to create inclusive rules of legality are to be seen throughout the Medieval period. Indeed, those Europeans had a genius for such creations. The fashioning of the law merchant is the most dramatic example of this. For what the legists set out to do was to create a system of rules and regulations that applied to all commercial and maritime traders whatever their nationality, whether they be Italians, Frenchmen, Englishmen, Spaniards or whomever (Berman 1983, Chapter 11). Hence there was a powerful *universalization* and *internationalizing* of legal rules and procedures originating in the Renaissance of the twelfth and thirteenth centuries that has been carried forward intermittently to the present era of globalization.

An examination of the great revolutionary reconstruction of Western Europe in the twelfth and thirteenth centuries shows that it witnessed sweeping legal reforms, indeed, a revolutionary reconstruction, of all the realms and divisions of law—feudal, manorial, urban, commercial, and royal (Berman 1983)—and therewith the reconstitution of Medieval European society. It is this great legal transformation that laid the foundations for the rise and autonomous development of modern science, but also the rise of parliamentary government, the very idea of *elective representation* in all forms of corporate bodies, the legal autonomy of cities and towns, and a vast array of additional legal forms unique to Western law. It is here in this legal reconstruction that we find the first effort to a separate Church and State. The idea was that just as the Church would follow its own law (i.e., Canon law), so too secular authorities would establish their own civil laws and courts.

At the center of this development one finds the legal and political principle of treating collective actors as a single entity—a *corporation.* The emergence of corporate actors was unquestionably revolutionary in that the legal theory which made them possible created a variety of new forms and powers of association that were unique to the West, since they were wholly absent in Islamic as well as Chinese law. Furthermore, the legal theory of corporate entities brings in its train *organizational* principles establishing such political ideas as *constitutional government, consent in political decision making,* the *right of political and legal representation,* the *powers of adjudication and jurisdiction,* and even the power of *autonomous legislation.* Aside from the scientific revolution itself, and perhaps even the Reformation, no other revolution has been as pregnant with new social and political implications as the legal revolution of the European Middle Ages. By laying the conceptual foundations for new institutional forms in legal thought, it prepared the way for the two other revolutions.

Moreover, since 1993 when this author first pointed out the powerful sociological significance of the corporate form of legal autonomy and its absence in Islamic and Chinese law (Huff 1993/2003, pp.129ff and 143–46), other scholars have shown in detail how this absence was a major cause of the economic divergence leading to a very long period of stagnation in the Muslim world (Kuran 2005; 2010).[3]

THE TRIPLE REVOLUTION

With this summary we see that the unparalleled legal revolution in Europe of the twelfth and thirteenth centuries implied the creation of numerous *innovative conditions* in three major spheres of societal development: the emergence of representative democratic politics, the creation of new economic rights and actors (corporate entities of many sorts) and scientific development. Behind all this was the Papal Revolution of the late twelfth and early thirteen century that had set off this legal reconstruction that radically altered the foundations of virtually all social, political and economic relationships. In thinking about this, the comparison should not be made with current democratic institutions and their practice, nor contemporary science and its ideals, but with the *contemporaneous institutions and practice of the early modern era of the three other civilizations* (China, India, and the fledgling Ottoman Empire).

The theory of corporate existence, as understood by Roman Civil Law, and refashioned by the Canonists and Romanists of the twelfth and thirteenth centuries granted legal autonomy to a variety of corporate entities, such as cities and towns, charitable organizations, and merchant guilds, as well as professional groups represented by surgeons and physicians. Not least of all, it granted legal autonomy to universities. All of these entities were thus enabled to create their own rules and regulations, and in the case of cities and towns, to mint their own currency and to establish their own courts of law. Nothing like this kind of legal autonomy existed in Islamic law or in Chinese law or Hindu law of the era (Huff 2003 chapters 4 and 8).

As concerns both the economic and political spheres, the theory of corporate existence, as understood by Roman civil law, distinguished between the property, goods, debts, and liabilities of the corporation and those of individual members. A debt owed by the corporation was not owed by the members individually. Likewise, *ownership* of property by the corporation was not equivalent to *jurisdiction* by the head of the corporation because those empowered to adjudicate within or for the corporation were distinguished from the *owners* of the property. Most importantly, the allegiance of the individual members was said to be to the corporation, not to other members of the corporation personally. These ideas served to create a foundation for a *public* versus a *private* sphere of action and commitment—a distinction still lacking in many parts of the world (Huff 2003, pp. 145f).

In short, the theory of corporate existence that was worked out uniquely by the twelfth and thirteenth century Western European legalists (but neglected by the Byzantines) created a whole new *bundle of rights*. These included the right to own property, to have representation in court, to sue and be sued, to make contracts, to be consulted when one's interests were affected by actions taken by others, especially kings and princes following the Roman legal maxim, "What touches all should be considered and approved by all" (Post 1964, chapter 4; Berman 1983, p. 608 n54).

Altogether these breakthroughs created foundational conditions for new forms of citizenship and for economic development. As a Dutch economist put it, this revolution set the stage for the "long road to the industrial revolution" (J. L. van Zanden 2008, pp 200f and 294; and also de Moor 2008).

Of course it was a slow process putting all these new ideas into action, but the door had been opened, a new legal framework had been gestated, and in the centuries to come, those ideas and forms of legality were to be transported across Europe, around the world, and now are found among the implicit assumptions of international law and globalization.

SCIENCE, UNIVERSITIES, AND THE AXIAL SHIFT

But let us not lose sight of the breakthrough conditions that enabled the scientific revolution to flower in the 16th and 17th centuries. In conjunction with the great translation movement of the eleventh and twelfth centuries a vast new cargo of Greek natural philosophy (enriched by Arab philosophers and commentators) arrived that Europeans, unlike their Muslim counterparts south of the Mediterranean, quickly made the center of their university curriculum.

The Europeans refocused the curriculum of the universities around the three philosophies—natural philosophy, moral philosophy, and metaphysics. Then they placed at the center of this new curriculum the natural books of Aristotle. These included his *Physics*, *On the Heavens*, *On Generation and Corruption*, *On the Soul*, *Meteorology*, *The Small Works on Natural Things*, as well as biological works such as *The History of Animals*, *The Parts of Animals*, and *The Generation of Animals.* It is with these books, Ed Grant observed, that we find "the treatises that formed the comprehensive foundation for the medieval conception of the physical world and its operation" (Grant 1984, p. 78). The study of these was indeed a core experience that was essentially scientific. Put differently, the Europeans *institutionalized* the study of the natural world by making it the central core of the university curriculum.

On the other hand, Islamic madrasas were concerned first and foremost with the "transmitted sciences" or "religious sciences." These were based on the Quran and the traditions of the prophet Mohammad (hadith studies) as well as Islamic law. None of the natural sciences as understood by Aristotle were taught therein (see Huff 2010, pp.153–58).[4] Furthermore they were pious endowments, not legally autonomous entities.

Moreover, the universities developed a *method* for taking up and disputing all sorts of naturalistic questions. It was a product of the lectures that the masters gave. It was based on compilations of the summaries of major physical questions, as well as original treatises. This literature and its use resulted in a concerted form of skeptical probing of a large set of questions in the natural sciences—physics, astronomy, cosmology, mechanics, and so forth. These probings included questions about the ultimate constitution of nature and the conditions of its transformation. Questions were asked about whether the world is singular or plural; whether the earth turns on its axis or is stationary; "whether every effecting thing is the cause of that which it is effecting"; whether things can happen by chance; whether a vacuum is possible; whether the natural state of an object is stationary or in motion; whether luminous celestial bodies are hot; whether the sea has tides; and so on for virtually every known field of inquiry (Grant 1974, pp. 199–210). And what is more, the European universities were unique possessors of Aristotle's canon of logical argumentation unknown in China. It is hard to imagine a more concentrated diet of scientific questions about the natural world and how it works.

In short, European intellectual inquiry was on an altogether different path than was to be found in the Islamic world or China. By digesting the corpus of the "new Aristotle" and its methods of argumentation and inquiry, the intellectual elite of medieval Europe established an *impersonal intellectual agenda* whose ultimate purpose was to describe and explain the natural world in its entirety in terms of causal processes and mechanisms. This disinterested agenda was no longer a private, personal, or an idiosyncratic preoccupation, but a publicly shared set of texts, questions, commentaries, and in some cases, centuries old expositions of unsolved physical and metaphysical questions. These inquiries set the highest standards of intellectual investigation. By incorporating the natural books of Aristotle into the curriculum of the medieval universities, a *disinterested agenda* of *naturalistic inquiry had been institutionalized*. It was institutionalized as a curriculum, a course of study, and it was this curriculum that remained in place for the next four hundred years in the European universities. It thereby laid the foundation for the breakthrough to modern science. It did so by *instilling a profound skepticism* about the sources of knowledge and *a deep curiosity about the workings of nature*. It inculcated a spirit of scientific curiosity that was unmatched anywhere in the world, as we shall see.

It may be correct, as Nelson suggested, that various intellectual breakthroughs occurred in "the logics of thought and action" in this axial shift, and helped transform European *naturalistic thought* permanently. At the same time, it was the decisive legal advances and their incorporation into European juristic practice that facilitated the institutionalization of new modes of thought that paved the way to modern science as well as modern economies.

MODERN SCIENCE AND INTERCIVILIZATIONAL ENCOUNTERS

Let us push this comparison one step further. As Nelson suggested, one of the most powerful comparative historical tools one can employ is that of intercivilization encounters. If we can find episodes in which a segment of European culture was intruded into a non-Western civilizational context, we would be in a strategic position to scrutinize the reactions of others to what Europeans apparently value and achieved. Likewise, one gets a better sense of what non-European thinkers valued and wished to preserve.

One fascinating window for exploring such an encounter concerns the telescope that I call *a portable laboratory* and *discovery machine.* As is well known, the telescope was invented in 1608 by a Dutch spectacle maker by the name of Hans Lipperhey. When Galileo heard about the invention but could not acquire a copy of the device for himself, he made his own instrument in the summer of 1609. During that fall he made a number of major astronomical discoveries. These included the discovery of the rough cratered surface of the Moon, the four satellites of Jupiter, the phases of Venus, and the "handles" of Saturn that turned out to be *rings* of ice and debris. Galileo also deserves credit for seeing sunspots and claiming that they are in fact eruptions on the surface of the sun. When most of these discoveries were announced in 1610 by Galileo in his little book titled *The Starry Messenger* (Galileo 1610/1989), they set Europeans on fire with enthusiasm for astronomical exploration. Consequently, secular and religious scholars, ambassadors, and merchants all attempted to acquire telescopes and to join this new inquiry, hoping to prove or disprove Galileo's claims. It also stimulated long-term efforts to built bigger and more powerful telescopes. In a word, the new telescopic device set Europeans on fire with curiosity and during the 17th century helped to transform astronomy from a slow plodding field into a new exploratory science constantly looking for new discoveries.

THE TELESCOPE GOES ROUND THE WORLD

For present purposes, we need only to ask ourselves what might have happened if the telescope were taken to China, Mughal India, and the Ottoman Empire during this same period of time, as indeed it was. Briefly the story tells as follows.[5]

The earliest certain transmission of the telescope to Asia appears to have occurred in 1613 when a Dutch sea captain brought it to Japan. In 1615, the British Ambassador, Sir Thomas Roe, presented a telescope to the Mughal Court of Sultan Jahangir at his headquarters in northern India. In the same year Chinese scholars could read a preliminary account of Galileo's celestial discoveries written and translated into Chinese by Portuguese Jesuits. By 1618 a newly designed "Keplerian" astronomical telescope arrived in China with a new batch of missionaries. By the 1630s the Chinese had a dozen or more treatises written in Chinese by the missionaries that explained the construction and use of the telescope, along with extensive explanations of Western astronomy.

In the Ottoman Empire, soldiers and merchants rubbing up against Europeans may have encountered the spyglass early on, but certain knowledge of the transport of the telescope to Istanbul had to wait until the late 1620s and early 1630s, when we know for sure that telescopes were in use in Istanbul. We know this because a Venetian merchant was executed for training his spyglass on the royal harem. Likewise, we know that the Ottoman Sultan Murad IV (r. 1621–1640) and his brother Sultan Ibrahim (r. 1640–1648) both possessed telescopes that they used to spy on Europeans and their own subjects. Furthermore the Ottoman navy had telescopes in the hands of their commander by the 1650s.

In India, by mid-century the French Jesuits, along with British traders, brought more and more telescopes, some of them six and twelve feet long, to the Mughals. But neither the Mughal scholars nor the Hindu rulers made any use of the telescope for astronomical purposes. When the Hindu ruler, Jai Singh started building his extraordinary stone and masonry observatories in Delhi and Jaipur in the early 18th century, he made no use of telescopes despite advice from Jesuit missionaries. Consequently, the observatories using ancient Hindu techniques and symbolism (though modified by Islamic sources) were less accurate than what had been achieved by Europeans decades earlier.

In short, all three of these non-Western civilizations had possession of multiple European telescopes by the 1630s, yet during the 17th and early 18th century their scholars made no effort to use them systematically (with a notable minor exception in China). They made no additions to astronomical knowledge, no improvements in the design of telescopes, nor any contributions to the development of modern scientific theory (for details see Huff 2010, chapters 4 and 5).

From this multi-civilizational encounter we can draw many lessons, but the most important is that outside Europe there was little interest in the cutting edge of scientific research as it was unfolding in astronomy and would culminate in Newton's *Principia Mathematica*. The new science of mechanics set out there, governed by the law of universal gravitation, was entirely made in Europe. Whereas European university graduates and laymen of all sorts were filled with curiosity about the natural world, outside Europe that interest was entirely stifled. Moreover, it was not just with regard to astronomy that we find a lack of curiosity for the same applies to medicine and human anatomy, microscopy, hydraulics and pneumatics as well as electrical studies (see Huff 2010, chapters 7–12).

The foregoing discussion suggests that civilizational analysis has much to offer by way of elucidating our past. It provides indispensable insights into how our world got to be the way it is—and others quite different. Will the same frames of reference help to chart the way of the future?

"THE NEXT CONVERGENCE"?

Although the historical record clearly shows a great divergence between "East" and "West" (Pomeranz 2000; Bayly 2000; Kuran 2011) going back, not just to the 18th and 19th centuries but all the way to the Medieval period, Benjamin Nelson highlighted the processes of universalization and internationalization. The gaps between East and West might be great, yet his faith was in the good intentions of historical actors who would fashion new forms of "universal brotherhood." This was not to neglect the likelihood of wars, injustice, or inequality but the reflection of a faith in positive human agency.

In his new book, *The Next Convergence,* Michael Spence (2011) has given us a rich overview of global economic development since 1945. From his point of view the world is undergoing a convergence whereby developing countries (sometimes called "the rest" of the world [Amsden 2001]) are now growing at an unprecedented rate (7 percent or more decade by decade) and which if continued for the next forty years would place them in the charmed circle of "developed" countries. Their per capita incomes would reach about 20,000 or above, which is the range of currently "advanced" economies (Spence 2011, p. 19ff).

From the perspective of current globalization theory Nelson's work and the details I have filled in will appear as the "pre-history of globalization." Nevertheless, what was wrought represents indispensable aspects of the universalization and coordination of human action in many realms that underlie what may be a new "convergence."

An array of European legal innovations created the foundations for a new system of international law and vital linkages to the continuing globalization of our planet. Beginning not later than the twelfth and thirteenth centuries efforts to create the merchant law and a variety of new legal concepts paved the way for both self-governance and interstate cooperation across Europe. New and indispensable concepts were forged that laid the foundations for legal autonomy, for a hierarchy of courts and legal jurisdiction, sovereignty, and democratic representation based on the idea of what touches all should considered and decided by all. All of these ideas created the basis for parliamentary government, something missing in other parts of the world, even in the 21st century. Whether it be a business organization, a governmental organization, a charity, professional association, or a municipal entity, it is clear that the fictive idea of a corporation, a legality autonomous entity, has been and will continue to be indispensable for political and economic action around the world. These foundational ideas did spread across the world and have provided a basis for greater economic and political cooperation. In Nelson's phrase, Westernization has become internationalization.

In the 20th century the fundamental ideas of Western and international law were put forth as the *United Nations Universal Declaration of Human Rights* (1948). Eleanor Roosevelt, who was so instrumental in getting this universal accord created and adopted by the United Nations, believed that it provided continuity back to the *Magna Carta, The French Declaration of the Rights of Man and Citizens* (of 1789), as well as the American *Bill of Rights* (Glendon 2001, p. 173; also Donnelly 1989). Although this was a Western initiative, representatives of the fifty UN members participated in the drafting of the document so that the vast majority of countries around the world today, including China, Muslim, and Middle Eastern countries, are signatories to the UN Declaration and its extended protocols.

Among the elements of that declaration we find freedom of conscience, of religion, of expression and association, as well as freedom of movement and many others. But even before this declaration, many international non-governmental organizations had come into existence with the goal of more broadly instituting international legal and scientific standards (Boli 2000; Driori, et al 2003, Chapter 3).

KNOWLEDGE ACCUMULATION AND TRANSFER

On the other hand, scientific discovery (and technological innovation) has been one of the great motors of historical change. The huge intellectual capital that Europeans built up during and following the scientific revolution of the 16th and 17th centuries must be credited with creating the possibilities of a great many technological innovations. These include the steam engine (not possible without the discovery *that air has weight*), electricity, electrical lighting, and the electric motor. None of these would have been possible without the scientific advances of the 17th and later centuries, and which had no counterpart outside Europe.

Seen from this comparative and civilizational point of view (and not a state-centered perspective), the industrial revolution and its sequel would not have been possible without the foundations laid by Western universities and the scientific outcomes that they made possible. In contrast to the globalization of so many aspects of Western and international law that are now taken for granted, the creation of first rate universities outside Europe and the West has been problematic. Even at the end of the 20th century, there were no world-class universities outside the West, not in China, India, or the Middle East.[6]

Accordingly, one of the questions that will need to be studied more intensively is that of the transplanting of universities outside the West, along with the inculcation of the vital spirit of scientific inquiry and its supporting institutional mechanisms. This is a realm where major developments emerge only very slowly and yet have profound effects.

Great hope has been placed on China and India as sources of major new technological innovations, yet so far that promise is far from being fulfilled.[7] At the same time, it has been assumed that if India and China continue to grow at 9 to10 percent each year, they will soon "rule the world" (Jacques 2009) and all great economic advances will come from them. Just how important these inquiries are can be seen from the following and a parallel from the past.

THE WIRELESS WORLD

Scholars recently have been claiming that up until the 18th or 19th century, China was just as economically advanced as Europe (see Pomeranz 2000; Wong 1997; Goldstone 2008; and for a summary of this research, Zanden 2009, pp. 275–82). Since most of those scholars ignored the fact that Europe had built up an extraordinary surplus of human, intellectual and scientific capital, along with a revolution in literacy, it comes as a great surprise that Europe took off in the18th century (or much earlier) and assumed an ascendancy of at least three hundred years (Buringh and Zanden 2009). Currently, "Real output per capita in Western Europe and North America is more than ten time that of many less developed countries" (Allen, et al. 2005, p. 1). In the meantime, China, India, and the Middle East did not fully close the literacy gap with the West until the end of the 20th century, and then not even in most parts of the Muslim world (see Tables E-1 and E-2 in Huff 2010, pp. 311-313; and UNHDR 2002).

For these reasons, it is important to carefully monitor the *differential* developmental paths of these contrasting parts of the world. The "Arab Spring" is a hopeful occurrence, yet the fact is that none of the Middle Eastern countries now going through this transformation is likely to be on the same rapid path of economic, scientific and technological advance that developing countries elsewhere will be experiencing. Turkey, not an Arab country, will probably be the Muslim exception in the Middle East. Likewise, Malaysia and Indonesia seem more promising outside the Middle East.

Furthermore, when it comes to transforming scientific discoveries into major technological and commercial breakthroughs, there is a formidable time lag. Consider for example, wireless communication. If we ask, who invented *wireless* communication and what scientific discoveries made wireless communication possible, the answer points to the late 19th and early 20th century. It was the German scientist Heinrich Hertz, working with the implications of James Maxwell's mathematical equations, who discovered that wireless signals are indeed transmitted by certain electric circuits (1888). Much more had to be done to get a practical application such as Marconi achieved in the 1890s.

If we jump forward to the present and ask what scientific breakthrough made wireless communication possible as we now know it, then the answer would be quantum mechanics. What scientists in the 1920s and 1930s discovered (along with the very existence of microparticles such as the electron), is that those particles travel in ways very different from ordinary size bits of matter, grains of sand and planets. The paths of microparticles defy the logic of classical physics. Nevertheless, it was discovered that under certain conditions and in certain materials, electrons can be knocked free producing a cascade of miniature communications. This was the idea behind the invention of solid state semiconductors.

In this case, the inventors of the transistor (Shockley, Bardeen, and Brattain in 1947) had been studying the latest findings in quantum mechanics (Weiner 1973). The upshot was the invention of the solid state transistor which would have all the functionality of a vacuum tube, and lots more. Once perfected (ca.1955–1957), it was realized that these solid state devices could hold many transistors. This heralded the invention of the microchip, or a computer on a thumb-sized piece of silicon (Handel 1967:80–92; Reid 1985/2001).

To shorten the story, it is only because of all these miniaturized circuits that we can have a computer small enough to fit into your cell phone and lots of other very small devices. The point is, however, that the whole computer revolution in its miniaturized form was highly dependent on the discoveries of quantum physics and a host of other technological breakthroughs. Of course, vacuum tube-based computers were made but they took up whole rooms, needed constant "debugging," used huge amounts of energy, and so on (Ceruzzi 1998). In a broader sense, the electronic revolution of the 20th century was a product of scientific discoveries whose real roots were in the 19th century or very early 20th century; yet their technological applications unfolded only during the second half of the 20th century. The lesson from this might be that neither China nor India, much less the Middle East, is likely to be the sponsor of a major scientific-technological revolution such as this in the 21st century. It is also probably true that the major scientific and technological breakthroughs that will shape the 21st century have already been made. We just do not know which they are.

CONCLUSION

Let us return to Michael Spence's thesis regarding a new convergence. From the point of view of a century long cycle of closing the economic gap between developed and developing countries that appears to have started after WWII, Spence's account seems highly plausible.

Nevertheless, Spence is aware that the likelihood of maintaining the high-speed pace of growth now seen among the developing countries and leading to economic convergence is both unprecedented and unlikely to be sustained without many detours. He also stresses the fact that getting to convergence requires far more than old fashioned "balance sheet" considerations. There are many unknown challenges. Both questions about ecological sustainability and the need for a constantly renewed knowledge base form critical parameters that must be dealt with in the next fifty years. The financial collapse of 2008 and its aftereffects suggest that the regulatory structures for the global economy are far from adequate and that trust in political leaders and economic advisers has greatly diminished.

Furthermore, the growth and development patterns around the world are very uneven. The Arab Middle East, for example, does not appear to be on this fast track at all. India has lagged behind China, but is still in the ball game and is expected to make sizeable gains in the next decade. Africa is probably somewhere in the middle, but suffers from major issues of corruption and self-governance. Still several African states have made remarkable economic progress recently.

Spence notices with approval that the G-7, the largely European group of countries that used to meet to guide world economic development, has quietly but very quickly been expanded to the G-20 as a result of the financial collapse of 2008.[8] This expansion, bringing many developing countries to the table of world economic leadership, can be seen as another example of universalization that Nelson often looked for.

Nelson predicted that the Protestant ethic would move to all parts of the world, though surely not massively, yet significantly. There have been many reports of a growing influence of Christianity and especially Protestantism in China. Weber's *Protestant Ethic and the Spirit of Capitalism* has been translated into Chinese and is often available in conspicuous places (Stückelberger 2005). Other reports suggest that even some officials have encouraged this "foreign import." Debate and discussions on this topic have taken place in universities and in high places. The book titled *Jesus in Beijing* (Aitkins 2006) makes strong claims about the growth of Christianity in China and its possible long-term influence (also see Spegele 2011), which of course remains highly problematic.

In Latin America many versions of Protestantism have appeared over the last several decades while the Pope's visit to Brazil in 2007 had overtones of trying to halt the spread of Protestantism and Pentecostalism there (Rohter and Fisher 2007). Even in the Muslim world a number of scholars find a variant of Weber's "worldly asceticism" making inroads and inspiring a new civic-mindedness (among others, see Ozdalga 2006; Yavus and Esposito 2005; Findley 2011).[9] In all these cases the new spirit of civic responsibility and diligence in the service of higher purposes may provide an extra impetus to growth and development. Still other Protestant life ways are spreading rapidly in some parts of the developing world (Gravois 2011).

The global community has in many ways stockpiled a great corpus of intellectual, scientific and legal capital that can be put to work integrating the various regions of the world economically, politically, and electronically. Surely the information technology needed for global communication at the highest levels is rapidly being deployed across our planet. Cell phone networks which are now so much less costly than landlines used to be, have radically altered the situation that we used to call the "digital divide" (Spence 2011, pp. 241–43).

Global politics, however, is no small matter and it remains to be seen whether international leadership can measure up to the challenges. There are world-scale problems with terrorism, international crime, immigration from underdeveloped to developed countries, insufficiently regulated markets, shortages of material assets and resources, including water, oil, and minerals, as well problems of climate change.

Will 2050 bring the great convergence, or will the accretions of centuries of differences of languages and religions, customs and alternative norms that crystallize into *civilizational congeries,* still play a major role in global politics and outcomes? Is China as well-positioned as pundits opine (Jacques 2009) to guide economic, political, and scientific development internally as well as internationally (cf. Suttmeier and Cao 1999; Cao 1999; Kerr 2007; McGregor 2010; Shambaugh 2008)? Will Chinese intellectuals and Party leaders seek to revert to an imagined past of greatness located in its ancient traditions, or will they attempt to remake China as a new liberal and progressive state fully compatible with international norms?

Will the "new Argonauts" (Saxenian 2006) working in Silicon Valley and similar locations in the United States be able to transfer high levels of entrepreneurship and cutting edge technological innovation back to India, China and other parts of the developing world, or will the United States remain the best place for them to thrive? These are questions for the future.

NOTES

1. Of course if it's true that Merton's own interests were broader as seen in his classic study, *Science, Technology and Society in 17th Century England* (Merton 1938/1970); and Merton (1973).
2. There is but one article on law in the *Globalization Reader*, and it is barely over four pages long. Its references do not go back further than the 19th century. See David Held (2000, pp. 165-671).
3. It is well worth citing Kuran's summary here:

   > In the 19th century, financial reforms in the Middle East included the legalization of interest, the establishment of secular courts, and banking regulations, all based on Western models . . . Islamic law blocked evolutionary paths that might have generated financial modernization through indigenous means. Sources of rigidity included (1) the Islamic law of commercial partnerships, which limited enterprise continuity, (2) the Islamic law of inheritance system, which restrained capital accumulation, (3) the waqf system which inhibited resource pooling (Kuran 2005: 593).

4. There has been such fundamental misunderstanding of this point that I should point out (as I have done in Huff 2010, pp. 13–14) that *neither* mathematics nor astronomy was considered a natural (or philosophical) science by Aristotle. Mathematics was subordinate to philosophy and astronomy was simply a mathematical discipline, basically geometry, concerned with making geometrical models. No one, including the Arab astronomers, until Copernicus and his followers, claimed to be proposing true models of universe. The citation of Galileo's uncompromising position on this noted earlier makes this radical turn unmistakable.
5. I omit here the many references and other details that can be found in (Huff 2010).
6. This is not a very well studied subject, though studies do exist. See Cole (2009). For the international study of university rankings: www.arwu.org/ARWU2008.jsp (retrieved August 8, 2011). For a sketch of university developments in the Muslim world up to the 20th century, see Huff (2003, Epilogue).
7. Some reports suggest that China hopes to make major progress in nanotechnology; (among see Appelbaum and Parker 2008; and Tom MacKenzie 2009).
8. This group now includes the following: Argentina, Australia, Brazil, Canada, China, France, Germany, India, Indonesia, Italy, Japan, Mexico, Russia, Saudi Arabia, South Africa, Republic of Korea, Turkey, the United Kingdom, and the United States of America. (www.g20.org/about_what_is_g20.aspx)
9. This movement inspired by Fethullah Gülen is truly impressive, so much so that many Turks and some commentators worry that it is becoming too influential, even in the United States (See Saul 2011; Kinzer 2011, especially pp. 38–39).

REFERENCES

Aikman, David. 2003/6. *Jesus in Beijing*. Washington, D.C. Regnery Publishing.

Allen, Robert C., Tommy Bengtsson, and Martin Dribe, eds. 2005. *Living Standards in the Past: New Perspectives on Well-Being in Asia and Europe*. Oxford: Oxford University Press.

Amsden, Alice H. 2001. *The Rise of "The Rest": Challenges to the West from Late-Industrializing Economies.* New York: Oxford University Press.

Appelbaun, Richard and Rachel A. Parker. 2008. "China's Bid to Become a Global Nanotech Leader: Advancing Nanotechnology Through State-Led Programs and International Collaborations." *Science and Public Policy* 35 #5: 319–334.

Arjomand, Said Amir and Edward Tiryakian. 2004. *Rethinking Civilizational Analysis*. Thousand Oaks, Ca: Sage Publications.

Arnason, Johann P. 2001. "Civilizational Analysis. History of." Pp. 1909–15 in *International Encyclopedia of the Social and Behavioral Sciences*. London: Elsevier. A longer version is in "Civilizational Analysis: A Paradigm in the Making." *Encyclopedia of Life Support Systems* (2002–2009), 34 pp. www.eolss.net.

———. 2003. *Civilizational in Dispute: Historical Questions and Theoretical Traditions.* Leiden: Brill.

Bayly. C. A. 2000. "South Asia and the 'Great Divergence.'" *Itinerario* 24 #3/4: 89–103.

Becker, Sascha O., and Ludger Woessman. 2009. "Was Weber Wrong? A Human Capital Theory of Protestant Economic History." *The Quarterly Journal of Economics 124* #2: 531–596.

Berman, Harold J. 1983. *Law and Revolution: The Formation of the Western Legal Tradition* Cambridge: Harvard University Press.

Boli, John. 2000. "World Culture in the World Polity: A Century of International Non-Governmental Organization." Pp. 262–268 in *The Globalization Reader* edited by Lechner and Boli.

Buringh, Eltjo and J. L. van Zanden. 2009. "Charting the 'Rise of the West': Manuscripts and Printed Books in Europe. A Long-Term Perspective from the Sixth through the Eighteenth Centuries." *The Journal of Economic History* 69 #2: 409–45.

Cao, Cong. 1999. "The Changing Dynamics Between Science and Politics: Evolution of the Highest Academic Honor in China, 1949–1998." *Isis* 90: 298–324.

Chase-Dunn, Christopher and Thomas D. Hall. 1995. "Cross-World System Comparisons, Similarities and Differences." Pp.109–134 in *Civilizations and World systems: Studying World-Historical Change*, edited by Stephen K. Sanderson. Walnut Creek, Ca: Altamira Press.

Cole, Jonathan. 2009. *The Great American University*. New York: Public Affairs.

Cole, Stephen and Jonathan Cole. 1967. "Scientific Output and Recognition: A Study of the Operation of the Reward System in Science." *American Sociological Review* 32: 377–390.

Ceruzzi, Paul.1998. *A History of Modern Computing*. Cambridge: MIT Press.

Donnelly, Jack.1989. *Universal Human Rights in Theory and Practice*. Ithaca: Cornell University Press.

Drori, Gili S., John Meyer, Francisco O. Ramirez, and Evan Schofer. 2003. *Science in the Modern World Polity: Institutionalization and Globalization*. Stanford: Stanford University Press.

Duchesne, Ricardo. 2011. *The Uniqueness of the Western Civilization*. Leiden: E. J. Brill.

Galileo Galilei.1610/1989. *Sidereus Nuncius or The Sidereal Messenger*. Translated with an Introduction, Conclusion and notes by Albert van Helden. Chicago: University of Chicago Press.

Glendon, May Ann. 2001. *A World Made New: Eleanor Roosevelt and the Universal Declaration of Human Rights.* New York: Random House.

Goldstone, Jack. 2008. "Capitalist Origins, the Advent of Modernity, and Coherent Explanation: A Response to Joseph M. Bryant." *Canadian Journal of Sociology* 33 1: 119–133.

Grant, Edward, ed. 1974. *A Source Book in Medieval Science*. Cambridge: Harvard University Press.

———. 1984. "Science and the Medieval University." Pp.68–102 in *Rebirth, Reform, and Resilience: Universities in Transition, 1300–1700,* edited by Pamela J. Transue. Columbus: Ohio State University Press.

———. 1996. *The Foundations of Modern Science in the Middle Ages*. New York: Cambridge University Press.

Eisenstadt, S. N. 2003. *Comparative Civilizations and Multiple Modernities.* Leiden: E. J. Brill.

———. 2004. "The Civilizational Dimension of Modernity: Modernity as a Distinct Civilization." Pp. 48–70 in *Rethinking Civilizational Analysis*, edited by Arjomand and Tiryakian.

Gravois, John. 2011."The Lions of Lagos, the Rotarians of Rawalpindi." *Washington Monthly* July/August. www.washingtonmonthly.com/magazine/julyaugust_2011/features/the_lions_of_lagos_the_rotaria030493.php?page=3

Findley, Carter Vaughan. 2011. *Turkey, Islam, Nationalism and Modernity: A History, 1789–2007*. New Haven: Yale University Press.

Hagstrom, W. O. 1965. *The Scientific Community*. New York: Basic Books.

Hall, Thomas D. and Christopher Chase-Dunn. 2006. "Global Change in the Long Run." Pp.33–58 in *Global Social Change. Historical and Comparative Perspectives*, edited by Christopher Chase-Dunn and Salvatore J. Babones. Baltimore: Johns Hopkins University Press.

Held, David and Anthony McGrew, editors. 2000. *The Global Transformations Readers*. Cambridge, U.K.: Polity Press.

Handel. S. 1967. *The Electronic Revolution*. Baltimore: Penguin Books.

Huff, Toby E. 1993/2003. *The Rise of Early Modern Science: Islam, China, and the West*. New York, 2nd edition. Cambridge University Press.

———. 2010. *Intellectual Curiosity and the Scientific Revolution: A Global Perspective*. New York: Cambridge University Press.

Jacques, Martin. 2009. *When China Rules: The End of the Western World and the Birth of a New Global Order*. New York: Penguin.

Kerr, David. 2007. "Has China Abandoned Self-Reliance?" *Review of International Political Economy* 14:1: 77–104.

Kinzer, Stephen. 2011. "Triumphant Turkey?" *New York Review* 58 #3 (August 18): 37–40.

Kuran, Timur. 2005. "The Logic of Financial Westernization in the Middle East." *Journal of Economic Behavior & Organization* 56: 593–615.

———. 2011. *The Long Divergence: How Islamic Law Held Back the Middle East*. Princeton: Princeton University Press.

Lechner, Frank J., and John Boli editors. 2000. *The Globalization Reader*. Malden, Mass: Blackwell.

MacKennzie, Tom. 2009. "China's Giant Step into Nanotech." *The Guardian,* March 26.

McGregor, Richard 2010. *The Party: The Secret World of China's Communist Rulers.* New York: HarperCollins.
Melko, Matthew and Leighton R. Scott. 1987. *The Boundaries of Civilizations in Space and Time.* Lanham, MD: University Press of America.
Merton, Robert K. 1938/1970. *Science, Technology and Society in 17th Century England.* New York: Harper & Row.
———.1965. *On the Shoulders of Giants: A Shandean Postscript.* New York: Harcourt Brace & World.
———. 1973. *The Sociology of Science: Theoretical and Empirical Investigations.* Chicago: University of Chicago Press.
Moor, Tine de. 2008. "The Silent Revolution: A New Perspective on the Emergence of Commons, Guilds, and Other Forms of Corporate Collective Action in Western Europe." *International Review of Social History* 53: 179–212.
Needham, Joseph, 1954–2004. *Science and Civilization in China*, 7 vols. New York: Cambridge University Press.
Nelson, Benjamin.1949/1969. *The Idea of Usury: From Tribal Brotherhood to Universal Otherhood.* Second enlarged edition. Chicago: University of Chicago Press.
———. 1974. "Max Weber's 'Author's Introduction' (1920): A Master Clue to his Main Aims." *Sociological Inquiry* 44 #4: 269–277.
Nielsen, Donald A. 2004. "Rationalization, Transformations of Consciousness and Intercivilizational Encounters." Pp. 119–131 in *Rethinking Civilizational Analysis*, edited by S. A. Arjomand and E. Tiryakian.
Ozdalga, Elizabeth. 2003. "Worldly Asceticism in Islamic Casting: Fethullah Gülen's Inspired Piety and Activism." *Critique* 17 (Fall): 83–104.
Pomeranz, Kenneth. 2000. *The Great Divergence: China, Europe and the Making of the Modern World Economy.* Princeton: Princeton University Press.
Post, Gaines. 1964. *Studies in Medieval Legal Thought; Public Law and the State, 1150–1322.* Princeton: Princeton University Press.
Reid, T. R. 1985/2001. *The Chip: How Two Americans Invented the Microchip and Launched a Revolution.* New York: Random House.
Ritzer, George, editor. 2007. *The Blackwell Companion to Globalization.* Malden, Ma: Blackwell.
Robertson, Roland and Catherine White. 2007. "What is Globalization?" Pp. 54–66 in *The Blackwell Companion to Globalization.*
Rohter, Larry and Ian Fisher. 2007. "Brazil Greets Pope but Questions His Perspective." *New York Times* May 9, 2007.
Sanderson, Stephen K., editor. 1995. *Civilizations and World Systems. Studying World-Historical Change.* Walnut Creek, Ca: Altamira Press.
Saul, Stephanie. 2011. "Charter Schools Tied to Turkey Grow in Texas." *New York Times* June 7, 2001, p.1.
Saxenian, Annalee. 2006. *The New Argonauts. Regional Advantage in a Global Economy.* Cambridge: Harvard University Press.
Spegele, Brian. 2011. "China's Banned Churches Defy Regime." *Wall Street Journal* July 28:1.
Spence, Michael. 2011. *The Next Convergence: The Future of Economic Growth in a Multispeed World.* New York: Farrar, Straus and Giroux.
Shambaugh, David. 2008. *China's Communist Party: Atrophy and Adaptation.* Berkeley: University of California Press.
Strayer, Joseph. 1970. *On the Medieval Origins of the Modern State.* Princeton: Princeton University Press.

Stückelberger, Christopher. 2005. "Why the New Interest in Max Weber's 'The Protestant Ethic and the Spirit of Capitalism' in Southeast Asia Today?" (China, Nanjing Union Theological Seminary Lecture, 26 April 2005). www.christoph stueckelberger.ch/dokumente_e/calvinweberchina.pdf.

Suttemeier, Richard and Cong Cao. 1999. "China Faces the New Industrial Revolution: Achievement and Uncertainty in the Search for Research and Innovation Strategies." *Asian Perspective* 23 #3: 153–200.

Snyder, Lee Daniel. 1999. *Macro History: A Theoretical Approach to Comparative World History*. 2 vols. Lewiston, New York: Edwin Mellen Press.

Turner, Bryan, editor. 2010. *The Routledge International Handbook of Globalization Studies*. New York: Routledge.

UNHDR. 2002. *United Nations Human Development Report: Arab Human Development* Report. New York: United Nations Publications.

Wallerstein, Immanuel. 1974. *The Modern World System, vol. I: Capitalist Agriculture and the Origins of the European World-Economy in the Sixteenth Century*. New York/London: Academic Press.

Walter, E. V., Vytautas Kavolis, Edmund Leites, and Marie Coleman Nelson, editors. 1985. *Civilizations East and West: A Memorial Volume for Benjamin Nelson*. Atlantic Highlands, NJ.: Humanities Press.

Weber, Max. 1904–1905/1958. *The Protestant Ethic and the Spirit of Capitalism*. Translated by Talcott Parsons. New York: Charles Scribers and Sons.

———. 1954. *Max Weber on Law in Economy and Society*. Edited and annotated by Max Rheinstein, translated by Edward Shils and Max Rheinstein. New York: Simon and Schuster.

——— 1961. *General Economic History*. Translated by Frank H, Knight. New York: Collier-MacMillan.

Weiner, Charles. 1973. "How the Transistor Emerged." *IIEE Spectrum* (January): 24–33.

Weisheipl, James A. 1964. "Curriculum of the Faculty of Arts at Oxford in the Early Fourteenth Century." *Mediaeval Studies* 26: 143–185.

Wong, R. Bin. 1997. *China Transformed. Historical Chances and the Limits of European Experience*. Ithaca: Cornell University Press.

Yavus, Hakan and John Esposito. 2003. *Islam, Turkey and the Secular State*. New York: Syracuse University Press.

Zanden, J. L. van. 2008. *The Long Road to the Industrial Revolution*. Leiden: E. J. Brill.

PART I
Introduction

ONE

On the Origins of Modernity: The Author's Point of View

I

The essays gathered in this book are a selection from the papers I have written over the last decade which are primarily addressed to the clarification of a perspective that I have been elaborating for a good many years now.[1]

From the first my work has been motivated by a need—comparable to that acknowledged about himself by Max Weber—to know how the modern and contemporary worlds have gotten to be the way they are—and are likely in their various ways to become as the earth turns. Thus, I early became intent on investigating the central social and cultural structures and processes of the complex civilizational communities and societies in the specificities of their diverse histories as well as their *extra-domestic* encounters, expansions, and interdependencies. So it is no wonder that the theories that I have been struggling to clarify present an evident contrast to those which are featured in our familiar surveys and texts. Without at any time abandoning an interest in studying the fabrics and textures of institutions and interactions from the so-called *micro*-sociological and *micro*-historical points of view, I have nevertheless become ever more aware as the years have passed of the critical importance of major processes and conflicts arising in *inter*-civilizational as well as *intra*-civilizational milieus for the general understanding of social action and cultural change.

Actually, I was drawn to this perspective as early as my first published work, *The Idea of Usury,* which had it first edition in 1949 (and its second enlarged edition two decades later). This work, which

---

Originally published as *"Einleitung"* ["Introduction"] to Benjamin Nelson, *Der Ursprung der Moderne: Vergleichende Studien zum Zivilisationsprozess* (Frankfurt am Main: Suhrkamp, 1977), pp. i-xvi. Published here in English for the first time.

will be found to bear the subtitle, "From Tribal Brotherhood to Universal Otherhood," purports to record the very extraordinary "transvaluation of values" illustrated in the exegesis of a text out of the Old Testament—Deuteronomy 23:19–20—and it seeks to explain how the double-edged commandment which bespeaks a society generally described as a tribal brotherhood was obliged to undergo opposed reconstructions, first in the medieval era and then in the days of the Protestant Reformers as a consequence of the insistently proclaimed commitment of Christianity and Christendom to the notions of universality and universal brotherhood.

That is where I first encountered in civilizational and intercivilizational contexts the differential workings of notion of universalization. It was critical to explain how it chanced in the history of the relations of societies, economies, and peoples in all of their interactions that as the range of enterprise expanded and the complexity of affairs in some way or other deepened, it got to be more difficult for people to see the world as it had been seen either by the men of the Hebrew tribe or the men of the Middle Ages. As it proved, a remarkable letter of Calvin to Claude de Sachins in 1545 came to seem a Magna Carta of a new form of Universal Brotherhood, one which I came to call *Universal Otherhood*, a Brotherhood in which, as I have written, "all men become brothers in becoming equally rather than differentially others."

Well before I completed *The Idea of Usury*, I came to feel that my work had a somewhat complex relation to the achievement and spirit of Max Weber. I sensed that the itineraries I was charting could be conceived as moments in the wider relations of the institutions and notions linked in the triangulated structure of *conscience, casuistry and the cure of souls.* My growing realization that there was no talking of one side of the triangulated structure without involving the others implied that I was resolved to give much more stress than did Weber to the critical importance of the reshapings of conscience, the inherited casuistries of opinion, and the claims of certainty and certitude by Luther, Calvin, Galileo, Descartes, Pascal and other reformers and early modern pioneers in science and philosophy for the understanding of the profound successions of revolutionary transformations in the structures of conscience and consciousness of the modern and contemporary eras.

In this spirit, I found myself doing special studies from these points of view of the meaning of the opposition of St. Bernard of Clairvaux and Abelard, Luther's attacks on the medieval organization of the Forum of Conscience and the ecclesiastical law, the trials of Galileo and the travails of Pascal.

These comparative historical sociological investigations into the cultures of conscience and consciousness, of opinions, proofs, and personal commitment proved to offer fresh ways to understanding decisive changes and new integrations of "cultural maps" and "symbolic technologies" which are now spreading across all areas of the world pursuing accelerated modernization and rationalization or their lively alternatives.

If I continued to explore what I perceived to be a "two-front struggle" to recast the fundamental rationales of both science and *praxis,* it was because I was proceeding from a strong sense that a key to our own cultural situation in the second half of the twentieth century might be found in the fusions and fissions of the myriad outcomes of the two great movements of the sixteenth and seventeenth centuries, the Protestant Reformation and the "Scientific Revolution."

At the crux of the issues I was seeking to develop there lay a far-flung postulate, the evidence for which I cannot assemble and debate now. In my view, the so-called Scientific-Technocultural Revolution through which we have been passing in our times is best understood not as a simple speeding-up or extension of the First Industrial Revolution, an economic-social revolution, however vastly magnified. Rather the so-called Scientific-Technocultural Revolution, of our days is an explosive Perspectival Technological-Organizational Revolution, one in which—as never before in recorded history—the fullest potentialities of the advanced sciences, mathematics and logic are finding expression in a sort of universal algebra (an algorithm) now being powered across the world by ramifying "Protestant" institutions, lifeways, and organizational systems.

Both of these movements, the Protestant Reformation and the Scientific Revolution—this point cannot be too strongly emphasized—involved fundamental critiques of the authoritative medieval integration of "conscience, casuistry and the cure of the soul," embodied in the *Court of Conscience.* In other words, the twofold changes in the fundamental "Moralities of Thought" and "Logics of Action" embraced in the earlier institution of conscience constitute the necessary foundations of the accelerated passage now in progress towards the universal symbolization—and automation—and collectivization of individual and social intelligence.

As different and even contradictory as they were in many of their views and motivations, there is one critical sense in which the pioneers of the Reformation and the pioneers of the Scientific Revolution converged. All the leaders—Luther, Calvin, Galileo, Descartes, Pascal—attacked the late medieval casuistry of conscience and probabilism of opinion at their very roots. Moreover, their attacks against every shade and grade of conjecturalism, fictionalism, and probabilism were put forward in the name of true knowledge, subjective certitude, and objective certainty. (The evidences for these views are offered in several papers in this volume.)

For the revolutions of the rationales of conscience to come to their fruition, the *Protestant Ethic* had to pass beyond Weber. The legacies of Luther and Calvin, Galileo and Descartes had truly to be linked together and become united. The new ethics, energies, and lifeways which had arisen in the successful Protestant Revolution had to be placed at the service of what contemporary scientists call "new world-models"—the natural philosophy, the mathematical physics, and the revolutionary algorithm of the new logics.

In short, I am proposing that the reshapings of the rationales of conscience compose fundamental phases in the makings of early modern cultures and that from the late nineteenth century forward, first mainly in Germany after an abortive episode in Napoleonic France, then in the United States and elsewhere, the by-products and the off-shoots of these makings were fused at great heats. Western civilization had to wait longer than many have supposed to arrive at the not yet believable take-offs and "exponential growth-points" of the contemporary era.[2]

As my work in this vein progressed, I became ever more firmly convinced that sociologists could not truly dispense with depth-historical inquiry into issues of this sort if they hoped to relate significantly to a series of on-going revolutions which they have so far been passing over much too lightly. An effort to repair this lack is made in the very first pages of my "Civilizational Complexes and Intercivilizational Encounters" (chapter 5) which surveys the following movements among others: the *scientific-technological-perspectival revolutions* which have been proceeding with ever more intensified pace, scope and impact since the beginning of the twentieth century; the *spatio-temporal revolution* associated with the above; the *rationalization revolutions* involved in the spread across the world of systems of instrumental and functional rationalization; the *world revolutions in the structures of consciousness and conscience* which includes conflicts within and among all strata and groups across the world in relation to each other over the by-products and issues of modernization, industrialization, Westernization and their diverse alternatives in "East" and "West" alike.

This stress suggests that, in my view, "forensic contexts" constitute a particularly strategic locus for systematic and comparative civilizational studies of transformations in consciousness and conscience. For now, "forensic contexts" may be taken to refer to the institutional settings and culture structures of disputes and conflicts, values, and rationales being litigated and fought over by whole societies, groups, and individuals. The very wide bearings of this stress on the vicissitudes of rationale-systems and decision-matrices of both theoretical proof and opinions relating to conduct and every form of praxis are illustrated in many of the essays which follow.

These strong interests in the polycentric, polyvalent, and the processual have led me to relate myself to the inheritance, including the inheritance of unresolved problems, bequeathed to us by Max Weber, Emile Durkheim, Marcel Mauss, Henry Sumner Maine, and the many present-day practitioners who have been moving toward a comparative historical, differential sociology in civilizational perspective. Having always wanted to know how and why at any given time in any given place any given set of people see—constitute—reconstruct—the world as they do, I have preferred to work as closely as I possibly could to the structures of "existence, experience and expression." A stress of this sort quickly leads one, as it has led me, to enlist the aid or Durkheim, Mauss, and the Durkheimians in the development of processual analyses of entire "civilizational communities," "whole societies," and "total social facts." I therefore have sought to do justice to recurring series of passages and reversals between varied states and structures of solidarity—between "collective" and "individual representations." Like Henry Sumner Maine and Louis Dumont, I am persuaded that a sociology or history which takes the *generic individual* as the point of departure of analysis readily falls prey to the snares of uniformitarianism, forfeiting critical opportunities to deal in a dynamic manner with complex historical accretions of structure and meaning in the central social and cultural institutions including the fabrics of law, science, literature, the arts and so on.

II

By now it must be apparent that I had come to feel that to understand developments of this scope in their depths we had to go beyond Weber's *Protestant Ethic and the Spirit of Capitalism* (1904–05) in the spirit of the new ground-breaking comparative historical studies which came from his pen between 1910 and his too early death in 1920. I refer especially to his studies of China and India in the *Economic Ethics of the World Religions;* his "Zwischenbetrachtung" or "Religious Rejections of the World and Their Directions"; and, above all, the last summing-up of his life's work, his "Vorbemerkung" ["Author's Introduction"] (1920) for the posthumously published *Collected Essays in the Sociology of Religion* (see Nelson 1973c). I felt the need to place increased stress on the role of the revolutions of the historical systems of "rationales" both of action and thought within the processes Weber himself described under the headings of rationalizations in East and West alike.

Aspiring to carry on case-historical research in the comparative civilizational manner, I repaired to Joseph Needham's great masterwork in progress titled *Science and Civilization in China* only to find there—and in his *Grand Titration*—challenges no less profound than those of Weber. The results of this encounter are reported in my essay on "Sciences and Civilizations, 'East' and 'West' " (chapter 10 below), which sets forth a number of my recent hypotheses on the distinctive trajectories so far exhibited in the civilizational histories of China and the West. I consider every one of these theses as a site for searching questions and further research:

Am I justified in contending, as I believe Max Weber did contend, that the crossings of peoples and patterns occurring in the midst of intercivilizational intersection issued in great numbers of extraordinary changes in the history of Europe, such as the distinctively autonomous "middle-class" Occidental city, different from cities elsewhere in the world; the unprecedented Occidental *university* (not college) which had few parallels elsewhere before the nineteenth and twentieth centuries; "natural" theologies that proved to contain the seeds of a universal mathematical physical science; a "moral" theology, moral philosophy and law, which were to issue in universalistic ethical and juridical structures and systems?

Does the evidence not indicate that from the twelfth century forward in Western Europe universalities and universalizations came to play an increasing part in all sorts of central settings and structures especially in the most decisive spheres of theology, philosophy, law, science?

Are we not safe in provisionally following the lead of Weber in his view that until now universalities and universalizations have achieved a unique degree and extent of universalization in the Western European World? Is it not the case that the distinctive features of civilizational structures which have characterized the modern West, for better or worse, rest upon special features of Western civilizational achievements of ancient Greece and Rome; the universalizing horizons, not altogether perfectly illustrated in the Hebraic world, and certainly the insistently continuing stress upon universality in Christianity?

To all these questions and to others in the same vein, I have replied in the affirmative without for a moment supposing that these crucial traits of the West have existed in themselves, for themselves, by themselves in some empyrean of concepts or heaven of ideals. This viewpoint hardly deserves to be called "idealist" because it allows that cultural and civilizational heritages play a notable part in historical process, nor will I allow that the postulate of distinctiveness of civilizational patterns is in any way an assertion of ethnocentrism or European triumphalism. Again and again in the papers which follow, I stress the *inter*civilizational character of civilizational and cultural production, the mutability—even precariousness of civilizational patterns, the endemic character of *intra-* as well as *inter-*civilizational conflicts, the shifts in intercivilizational ascendancies. The tangled vicissitudes of these universalities and universalizations in the more recent histories of the West and East alike are discussed below (in chapter 5).

To return now to the paradoxes proposed by Needham: As I read the evidence, the movement toward a universalizing science occurred in the West before it could gain momentum in China, this despite the fact that, as Needham shows so clearly, China attained a high level of technology before the West did. A particularly important key to the development was found in the differential responses of so-called "East" and "West" to the pressures in the direction of widened fraternity and expanded universality.

A critical element in this process was the emergence of patterns of interaction and communication which spurred recourse to universalizing logics of decisions in the intellectual-theoretical spheres as well as in the *practical-ethical juridical-spheres.* As I have already noted, the rationales related to these two spheres were grafted onto the same root, the notion of *conscientia,* on the base of which there evolved from the twelfth century forward an integrated set of teachings and institutions which came to play a central role in Western European medieval life and thought.

The striking idea of the individual relating to his conscience in all the spheres of thought and action was the pivotal assumption of this structure. The *concrete* individual was presumed to have access to *universal* norms of the utmost generality.

So far as one can now tell, the situation tended to have a different aspect in the "classical" histories of many non-Western lands. In China, for example, save for a few widely separated thinkers there was little continuing reinforcement for, or institutionalization of, universalistic notions. The Confucian stress on the maintenance of properties and pieties in the five sovereign relationships tended to restrict greatly the access to universalized norms.

Fuller research, reported herein and elsewhere, on these relations in comparative historical terms—especially in China, India and the West—further confirms this finding. A vigorous two-way flow—a sort of "double dialectic of double universalization process"—is not to be expected where rigid sociopolitical barriers resist passages to wider society and community and where the generalization of ethical and knowledge systems are strongly restricted by the stringencies related to the structure and employment of the symbolic technology.

The tangled relations among the processes of fraternization, universalization, and rationalization are reviewed in two of my latest essays: "On Orient and Occident in Max Weber"; and "Vico and Comparative Historical Civilizational Sociology" (Nelson 1976a,c).

III

A new essay by Jürgen Habermas—"Geschichte und Evolution" (Habermas 1976, pp. 200–59)—speaks directly to these issues and refers to my own writings in this context. I cannot here deal in detail with all the questions which are at stake in Habermas's wide-ranging discussion. I would, however, wish to clarify a few points where he appears to me to misread my intentions or to propose an alternative to my views which I am reluctant to accept.

Let me say first, however, that I am happy to subscribe to a good number of the principal points of Habermas's analysis. I would especially single out in this connection his strong emphasis on the principles of organization of societies. I, too, believe that the technological is only one of the basic dimensions and I concur in stressing the importance of (moral) regulation of interaction and symbolic structures which allow individuals and societies to attain and retain their identity.

I find myself differing with Habermas's emphasis, however, when he takes leave of the known and knowable historical structures out of preference for a general classificatory model describing the ontogenesis of knowledge-structures and role competencies. On this base he raises some problematical questions concerning the tenor of my essay on Needham and Weber.

It is not my personal predeliction but rather the evidence of the historical developments which requires the references to the differentially specific civilizational settings of later developments toward wider universalizations. The position developed there cannot by any stretch of imagination be called "idealist" nor can it properly be conceived to be Eurocentric.

Habermas takes exception to my calling attention to the oft-cited importance of the Greek and Roman and Judaic-Christian sources of European universality from the twelfth century forward. Every one who has sought to understand the different patterns of development of Chinese and Western science has had to ask about the relative significance of this fact.

Here, in my view, Habermas has been led to extremes by his

intense commitment to an evolutionary—as opposed to an historical—perspective. However, it hardly helps to refer to an antecedent *a priori* transcendental model to ground the claim that all "High Cultures" following the so-called Axial period have an equal potentiality to universality. On this score it is no easy matter to tell by what logic or with what evidence it is possible for him to describe and explain the diverse patterns of cultural and social structures at allegedly the same levels of evolutionary development in different parts of the world at different times.

Habermas misses some interesting turns in the way in which Jaspers describes the meaning of the changes wrought in the so-called Axial era (see Jaspers 1949, pp. 19-29). The fact that there were significant passages in high cultures toward a widened access to the difference between *mythos* and *logos* or a greater awareness of the reflective consciousness may not be taken to mean that the philosophical, scientific, juridical, and religious structures of China and India were the same as those that were developed in classical Greece. To treat these differences as irrelevant to the understanding of the subsequent civilizational patterns of "West" and "East" is to go against the evidence. I am forced to conclude that in the matter of the meaning and institutions of universality and universalization process Habermas is a "loose constructionist" and I a "strict constructionist." Latent universality and embodied and articulated universality are very different notions and realities.

In my view, an excessive price is currently being paid wherever theorists and researchers neglect comparative historical perspectives in favor of the ontogenetic epistemologies of Jean Piaget and Lawrence Kohlberg (see Kohlberg in Mischel 1971, pp. 151-235). The specificities of the value orders of different civilizational communities can hardly be deduced from abstract schemas of moral development or evolution.

The essays I have written in this sphere start from an assumption that Habermas supposes I am denying. I altogether believe that what he is describing as the universalistic potential is not a peculiarity of the Western tradition but, to use his terms, is the potential in all those world images that originated between 800 and 300 B.C. in China, India, Greece and Israel. I go beyond this and suppose that, in terms of pure potentiality, all peoples may be said to be possessed of a potential to engender universal structures and yet so far, for better or worse, few have gone very far in articulating and institutionalizing universalities in the spheres of social relations and cultural designs. To say without accompanying explanatory references that "we are not able to explain why the structures of consciousness, potentially usable in several places, have only been used in one place for the mastery of evolutionary demands and have been institutionally embodied" (Habermas 1976, p. 242), is to allow the impression that little or no research of value has been produced in this area. The alternatives proposed by Habermas himself for the explanation of the Western case prove schematic, falling short of attaining a requisite level of specificity or probability to allow detailed discussion here. He seems, indeed, to be propelled into unlikely conjectural histories by the very strength of his commitment to his transcendental model. His "equal potentiality" thesis is expressed so strongly at times as to raise the question whether from his general viewpoint the central sociocultural structures of all the high cultures—notably China—may, to all intents and purposes be taken to be the same as those that were developed in classical Greece. To treat these differences as irrelevant to the understanding of subsequent civilizational patterns of "West" or "East" would be to set a great deal of history aside.

The more one studies the new literature on social evolution the more one is tempted to conclude that "social evolution" gives the impression of a *post facto* simplified summary or model of how allegedly peak developments are said to have occurred. Willy-nilly, "social evolution" seems to incline toward *unilinearism*. The fact that the current views generally prefer to identify themselves with *multilinearism* does not alter the point here.

All of the particular values that historians and historical sociologists place upon specificities of historical experience are lost in the evolutionary reconstruction. The following regularly get washed out in the evolutionary summary of what has transpired: 1) the sense of history as project; 2) history as construct and construction; 3) the *experience* of history in the intentions of actors; 4) the imperative importance of historical memory for personal and group identity; 5) history as tragedy, as comedy, as farce; 6) history as tradition.

The evolutionary position as currently constituted substitutes the ontogenesis of knowledge structures and the notion of genetic epistemology for the careful study of actual structures of consciousness in all their diversities and complications. In the new evolutionary view little effort is made to attend to the specificities of social, cultural, political, intellectual history, above all, the half dozen (or more) particular *milieus* or *settings* of process, interaction and experience. Families, kindred lineages, localities, nation-states, societies, civilizational structures constitute our different but interpenetrating settings which need to be entered into our equations. Only by introducing these differential structures do we leave place for the perplexities of choice.

It is not easy to understand how a theory which describes itself as Marxist, however qualified, would lack for ways of giving greater weight to the specificities of actual histories of actual groups, struggling over divergent interests, contending for domination, evolving new structures of rationales, rationalities, engendering institutions, organizations, ideas and justifications.

The new evolutionary view seems to have relatively slight interest in conveying the character of conflicts, the transformations of consciousness and conscience which have occurred from one time to another in one place or another. The abiding interest of the new evolutionary view is to identify the forms of models which appear to be embodied in the evolutionary—not empirical—histories of men. The deep intentions of individuals and groups struggling to express value are set aside as being without relevance.

Some last words about these issues may be allowed at this point: The evolutionary position as now espoused by Habermas and his associates is not readily reconciled with the intentions or achievements of men like Weber, Durkheim, Marcel Mauss, Henry Maine, Joseph Needham and many others who have sought to serve a purpose related to my own. It is not without interest that these men needed to oppose the many misconstructions which regularly appear to follow from too freely and loosely expressed an evolutionary position.

In closing I would like to express the hope that these essays will be found to comprise a contribution to the disciplined study of what will one day—before too long—be found to constitute a "*new* science" of civilizational analysis. It is a science which is a structure of investigation having key anticipations in Max Weber, Maine, Durkheim, Mauss and others, but achieving its most interesting contemporary exemplifications in the work of a number of men who have given us outstanding studies on many historical civilizational complexes and patterns in the past two decades. My list of major contributors would include the following: Joseph Needham, Gustav von Grunebaum, Joseph Levenson, Etienne Balazs, Marshall Hodgson, Clifford Geertz, Louis Dumont, Lynn White, Jr., and others named in the following pages.[3]

It will remain for the days ahead to spell out precisely what relations obtain between civilizational analysis or civilization-analytic sociological research as I and others conceive it and the more established types of sociology. Already the evidence is overwhelming that many of the main challenges confronting sociology in our days cannot be met without recourse to civilizational and intercivilizational perspective.

With the help of the aforementioned sociological—and other historical—pioneers, I have been striving to carry out detailed studies in the comparative historical, differential sociology of civilizational patterns and intercivilizational encounters. With these frames as my reference points, I have sought to throw light on the differences in the passages to—and from—modernity in "East" as well as "West" in the spheres of law, conscience, consciousness, science and in the images people have regarding state, society, community, authority, individual, future, freedom, and so on.[4]

NOTES

1. A number of points considered in this Introduction are broached in a conversation I shared with Niklas Luhmann at the Graduate Faculty of the New School on December 11, 1975; see Nelson (1976f).
2. I have drawn freely here on my essay, "Scholastic *Rationales* of 'Conscience,' Early Modern Crises of Credibility, and the Scientific-Technocultural Revolutions of the Seventeenth and Twentieth Centuries" (Nelson 1968a). A related statement appears in chapter 3 below.
3. Some of my reasons for not including the names of renowned civilizationalists in the above list will be found in my paper, "*De Profundis . . .*: Responses to Friends and Critics" (Nelson 1974d [this volume, chapter 13]).
4. I would like to thank Professor Habermas and his colleagues at the Max-Planck-Institut zur Erforschung der Lebensbedingungen der wissenschaftlich-technischen Welt in Starnberg, especially Wolfgang Lepenies, Wolfgang Krohn and Gernot Böhme, for graciously inviting me to participate in lively discussions at Starnberg.

   I am grateful for the aid and counsel of many friends, students, and research assistants. Of these I would now name only a small number: Charles Lawrence and Jerome Gittleman of The Graduate Faculty of the New School for Social Research; Arthur Smith and Wallace Davis of the State University of New York at Stony Brook.

PART II

Conscience, Cultural Systems, and Directive Structures

TWO

Cultural Cues and Directive Systems

It has been by custom and law determined
for the eye, what it is supposed to see and not to
 see,
for the ear, what it is supposed to hear and not to
 hear,
for the tongue, what it is supposed to say and not
 to say,
for the hand, what it is supposed to do and not to
 do,
for the foot, whither it is supposed to proceed and
 not to proceed,
for the heart, what it is supposed to want and not
 to want.

ANTIPHON[1]
(480–411 B.C.)

INTRODUCTORY

Cultures beg to be understood as symbolic designs—complexes of dramatic, defensive, directive, and economic systems.[2] Living actors take, play, make, and are made by roles. The social-cultural-role settings in which we chance to locate largely establish the repertoire of parts we come to perform in the course of our lives and the "batteries" of cues we encounter. If we are to be able and willing to act our assigned and elected parts, our directors have to want and know how to assist in the development of directive programs which are felt to possess a favorable balance of stability, consistency, and congruity with our lived experience, feeling of fitness and sense of "reality."

Originally published as "Actors, Directors, Roles, Cues, Meanings, Identities: Further Thoughts on 'Anomie,' " in *The Psychoanalytic Review* 51, 1 (1964): 135–161. An earlier draft was read at a panel of Section H (Anthropology) of the American Association for the Advancement of Science, meeting in Philadelphia December 30, 1962, under the title "Social Structure, Cultural Process, Personality System: Boundary Paradigms." Reprinted with permission.

Whenever these frames of existence and reference go out of phase, rhyme and reason are felt to vanish from the world. Readily available script samples cease making sense. *Alienation* and *anomie* reign. In their perplexity and rage, actors cast about for new plays with new directors, roles, and rewards. The show goes on somehow. . . .

Such is the Round of Life viewed as Theatre and Game under the greater or lesser dominion of the Wheel of Fortune. Now to translate our central themes into the more formal idiom preferred by those—sociologists, anthropologists, and psychologists alike—who are counted specialists in the study of the relations of social structures, cultural processes and personality systems:

Cultural signals and emotional attitudes figure as "inputs" and "outputs" in every instance of social action and social exchange. These communications—their patterns, "meanings," "circuit flows"—have yet to be precisely charted in the light of available "grids of analysis," both old and new.

Toward this end, I draw freely in what follows, upon a host of sources: Durkheim, Freud, Talcott Parsons, Kenneth Burke, Robert K. Merton, Erik Erikson and other renowned explorers of the central social and psychological frames of reference, including contemporary learning theorists, information theorists, socio-cultural analysts of dissociative process, and others.

The celebrated schemas of Merton (1938, 1959) and Erikson (1956), which are currently setting the patterns for studies on the causes of anomie and "crisis of identity," are examined in the first section of our paper (I) and are found to need relocation in a larger context of an evolving framework more congruent with the original properties of Durkheim than either Merton or Erikson are discovered to be.

A more persuasive *general theory* than is now available depends on the development of sound guidelines for fresh research. The suggestive models of human action proposed by Kenneth Burke and Talcott Parsons are reviewed by way of introduction (II). The following new themes are then treated in sequence:

- III: Cultures as symbolic designs, especially as symbol economies and directive systems.
- IV: The six classes of cue–systems embraced in the directive systems.
- V: Varieties of the system-states assumed by the directive cues when these are analyzed for the measure of their relative consistency (emotional as well as logical), their stability, and their congruity with the perceiver's sense of fitness and "reality."

Our closing pages seek to express the ways in which alienation, identity, meaninglessness and *anomie* appear in the light of the new perspectives.

I. ANOMIE AND IDENTITY: MERTON AND ERIKSON

Professor Merton's renowned essay "Social Structure and Anomie" (1938, 1959) has served as a platform to sociologists for over two decades now. The reason is not hard to find. Aiming at a distinctively *sociological* approach to *anomie* and "deviant" behavior, Merton pitted Durkheim against Freud and appeared to succeed in bracketing psychological factors as "residual categories." In the view of the present writer, this outcome represented neither of the alleged antagonists, Durkheim or Freud, with faultless fidelity.

Two features of the American social and intellectual environment of the 1930s seem, especially, to have stirred Merton to develop a fresh point of departure. The prevailing ways of explaining social phenomena were transparently defective, a hotch-potch of simpleminded biological and psychological notions. Even professional social scientists could not be counted on to speak sensibly about the sorts and sources of deviance. In place of all this, Merton worked to direct attention to the intrasystemic mechanisms of social and cultural deprivation. These systemic structural strains *rather than* intra-psychic fantasies or intra-organic developments needed to be identified as the conditions which prompted men in their different ways to enter upon deviant paths.

Deviant behavior, Merton added, needed to be recognized as *socially structured* behavior. Ours was a society in which everyone was directed to one clearly prescribed cultural goal, Success. The approved opportunities, however, for the achievement of success were differentially and invidiously distributed. In other words, the varying rates and sorts of deviance traced to the notable discrepancies between the fully institutionalized success norms and woefully under-institutionalized legitimate means for the fulfillment of these goals.

The response of sociologists to Merton's new orientation was both quick and enthusiastic. "Pure sociology" was now new born. Its professors had now acquired a fresh starting point, a master-hypothesis which seemed capable of generating scientific investigations of previously inchoate masses of data without the necessity of resorting to the extraneous principles of biology, psychology or physical anthropology.

Until recently it was only rarely noticed that Merton's schema represented a translation of Durkheim's perspective into terms which expressed the urgencies of America in the 1930s. The awareness of this aspect of Merton's paradigm is now becoming more general. We freely admit that a prime element in our own impulse to restate this problem is the sense that these frames do not seem to do justice to the social and cultural experiences of the 1950s and 60s. The strains of the recent period carry us beyond the image elaborated by the Mertonian school of newer models of analysis. Some of these, it will prove, involve the recapture of slighted perspectives in Durkheim's original terms of reference.

In current sociological parlance, Durkheim thought he had proved that rates of anomic suicide—not quite the same as Merton's cases of *antinomic* deviance—varied inversely with the *measure* of *integration within* as well as *among* the central coordinating systems—the cultural, social, and personality systems. *Anomie* for Durkheim refers to at least three related sets of conditions, three associated system-states:

(a) the absence of a determinate structure of compelling goals;
(b) the lack of effective order in the regulative mechanisms of society;
(c) confusion in the individual's sense of experience and self.

Against this background, Merton's influential schema may now be recognized as a foreshortening of Durkheim in the interest of consolidating a *purely sociological* model of explanation.

A fresh (and promising) start in a desired new direction was luckily made by Erik Erikson. Coming from a broad background of psychoanalysis, anthropology, and cultural history, Erikson fastened on a theme neglected not only by Merton, but strangely by Freud himself. Indeed, Erikson's new frame of reference carried Freudian notions far beyond the original limits of the libido theory, offering extremely valuable cues for charting the sequence of phase-specific challenges to the sense of identity encountered by individuals in their passage through the psycho-socio-cultural—as well as the psychosexual—stages of the life-cycle.[3]

Yet it will be observed, Erikson referred hardly at all either to the structural determinants of "crises of identity" or the sociological theories of *anomie*. His neglect of the research inspired by Durkheim and Merton posed a mark of the weakness of communication at the frontiers of the social sciences.

The paradigms presented here constitute an effort to repair this unhappy outcome of disciplinary isolationism. By way of introduction, we begin with a familiar pair of models of human action proposed by Kenneth Burke and Talcott Parsons.

II. MODELS OF ACTION: BURKE AND PARSONS

Kenneth Burke (1946a, 1946b, 1961) has proposed an ingenious dramatistic model in various writings under the name of the *pentad,* a cluster of five terms naming the elements of every dramatic representation of action. This *pentad,* which irresistibly calls to mind the title pages of theatrical playbills, appears to Burke to be a compelling point of departure for the *natural historical* study of human action in everyday settings.

In this schema, we begin with the ceaseless flow of (1) *Acts* impinging upon the senses, demanding to be understood and explained.

Four additional notions, Burke continues, arise in any examination of the context and meaning of acts:

Every act occurs upon a sort of stage, which can be defined as a (2) *scene.* A scene, in the view of the present writer, needs to be thought of as a place and a time with its setting and profile of resources and stringencies, non-material as well as material. (A quick way of indicating the suggestiveness of this model would be to observe that monocausal climatological theories, for example, represent claims that *acts* are simple functions of *scenes.*)

Upon this scene we always find an (3) *actor* (agent) or *actors* (agents) inevitably acting in relation to some other subject *(alter),* communicating through symbols, using certain sorts of (4) *agencies* (implements, artifacts, vehicles of various sorts), doing all this with some end in view, for which we may use provisionally the troublesome word, (5a) *purpose,* or under the influence of some set of antecedent circumstances, for which we may use a no less vexing word, (5b) *cause.* (There is no need at this time to trace out the associated notions of intention, motive, function, etc.)

I do not see how anyone can map any system of action without referring to all of the aforementioned notions: Act, Scene, Agent, Agency, Purpose, Cause. Burke's own analysis moves forward from this beginning to a discussion of the combinations and permutations of the "ratios" of his pentad of main terms: act/scene, scene/agent, scene/agency, agent/agency, etc.

We will not follow him here. We turn, instead, to the *four-systems model* of Talcott Parsons, which lends itself more readily than does Burke's convergent schema to *marking out the boundaries of the several sorts of socio-cultural analysis.*

In the view of Professor Parsons (1957), one who seeks to explore the determinants of behavior of any individual would be wise to name or make reference to components of *four major systems:*

1) The first system is the biological-physiological *system specific for the human species* and the behaving organism. Here, in my view, one must be careful to stress the fact that an individual's physiological system has both generic and unique features. To emphasize the former at the expense of the latter is to blight our hope of understanding human behavior at the very outset. Each human being is possessed of biochemical individuality and idiosyncratic as well as species-specific patterns. Nature no less than nurture is a source of individuated behavior patterns.[4]

2) The *personality system* denotes the behavior-disposition pattern of the individual. For the present purpose, we may highlight the fact that a given individual with a given physiological system develops his *character* in response to the cultural cues that are mediated to him by representatives of the social system.

3) The *social system* is referred to when we mean to designate a network of individuals and collectivities, the clusterings of *egos* and *alters* in organizational frameworks defining statuses and roles.

4) The term *cultural system,* in the restricted sense here intended, names units of another sort, that is signs and symbols. (Material artifacts, techniques, and above all, learned patterns of shared habituations that form part of the cultural heritage are reserved for treatment elsewhere as categories on the boundaries of the social and cultural systems.)

For our present purpose, the units of the social system are individuals, groups, social relationships; the units of the cultural system—always superorganic—are signs and symbols (Kroeber and Parsons 1958). *The culture system is mediated to the individual through the social system.*

From this point forward, I shall focus on neglected aspects of the "culture system."

III. CULTURE AS SYMBOLIC DESIGNS AND DIRECTIVE SYSTEMS

Vigorous discussions over the past decades by Kluckhohn (1952b), Kroeber (1952a, 1952b, 1958), Parsons (1951, 1959a), Hallowell (1959) and others emphasize the need for renewed exploration of the concepts and contents of culture. Happily, one-sided reductionist accounts of culture as projective responses to the social relations of production or the mothering patterns are passing out of fashion in favor of approaches doing greater justice to the regulative functions of culture. There is a fresh readiness to defer the fixing of causal explanations until due attention has been paid to the horizons opened by other perspectives which may be described under the names of the intentionalistic, cognitive, and the configurative approaches.

Four among the numerous connected ways of interpreting the more or less articulated structures or "universes of meaning" more familiarly known as "symbolic forms" (Cassirer 1953, 1957; Benedict 1934; Langer 1942) which culture comprises, seem especially relevant to the present purpose. I shall refer to these central perspectives in the following ways:

a. Culture as a *Dramatic Design,* serving to redeem time from the sense of flux by investing passage and process with the appearance of aim, purpose, and historical form.

b. Culture as *Defensive System,* comprising an array of beliefs and attitudes which help to defend us against vexing doubts, anxieties and aggressions.

c. Culture as a *Directive System,* that is, as complexes of instructions charging us to perceive, feel, think and perform in desired ways.

d. Culture as a *Symbol Economy,* that is, a value enterprise-organization whose primary resources and net incomes are symbols of varied worths. Society is here perceived to constitute an inevitably scarce supply of coveted symbols.

Reserving the last mentioned approach for the closing pages, I shall briefly characterize the first three perspectives in turn—I make no claim that I escape overlap:

(1) Culture always cries out to be regarded as symbolic form translating experience as *dramatic design.* Depending upon one's perspective, mood or philosophic tradition, the design is either celebrated as the ultimate revelation underlying all appearance or exposed as sheer convention barely concealing the void of chance. On this view, culture in the sense of form is man's supreme, albeit most ambiguous, discovery. Were it not for the intervention of human concern, the flux of nature and time would seem without distinction or direction. Events intrinsically empty of meaning or at best agonizingly equivocal in implication achieve the status of a representative symbol; come, indeed, to constitute a higher Truth through the human device of Postulation and the human production of consensus induced by postulation (Nelson 1963a).

To study culture in this spirit is to study the complex processes connected with the invention, attribution, coordination, and action of meaning. "In the Beginning," we are reminded, "was the Word," and by the power of the Word (Burke 1936, 1961) the chaos of Existence is converted into a cosmos of culture. Forever after, Nature is left to imitate Art and Illusion to interpenetrate "Reality."

This sense of culture as form has been expressed in radically different ways. On the one hand, there have been Plato, Kant, Hegel,

Cassirer (1953-7), Huizinga (1924), Santayana (1905-6, 1922), Whitehead (1933), Suzanne Langer (1942). On the other hand there have been the young Hegelians (Loewith 1941), Kierkegaard, and the Existentialists, Vaihinger's *Philosophy of "As If"* (1911), Pirandello, Sartre (1953), the contemporary leaders of the Theatre of the Absurd (Esslin 1961; Nelson 1963a). In our own country, a related outlook has been elaborated by the "symbolic interactionists" building on the foundations of George Herbert Mead and his followers (A. Rose 1963). Very recently a fresh effort has been made to do justice both to the natural origins and the ideal possibilities of this two-sided "precarious vision" (Berger 1961; Becker 1962; Burke 1954).

(2) The second way of talking about culture—as a *Defensive System*—owes its recent accent to Freud (1930) and Malinowski (Parsons 1957), who tended to emphasize the defensive function of all cultural elements. On this view, we never truly understand what any cultural element comes to mean until we recognize the bio-social framework (Lasswell 1932). Every sphere of culture proves to contribute to the commanding task of making men more at home in the only world they inhabit. The anxieties, fears, aggressions generated in the lives of individuals and societies are perceived to be mitigated by culture's protective forms.

Although I have spoken above of the Freudian tinge of this approach, I ought in fact to say that the intimations of this view may be found among the spokesmen of the so-called Existentialist tradition—St. Augustine, the French moralists, Pascal, Kierkegaard, Heidegger, Tillich, and others (Bochenski 1947; Heinemann 1958; Kaufman 1956; Nelson 1962c, d, e, and 1963a; Passmore 1957).

(3) Under the heading of Culture as a *Directive System* I shall be speaking in what may seem to many to be the least problematical way of treating the matter, one which comes closest to the models of learning theorists, and which perhaps is more readily susceptible to the kind of articulation and research which experimental psychologists prize. I shall simply talk of culture as *streams of cues* which are mediated to individuals with a view to charging and aiding them to define and respond to any possible point of reference—sign, symbol, object, event, person, situation—they might chance to encounter or fancy. In short, the shared solutions, the learned roles and the imposed constraints shall here be reconceived simply as cues of one or another sort which are mediated in different fashions by social groups.

Before discriminating the cues into their categories, I wish to speak briefly of the process of transmission, which I have been calling mediation. Whatever society one studies, one discovers that the cues are mediated by a motley army of authorized and unauthorized groups and individuals who can collectively be described as the influential others—familial paradigms, extra-familial supervisors and cynosures, cultural paragons, mediatorial elites—The Grand Army of Officers and Aides (commissioned and non-commissioned alike) who have been trained with responsibility for the processes of induction. Whatever we call these officers and aides, they are responsible for the defense of the interests of the governing powers. They are authorities in the interpretation of scripts and the establishment of the directive programs. At any given time these officers and aides have varying degrees of *formal authority, indirect influence* or *effective power* (Bierstedt 1950) in respect to the operation of the mediation process.

Parents and parent-surrogates are authorized, formally and explicitly, to serve as officers of induction and therefore have prior significance in the mediation of the cues of the culture. To parents can be added kinsfolk, elders, teachers, public officials and all other persons who have some sort of acknowledged authority to mediate cues and develop directive programs. For the present purpose we may temporarily defer enumerating the great number of agents (including institutions) of induction who achieve power or influence in respect to the mediation process.

In elaborating this paradigm, it will be noted, I deliberately continue to include a variety of concurrent perspectives, namely officers of induction, the characteristics of the cue arrangements, the different kinds of rewards and penalties, and so on. It is of the utmost importance, in my view, to keep all of these in focus at the same time. A great many of the conflicts which produce the sense of quandary, inextricable entanglement, alienation, apathy, *anomie* and so on result from incompatibilities among these numerous points of reference. Among these incompatibilities, discrepancies within each system and among the various systems loom very large.

IV. CLASSES OF CUES AND THE DIRECTIVE SYSTEM

Culture may surely be construed as a repertoire of cues, nonverbal as well as verbal, a "directive system"[5] serving to move individuals and groups to perform in accordance with desired norms. At least six classes of cues may be discriminated. For the sake of economy of diction, I use Latin gerunds to name them:[6]

1) *Percipienda* cues—this first and most embracing class of cues comprises directives which charge us to perceive any possible object, person or occasion in one or another way.

2) *Sentienda* cues—this second set of cues directs us to have one or another feeling in relation to any possible person, object, event or

situation. (After extended analysis it has seemed prudent to consider the *sentienda* cues as a second category rather than as a subclass of the *percipienda*.)[7]

3) *Agenda* cues—this third set of cues charges us to perform or not to perform one or another act on sanction of penalty or promise of reward.

4) *Credenda* cues—are those signals or symbols which tell us what or how we ought to believe or not to believe.

5) *Miranda* cues—are those directives which define what or whom we ought to hold in awe, what or whom we ought to marvel at.

6) *Emulanda* cues—this sixth set of cues influences us to emulate persons or imitate behaviors of those presented to us as role models, social paradigms, or cynosures.

Let us now talk in turn of these various classes of cues which are likely to be recalled easily enough under the mnemonic PSACME:

1) *Percipienda:*

The first class of cues, which I am calling the *percipienda*, comprises the central frames of reference, that is, the ways in which we are instructed or compelled or persuaded to see anything at all. They constitute our generic modes of perceiving and interpreting anything of interest. Hardly an hour passes in the socialization process without the child being told what and how to see, hear, feel, smell, touch and construe.

A very considerable part of the child's upbringing consists of his being influenced to avoid or not see what is before his very nose. The withholdings of recognition until the taboo is ended are of a piece with the denials, isolations, suppressions and other mechanisms of defense practiced throughout our lives. How else explain the fact that societies and individuals do indeed have such vastly different general frames for seeing external nature?

Proponents of naive realism have a tendency to suppose that what is "out there" is seen by everybody in the same way at all times, but it is surely clear at this late date that culturally-determined elements enter into the organization of anything and everything which is perceived. Under the *percipienda* one would be tempted to include the central themes of the categorial structures of societies of every description. The pioneering—but not always conclusive—efforts in this domain are associated with the names of Dilthey, Nietzsche, Scheler, Karl Mannheim, Max Weber, O. Spengler, B. Whorf, C. Lévi-Strauss, A. O. Lovejoy, L. Spitzer, S. Pepper, M. Dvörak, and others, including Clyde Kluckhohn, Florence Kluckhohn, Mircea Eliade, and Meyer Schapiro.

2) *Sentienda:*

Recent researchers have explored many impressive sorts of inconsistency, where the explicit directives—*percipienda, agenda, emulanda,* etc., point one way, the nonverbal *sentienda* cues point another way, the reward incentives remain ambiguous, and so on. Under the heading of the "double bind," Bateson, Jackson, Haley and Weakland (Bateson and others 1958) have described the price paid by individuals and societies when schizophrenogenic mothers, families, or for that matter, any other agents of induction, continually emit discrepant signals, especially *sentienda* signals, composed of contradictory or ambivalent suggestions.

Particular credit on this score is due to Breuer, Freud, and the psychoanalysts for having established the large part played by conflicted feelings and contrary *sentienda* cues in the development of personality.

3) *Agenda:*

Everyone functioning in society necessarily comes into contact with imperatives. Groups inevitably discriminate among acts requiring to be performed or avoided establishing positive or negative incentives mediated by the officers of induction with varying resorts to power or persuasion. Who has not heard of the *agenda* of a club or committee meeting?

It needs to be noted that we are often directed to do different things by the same as well as by different officers of induction. A selected sub-class of conflicts of this kind has been described as "discontinuities of cultural conditioning" (Benedict 1938). Another notable type of conflict arises from the felt discrepancies between the widest frames of encompassing principles and the seemingly *ad hoc* concessions made in particular cases on the grounds of utility, expediency, mercy and the like. This conflict has traditionally been described as the struggle between conscience and casuistry (Nelson 1963b).

4) *Credenda:*

Our fourth class of cues we are calling the *credenda*. There is no social system which does not require individuals to affirm a belief of some sort. In our own society we are hardly ever certain what it is we are supposed to believe; only rarely are beliefs explicated clearly. In any number of other societies, however, we find official philosophies stipulating explicit assent to a fixed system of doctrines, canons, or persons. Citizens of the "Free World" have a tendency to suppose that it is insufferable to be obliged to give assent to an explicit *credenda* system. As a matter of fact, there is a great deal of evidence that in many societies it is exactly the lack of explicitness that is the source of the greatest confusion and conflict. I think it apparent that there can very well be a great many sorts of incompatibilities of disharmonies between the directives of the *credenda* system and the directives of the *agenda* system. In other words, we may be called upon to believe certain things which are, in fact, contradictory to the things we are told to do; this is a commonplace in the socialization of any child.

It is impossible to exaggerate the importance for the history of world-wide political and religious affairs since the 16th century of the struggles between systems which require knowledge and faith to be explicit, e.g., Protestantism, and those which only require them to be implicit, e.g., Catholicism (Capéran 1934; Kirk 1925).

5) *Miranda:*

The fifth set of cues may be called the *miranda* cues, that is the signals directing us how to respond and act in the presence of the awesome (Otto 1917). *Miranda* refers to those persons, symbols, acts, objects or occasions we are charged to marvel and wonder at, deem sacred, that is, hold in awe. An elementary distinction needs to be maintained here—awe, marvel, and wonder involve an intensity of response, a passionate, non-reflective commitment, which may not be explicitly demanded by the *credenda* cues. Normally, in a complex culture the prescribed beliefs are subject to a certain degree of analysis. The *miranda* cues rarely admit of discussion; they lose their force and attraction if they are excessively scrutinized. Here again, as in the case of *credenda* cues, we have further possibilities of incompatibility, since at any time that any cue is mediated to us there is a concurrent message which is *invidious* in character. Some subject is set before us as an object of reverence by contrast to others not deemed entitled to any notable marks of esteem.

6) *Emulanda:*

The sixth class of cues may be named the *emulanda*. Here we refer to those persons or objects we are charged to emulate. In many societies there is no question whatever as to who are the proper persons to be in some way or other emulated. One of the critical features of contemporary society is that there are great confusions on everybody's part as to who are deserving of emulation (Nelson 1954).

Here is where we find the basis for the consideration of the role-model or, as I prefer to say, the paradigmatic element (Coleman and Nelson 1957a) in all behavioral relations. At any time, wherever there is mediation of any cue, a person who is engaged in the act of mediation is, as I say, either presenting himself as the one to be imitated, or else presenting someone else as the model for imitation. It is the paradigmatic element in all interactive situations which proves to be so critical because we absorb the cues through persons, and therefore we interpret the *agenda* and *credenda* cues through the way in which we connect with those whom we undertake to emulate.

V. CULTURE AND PERSONALITY: SYSTEM-STATES

I devote the penultimate section of this paper to a survey of the possible outcomes of and responses to the mediatorial performances. Bracketing for a moment all of the deficiencies in the performance of assigned roles which result from constitutional or learned incapacities, we can identify many possibilities. At the present time, I shall name only three of them:

(a) The cues of any system, or the cues of all the systems taken together, can be viewed in terms of their *consistency* or *inconsistency*. The shades and grades of consistency can be ranged on a continuum, all the way from perfect consistency to absolute inconsistency. It is clear that if an actor is to be able to function in relation to cues, the cues have to have a certain degree of consistency. It will be noted that I do not write "complete consistency." Just how much and what sorts of consistency may be required to cue action cannot be decided abstractly. Certainly, however, we know that complete inconsistency in the cue system, and among the cue systems, produces a sense of the absence of any kind of rhyme or reason in the directive system.

(b) Secondly, we may think of the cues as exhibiting *stability* or *instability*; that is, the cues seem to remain stationary or to undergo change at varying rates of speed. Here the pace, the character, intensity, direction of change are all critical to the picture. As a moment ago, I raised doubts concerning the need for "total consistency," I wish now to add that we must be careful not to assume that it is only where the cues are altogether stable that the actors are altogether happy. I think this is an error that has too often been made. In this model, it will be noted, the subject is not assumed to be a mechanical product of the cultural cues directed at him by any particular group. Being individuated in nature as well as by nurture, the individual has unique responses to experience as well as unique configurations of experience. On this view, we have the possibility that for an individual the cues may be excessively stable, or they may be excessively unstable.

(c) Finally, I introduce the notion of the *degree of congruity of the cues with the actor's sense of fitness of things and "reality," however derived.* Here again, it is imperative to avoid simple equations of variabilities in the cues with a feeling of incongruity. The experience may be derived from intrapsychic influences, from the proddings of the physiological system, or any other stimulus, but there is always some extra-social cultural component in the development of the sense of experience. Even though cues may be changing rapidly they may be quite congruent with the experience of the agent. In the same sense, cues may be relatively inconsistent without offending an agent's toleration. If he has come to expect a certain measure of inconsistency, he will not be too shocked.

But there is a situation which we find ourselves confronting when we encounter people who declare themselves, or who are reliably reported, to be confused as to what they see in the way of central perspectives, what or how they feel about anything or anyone at all, what is expected and so on. The total complex of cue systems mediated by the society, that is, all the cultural cues taken here together, are here experienced as *excessively unstable, excessively inconsistent and excessively incongruous with their own sense of fitness and "reality."* Where, as in this case, the "meaning systems" are no longer directive; where each is faced by what only poets and European philosophers are allowed to call an "abyss of meaninglessness"; there, power (or desire) to develop adjustive behaviors which are at all operative in the prescribed environment may be excessively weak, diffuse, intermittent, slight or altogether lacking.

Is this not the sort of situation or sense, which we have come to refer to as *anomie,* that is, a situation or sense marked by the absence of structured order, whether *objective* or *subjective* or *both?* Are these not the situations in which severe "crises of identity" are likely to occur? The modalities of the experienced feelings are exceedingly complex, extending over the entire range embracing apathy, indifference, identity disorders, dereism, depersonalization, dissociation. That the cultural cues are seriously askew these days in the sense of being excessively unstable, inconsistent and incongruous with the experiences and expectations of actors would seem to be a large part of the explanation for the fact that so many of those who are coming into psychological treatment at the present time exhibit severe confusions over norms and crises of self-reference and identity. The current psychiatric terms, "borderline," "divided selves," "existential vacuum"; the current literary terms, "beatnik," "hipster," "angry young man"; the current sociological terms "alienation," "bureaucratization of the charismatic," "heterogeny of ends" tell an important story which has yet to be generally appreciated.

SUMMARY

The present essay is part of an effort to understand the predicaments of contemporary society, culture, and persons in so far as these are fairly described by the words "alienation," crisis of identity, and *anomie*. A preliminary survey of available patterns of analyses in this area revealed an unexpected gap. Neither Merton's epochal "Social Structure and Anomie" (1938) nor Erikson's brilliant "Problem of Ego Identity" (1956) seems to range far or deep enough to constitute a *general theory* needed for the understanding of our present situation.

The way forward seems to require the recovery of certain lost accents of Durkheim's original design which had connected *anomic suicide* with lacks of integration within *as well as* among the central coordinating systems—the cultural, social, *and* personality systems. In this spirit we found ourselves construing cultures as dramatic designs, defensive elaborations, directive systems and symbol economies. Within this framework we placed our stress on the schisms in the central definitions of values and the discontinuities in the processes of articulation and communication of the various classes of cues which culture comprised. Centering mainly on cultures as directive systems and symbol economies, we observed that the directives fall into six categories: the *percipienda, sentienda, agenda, credenda, miranda,* and *emulanda*. Although it might be said that signals fusing all these directive instructions are always emitted by all actors relating to one another in any situation, it is also the case that symbol economies function as parts of stratified social and political organizations.

Throughout the paper suggestions have been developed for the ways in which the analyses of the shapes and ranges of the "cue systems" are affected by social arrangements and interests of the cadres of *influential others*—the "paradigmatic figures" (Coleman and Nelson 1957a), "officers of induction," "mediatorial elites" (Nelson 1962e), who come to play prominent parts in defining situations and affecting the life chances of others. All groups aspiring to preeminence of authority, power, influence, prestige necessarily contest with one another for control of the symbol economy. A full discussion of the bearings of the social organization of the directive systems upon the institutional processes of symbol production and distribution is reserved for a separate paper.

A FINAL WORD

However else we may regard them, we dare not forget that Human Action and Culture are Drama. If in "real life" we are not in truth "characters" anxiously in search of our Author, we are surely in search of the scripts and cues we require for the definition of the roles and the parts we are expected or disposed to play. To act at all, we have to learn to mean something to somebody. This is impossible without a meaningful script and another actor willing and able to understand us. To endure the game over long stretches of time, we need to feel that we can possess and transmit meaning, we must have title to our faces or at least have access to masks which we can wear with a modicum of assurance; we must have approved resort to identities which are not beyond the pale; the rewards and promises—"eternal" as well as "temporal"—must seem somewhat more than absurd.

Every actor necessarily requires to get hold of some sort of script if he is to perform some sort of role with a hope of being understood and validated by others. The capacity of an individual to evaluate his identity in the established "Nomenclature," as Jean Genet puts it, will vary with the states of the cue-systems and the measure of his control over processes of the definition of the situation. If individuals in sufficient number fail to achieve a favorable balance of stability, consistency, sense of fitness and "reality"—if the sense becomes widespread that those at the controls are indifferent to their plight or actively opposed to their justified claims—feelings of alienation and *anomie* make their presence felt everywhere.

Soon there is a cry from every throat: "*Things* are in the Saddle! *Whirl* is King!" The time of reckoning has arrived. "Moral chaos is upon us"; we are at the abyss of "meaninglessness"; we are on the way to some fateful new order.

One familiar scenario of our time runs as follows: The bearded prophets are the first on the set, the inevitable heralds of the new turn. Before long we are grovelling before the Hangman. When the Commissar arrives on the scene we hold our breath. But finally, after many great reckonings, we are relaxed; a new Joviality reigns. The final scenes find us laughing with the Clowns, the Acrobats, the Dancing Girls, the Puppets, and the wonderfully genial Comrade, The Master of the Revels.

The show goes on . . . somehow. . . .

NOTES

1. The above-cited Legend from Antiphon was discovered in Z. Barbu (1950, p. 19) just as these pages were going to press.
2. The professional reader will readily detect the omission here of the one perspective which has been most intensively discussed by technical anthropologists, the notion of culture as *adaptive* response. Consideration of this approach is deliberately set aside for treatment elsewhere. The current terms of *adaptational* analysis seem too imprecise to enter fully into our present discussion.
3. For this Erikson may be counted a pioneer, along with Heinz Hartmann, Ernst Kris, R. Loewenstein, and others (Rapaport 1961) in "psychoanalytic ego psychology." In his chosen domain, indeed, Erikson seems to me without peer.
4. I stress the duality in the individual constitutional endowment chiefly because there is too great readiness on the part of many writers to refer to the constitutional endowment as the explanation of the generic or common habits. Bio-chemical individuality needs to be in our picture throughout our investigation.
5. Nothing in this paradigm is intended to imply that agents (ego, alter, etc.) fully intend or know what occurs in their interactions. As Freud understood, it is hard to exaggerate the range and power of subliminal "unconscious" communication. The same awareness is not always found in current renderings of learning theory, including social learning theory, where agents are frequently assumed to know and intend their meanings. Whatever our assessment of the processes of conveyance, however, the directives may be found to fall into classes of cues.
6. Points of departure for the present schema will be found in Charles Merriam's suggestive study, *Political Power* (1934).
7. For the nonce, the wants *(desideranda)* of individuals and groups, about which social theorists from Plato to Durkheim have written with so much insight and passion, may be placed under this rubric.

THREE

Self-Images and Systems of Spiritual Direction

I. INTRODUCTION

In the midst of their incessant stresses and perplexities, men have everywhere struggled to win relief from the pangs of indifferent Fortune. Resources without number have been pressed into service in the effort to still the agonies of their spirits and, indeed, to gain lasting assurance of justification in their own eyes and in the eyes of their Divine Rulers. Among the most important of these resources have been the symbolic designs through which they have learned to express and enact meanings: images of the cosmos and universal destiny; schemes of group and personal identity; systems of self-culture and spiritual direction. It is of these last-mentioned patterns that the present essay mainly speaks.

Although we cannot now—and may, indeed, never be able to—say exactly how well the diverse systems of mental healing fulfilled the hopes and needs of the different peoples they were intended to serve, we may venture a number of preliminary observations.

At the core of all the different systems, there is discoverable a set of common concerns and procedures (Dill 1904; Eliade 1960; Harrison 1912; Linton 1956; McNeill 1951; Opler 1959; Pettazoni 1929). Nonetheless, each system of mental healing has its own metaphysical commitments; its own way of classifying the various sorts of passions and infirmities which men experience; its own roster of accredited dispensers and techniques of cures (Dodds 1951; Eliade 1959; Zilboorg 1941).

Rather than attempt to elaborate cross-cultural comparisons of these varied institutions at this juncture (Eaton and Weil 1955; Field 1961; Opler 1959; Sachs 1941; Wallace 1959), I would simply begin by

---

Reprinted from Samuel Z. Klausner, ed., *The Quest for Self-Control: Classical Philosophies and Scientific Research* (New York: Free Press, 1965), pp. 49–103. Reprinted with permission.

noting that the stresses and burdens we are called upon to bear seem to us today to stem from a variety of sources. We are prone to distinguish the following sorts of suffering.

1. Those felt to be universal experiences of every biological organism. For example: hunger, pain, sexual desire, illness, onset of death.

2. Those ascribed to the workings of central institutional structures (the economic, social, political, legal organizations).

3. Those which are felt to be interpersonal in nature and which are generally charged to the malice or ignorance of other persons—injustice, deception, treachery, privation, enforced isolation, enforced contact, loss of trust, punishment, etc.

4. Those generally ascribed to the incursions of the unconscious in the intrapsychic sphere. For example: anxiety, shame, guilt, obsessiveness, loneliness, inability to love, inability to work, feelings of meaninglessness, persecutory fantasies, convictions of omnipotence, homicidal impulses, incestuous desires, etc.

The operations of psychoanalysis may be cited as a preliminary illustration at this point. When compared with the other familiar systems, it does not in the first instance concentrate on reinforcing abilities to tolerate sufferings located in the first two categories above. There are surely many Yoga exercises (Evans-Wentz 1949; Eliade 1958) which prepare one more directly (and indirectly) than does psychoanalysis to tolerate hunger, pain, privation, disease, sexual desire. Psychoanalysis seems to concentrate on developing the power to bear burdens located in the two latter categories (the interpersonal and the intrapsychic spheres). Throughout the entire Freudian schema, emphasis is placed on coping with frustrations and anxieties arising from unconscious repressions of impulses to perform forbidden acts in relation to inappropriate objects, notably the members of one's own family. For Freud, experiences in the earliest years of life in the bosom of one's family are the paradigms (Coleman and Nelson [Nelson 1957a]) of all subsequent development: they are discovered to influence the shape and character of all successive extrafamilial adult contacts even in the seemingly most rational impersonal environments.

Freud also places great emphasis on the therapeutic importance of a maximum power to know and acknowledge one's own fantasies. Acceptance of one's inner demons is one matter, says Freud; uncontrollable compulsions to act on their every call is another. Here he breaks with those previous traditions which assimilated evil thoughts to crimes.

The implications of Freudian psychoanalysis for the distribution of values located in the second category cannot be stated unequivocally. Freud does not summon men to address themselves directly to the collective remaking of their institutions by planned political action. For this reason, social and political critics (Brown 1959; Rieff 1959; La Piere 1959) have often accused Freudians of conservative and even reactionary leanings. Yet a more disinterested view will suggest that a particular affinity for Freudian views will be found among the mobile metropolitan populations of the advanced industrial societies, which strongly emphasize the consensualistic motifs in their universalistic creeds (Kardiner 1957). It may also be noted that official Communist criticisms of Freudian psychonalysis emphasize its objectionable stresses on individualism, as evidenced in its encouragement of personal pursuit of current gratification in disregard of eventual realization of society's collective goals (Laqueur and Lichtheim 1958). Yet the incessant charges that, as opposed to the creative "freedom" ascribed to man in Soviet psychology and philosophy, "bourgeois" psychoanalysis and sociology promote "idealistic fatalism" (Bauer 1952), confirm substantial evidence from other quarters that the Soviet leaders have no desire to spread favorable attitudes to the sorts of *analysis,* whether in psychological or sociological spheres, which are favorably regarded in the so-called "Free World" (Nelson 1962e).

These preliminary references to psychoanalysis should suffice to indicate that there is no one superior medicine for every occasion. Every scheme of training has its built-in defects. We never acquire capacities without becoming unfitted for some other task. This holds as true for psychoanalysis as for any other system of self-direction.

Evidence gathered from prisoners of war and concentration camps (Beck and Godin 1951; Bettelheim 1943; Schein and others 1961; Lifton 1961) seems to indicate that three groups fared unusually well in maintaining their equilibrium in extreme situations: fervent devotees of sectarian movements, such as Jehovah's Witnesses; specialized intellectuals, notably mathematicians (Weissberg 1951), practiced in detaching themselves from external circumstances; criminal psychopaths impervious to dominant moral codes. None of these groups characteristically frequent psychoanalysis.

Indeed, we would be remiss if we failed to recognize the extent to which the accredited healers of a given society come to act as a privileged group offering their services on their own terms. The history of conflicts within and among different groups of directors needs to be seen as an illustration of the perennial conflict of mediatorial elites (Nelson 1962e, p. 18). It is, therefore, not surprising that accredited healers have seemed from time to time to be more concerned to reinforce their status-income-and-power claims than to expand the ability of their charges to direct themselves under stress.

II. CULTURE AS A DIRECTIVE SYSTEM

[If we adopt the distinctions set out in the last chapter between culture conceived as a *Dramatic Design*, a *Defensive System*, a *Directive System*, and a *Symbol Economy*, we may find a point of departure in the stream of *cues* comprising the *Directive System*. These we observed fall into six categories: the *percipienda, sentienda, agenda, credenda, miranda,* and *emulanda*.]

It hardly needs saying that (1) these six classes of cues are directed at us by agents of induction variously located throughout the complex of linked social systems: the family system, the educational system, the religious system, the political system, etc.; (2) the opportunities for discontinuities and contradictions in the communication of cues are endless (Benedict 1938); (3) The effect of these signals is notably influenced by the context, style, mood of the communication.

Specialists in the systems of spiritual direction are called in when subjects go aground in internalizing and institutionalizing these cues (Parsons and others 1961) and other experiences into workable relations to "realities." The possibilities of mishap are countless, varying from one society to another. The cues of any class or all six classes of cues may be experienced as intolerably stable or unstable, or intolerably incongruous with the subject's sense of experience (see chapter 2, pp. 29f). The agents of induction may be felt to be arbitrary in their provision of rewards and penalities. The task of the spiritual directors is made especially difficult when a state of normlessness *(anomie)* or crises of meaning and identity prevails (Durkheim 1933, 1951; Whellis 1958; Erikson 1959).

As the sympathetic reader will readily sense, the tell-tale terms in our last sentence have returned us to our original starting point. The remaining sections of our paper will lightly survey representative aspects of the self-images and systems of spiritual direction which have evolved in the history of civilization. The following plan describes our prospective itinerary:

III: "Occident" and "Orient": Some Similarities and Differences.[1]
IV: Socrates, the Stoics, St. Augustine.
V: The Middle Ages: Conscience, Casuistry, and the Cure of Souls.
VI: The Transmoral Conscience: From Luther to Freud.
VII: Psychoanalysis and 20th Century Culture.

Throughout we shall be concerned to observe the ways in which acts and attitudes are defined, meanings ascribed, identities attained, anxieties allayed. Do we need to add that we make no pretence to completeness of coverage?[2]

III. OCCIDENT AND ORIENT

It will not do, as Rudolf Otto has so brilliantly shown in his *Mysticism, East and West* (1957), to fall into the error of supposing that the images of the soul's illness and recovery vary entirely with changes of time and place. All mysticisms, he observes, have certain elements in common along with their differences. On close textural comparison, Sankara and Meister Eckhart often seem to be speaking the same idiom.

A similar observation applies to the innumerable writings on mental hygiene which succeed one another in the history of mankind (Jung 1938). It is surprising to note how many important assumptions they seem to share with respect to:

1. the origin of what we may be allowed to regard as the nuclear traumas of mankind;
2. the characteristics of what are here being called "the madnesses" of men;
3. the methods proposed for the overcoming of these madnesses;
4. the roles accorded to or claimed by spiritual directors or masters.

Yet we would be remiss if, in our concern to establish certain underlying unities of expression and attitude, we failed to observe very notable differences. By comparison to Sankara, Eckhart is inextricably Western, indeed in Otto's terms, Gothic and Faustian. Otto writes:

> His [Eckhart's] mysticism is quiveringly *alive* and of powerful vitality, and therefore far removed from "Abstraction." It is therefore also very far from Sankara and Indian mysticism, and the reason for that difference lies in the foundation from which it rises (1957, p. 168).
>
> That is indeed numinous rapture. At the same time it is subtly different from that of Sankara. This difference moreover is connected with what we have already described as the Gothic element in Eckhart's conception of God in contrast to Sankara's static Indian conception of Brahman. This distinction between the Gods occurs again in the emotions with which they are sought, striven after, experienced. For Sankara when the soul (ātman) has "come come" to the eternal Being (Ātman) it is there, it has arrived (āpta), it is at rest and fully content (sānta). But Eckhart is, in truth, never "there," never in a final static rest . . . (1957, p. 185)

Similarly, if one were asked to sum up in a phrase how Hindu and Western Christian schemas of spiritual direction differ, one might venture to say that classical Oriental methods (notably Hindu and Buddhist) seem to be directed at the overcoming of the anguish of the individual ego undergoing pain and privation in a remorseless world incapable of being notably ameliorated, to say nothing of being redeemed (cf. Weber 1946, pp. 323–359). Ultimate hope is placed in the escape from this burden by the return to the primal, undifferentiated ground where all oppositions vanish.

The Western Christian image of man's Fall from Grace (Williams 1927) begins with the same sequence: the original unity; the unclear trauma; the loss of paradise; the separation from the source of all goodness, truth and virtue; the haunting sense of alienation and estrangement. As in the East, alienation exhibits itself intermittently as:

1. over-attachment to irrelevant ideas and values which will be of no account in regard to man's eternal life;
2. under-attachment to the unfailing source of joy or peace;
3. infestation of one's spirit and the world by alien powers and noxious thoughts.

But here the accent dramatically changes and the difference from the East emerges. In contrast to a number of the ancient Oriental schemas and their neo-Gnostic expressions at explosive junctures in Western history (Grant 1961; Jonas 1958, 1960; Voegelin 1952; Cohn 1957; Nelson 1954), the world is described as good, the creation of a *good* God (Trémontant 1955). The value of the individual soul receives the strongest confirmation. The method of mitigating estrangement and madnesses are in the first place "a way *back*" to the primal undifferentiated ground. And then once more the new elements assert themselves strongly. The way back assumes the character of a "way forward," forward to the struggle for mastery of self and the world (Weber 1946, pp. 323ff).

Oriental schemas of self-direction are, in the ultimate sense, acosmic. The immense machinery set into play to liberate us from attachments to the world of passing illusions have as their primary purpose the overcoming of any sense of apartness from the One, which is all-encompassing and unchanging. Western teachings, even when they sound alike or are influenced by the Oriental works, preserve their peculiar flavor. They are, in the end, activistic. (Professor Otto's translators have written "actualistic," which seems to me not quite what is needed to express the situation.) To be sure, acosmism is a recurrent motif in all schemas of self-direction, as it is a phrase in every pilgrimage of the spirit, but quietism has never managed to assert itself as a dominant doctrine in the West (Nelson 1954).

It is instructive, in this connection, to study the details of the controversy over Quietism and Disinterested Love connected with Fénélon and Madame Guyon. Quietism was eventually condemned as a heresy for it appeared to contradict the commandment to love one's neighbor as one loved oneself. Holy egoism, the prescribed sacred love of oneself, was too strong in the Western world to be dislodged by Quietism (Brémond 1928; Kirk 1937). Mystical individualism feeds into the instrumental activism of the modern era (Weber [1904–05] 1958a).

Eastern treatises on the direction of self and others are likely to contain extremely detailed prescriptions and recommendations for the achievement of desired effects. One has only to look into any of the countless Yogi manuals (Eliade 1960), *The Tibetan Book of the Dead* (Evans-Wentz 1949), the Zen treatises (Suzuki 1956), and other works of spiritual hygiene (Jung 1938). The methods to be employed range from auto-hypnotic trances to extremely intricate sets of physical exercises intended to demolish what are sometimes called the "body armor" or the somatic resistance. There is relatively little of this in the West.

Could one reason for the contrast be the limitations placed on magic and the magical viewpoint in Western thought (Weber 1951, 1958a)? Perhaps Western postclassical, Judaeo-Christian culture is simply more generally philosophical and psychological. In truth, we have no explanation. We simply have a fact. The Western treatises go on the assumption that the "way back," which is also the "way forward," involves the journey through the "clouds of unknowing," in search of the new self.

Now and then a particular master or theorist will recommend procedures which have the ring of auto-hypnosis. Thus, for example, in the *Spiritual Exercises* of St. Ignatius Loyola (d. 1556), we have an extraordinary systematic arrangement of meditations which are aimed to lead the believer to arrive at an ultimate and irrevocable decision to be a soldier in the ranks of the good Lord and an enemy to death of the Devil and all his works. The exercises of Loyola involve repeated reflection with the mind's eye on the terrors of hell, on the sufferings of Jesus on the Cross, on the fires of purgatory, on the dread of damnation (Fueloep-Miller 1956). All else is to be eliminated from awareness in order that the experience of horror might be complete and the need for redemption might be experienced in the depths. It is no secret that Muslim elements have been detected in the extraordinary symbolism of Loyola.

In closing this section, one is again compelled to cite Otto:

If we turn again to Sankara, we can measure in full the distance between the two masters. Sankara knows the ātman in us, but this ātman is not the soul in the Christian and Eckhartian sense: it is not "soul" as identical with "Gemuet," infinitely rich in life and depth, a place of ever fuller experience and possession, an "inward man" with the characteristics of the biblical conception of this word. Least of all is his āman, "soul" in the sense of religious conscience, which "hungers and thirsts after righteousness," and for which "to be" is to be righteous with the very righteousness of God. Sankara's mysticism is certainly mysticism of the ātman, but it is not soul-mysticism as *Gemuetsmystik*. Least of all is it a mystical form of justification and sanctification as Eckhart's is through and through (1957, p. 206).

Eckhart thus becomes necessarily what Sankara could never be: the profound discoverer of the rich indwelling life of the "soul" and a leader and physician of "souls," using that word in a sense which is only possible on a Christian basis. Upon Indian soil there could never have developed this inward unceasing preoccupation with the soul's life as a life of Gemuet and of conscience, and therewith the "cura animarum" in the sense which is characteristic of, and essential to, Christianity from the earliest days. It is upon this calling as a *curator animarum* (shepherd of souls) that finally everything which Eckhart has said or done as a preacher, as a simple Christian or as profound Mystic, depends. (1957, p. 215)

And now to Greece and Rome.

IV. SOCRATES, THE STOICS, ST. AUGUSTINE

In the ensuing discussion I shall refrain from dealing with the intricate developments in the sphere of spiritual direction in classical antiquity. The few remarks I will permit myself in this connection bear upon one issue of particular interest to the argument of this paper, the recurrent rivalries among different sorts of spiritual directors throughout the history of the West.

I have elsewhere sought to show that from the time of Socrates to our own day, philosophy and psychiatry have been in a relation of antagonistic cooperation. Philosophers (for example, Epicurus, Descartes) have been as prone to proliferate psychiatries as mental healers (for example, Jung, Binswanger) have been to proffer philosophies (Nelson 1962d). These crossings of the never well-defined twilight zones inevitably occur in times of troubles when men grope for help from every source. At such times, philosophy accentuates its concern with spiritual direction, presenting itself as the true hygiene of the straying mind. Logic and physics are treated as simply the first steps on the way to therapy. Ethics becomes the quest for consolation (Dill 1904).

The different faces of philosophy (Cushman 1958; Nock 1933, pp. 164–86)—science, self-examination, therapeutic conquest of ignorance, consolation—are all mirrored in the life and thought of Socrates. An increasing stress upon the cathartic function of philosophy as a purgative and therapy is apparent in the Hellenistic schools, in the Epicurians, Cyrenaics, Cynics, Skeptics, Stoics (Arnold 1911; Lovejoy and others 1935; Murray 1925).

Through philosophy we win our way to a holy apathy and detachment; a relief from the pains afflicting anyone who sets his heart upon unattainable ends or evanescent pleasures. The task of philosophy is to present an unassailable truth free from illusion or blandishment. The Stoics bid men to live in accordance with nature and to conquer every impulse which divided them from natural law. The Cynics emphasized release from attachment to complex products and modes of satisfaction. Innumerable stories of Diogenes the Cynic connect happiness with abandonment of vain imagining and futile pretence (Lovejoy and others 1935). The Skeptics also viewed philosophy as the criticism of illusion. Their minute examination of the traps of logic and epistemology were intended to free men from subordination to painful superstition from which they could win no joy. The stress on the therapeutic role of suspension of belief recurs all through the Skeptical tradition and is perhaps seen most clearly in the writings of Sextus Empiricus (d. *ca* 250 A.D.; 1933).

One of the clearest ways of seeing the distinction between classical and patristic Christian approaches to the direction of self and mind is by comparing two sets of meditations—the *Meditations* of Marcus Aurelius and the *Confessions* (or Meditations) of St. Augustine (Dill 1904). Marcus Aurelius strives to present himself as philosopher King, a perfect Stoic. The net effect he seeks to convey is that he is possessed of an indomitable will to free himself of every infirmity and defect through his own exertions. Everywhere he looks, he sees shortcoming, pettiness and a failure to express the world spirit in action. His meditations are aimed to purge himself of every least unworthiness (Dill 1904; Oates 1940).

St. Augustine sounds a different note. He talks frankly of his boyish sins and manly passions; he humbly admits his metaphysical bewilderment and his recurring fear of meaninglessness. Feeling himself adrift at sea, he does not fear to avow his need for faith, hope, and love (Burnaby 1938; Dill 1898; O'Meara 1954; Gilson 1960).

The contrast between the Roman Emperor and the Christian Bishop has always seemed to me to have peculiar relevance for the understanding of the fluctuating sensibility of our own time.

V. THE MIDDLE AGES: CONSCIENCE, CASUISTRY AND CURE OF SOULS

A paradigm familiar to contemporary social scientists may help to express many central convictions of the medieval Christian consciousness. Freely adapting for our purposes a schema originally devised to classify the determinants of culture and personality, we may say that the Church viewed mankind and the world as follows:

Every man was in certain respects

1. like all other men,
2. like some other men,
3. like no other man.

(How odd the forthcoming details will sound to those who recall the expansions in the original essay [1948] by Clyde Kluckhohn and Henry A. Murray!)

1. *Every man is like all other men* in certain critical respects. All men are assumed to be sons of God. As such they share in the possession of reason and are answerable to God for the right use of reason. Thus all men are obliged to obey the moral law made available to them by the Law of Nature (Gierke 1900). As spiritual brothers, all men are obligated to the requirements of brotherhood (Nelson 1949a).

2. *Every man is like some other men* in the sense that they form historically separate communities within the universal brotherhood. Only Christians have received Christ and the New Testament. All Christians—and only they—are bound to preserve the true Faith, to obey the precepts of the Church and the Canon Law (Kirk 1925).

Yet Christians differ among themselves in many ways, most obviously estate and vocations. Warriors, peasants and monks have distinct "callings." Only the monks are called the "religious"; obey the counsels as well as the precepts; observe a rule; live by the triple vow of poverty, chastity, and obedience; may expect to qualify for the status of perfection.

3. *In the ultimate sense, no two men are alike.* Each is a unique person immediately responsible to God for the welfare of his soul and the well-being of his brother.

Who has not heard of the fateful medieval integration of the beliefs and sentiments stated in this paradigm? The extraordinary stress on the responsibility of each individual for the activity of his will and the state of his soul attained its height in the High and Later Middle Ages (Kirk 1927). Three sets of ideas and institutions—none entirely new in human history—were now fused into a single structure of spiritual direction never before (some will say never since) matched in complexity (Lea 1896). I refer to the beliefs and cultural arrangements embracing the determination of the individual *conscience* (Kirk 1927);

the realization of the dictates of conscience in the perplexing cases (or alternatives) in the here and now called *casuistry* (Lea 1896; Nelson 1963b); the management of errant, perplexed, and obsessively scrupulous consciences (McNeill 1951), the so-called *cura animarum* (care or cure of souls). All three of these bodies of ideas are found in many parts of the world—surely they were previously known in one or another way to the people of the ancient Orient, and, above all, to the Greeks and Romans (Thamin 1884)—but never before the Middle Ages nor after have they been so systematically elaborated in thought or so closely connected in practice. In the Middle Ages this imposing institution in its more generally known form came to be called the Forum of Conscience and the Tribunal of the Soul. It is this court which was later to become the source of the Jurisdiction of the Chancellor, "a Judge of Conscience," The Keeper of the King's Conscience, in Equity (Fifoot 1949, pp. 301–07, 321–29; Vinogradoff 1928).

Actually, conscience was the center of two related but separated institutions of spiritual direction to which different sorts of persons repaired for different reasons. Every Christian without exception was answerable to the Forum of Conscience for the sins he had committed and for the state of his soul. In this tribunal the presiding officer performed a complex of functions. He was a confessor, hearing or eliciting admissions; a judge, fitting the penalty to the crime; a physician, providing solace to the sinner without traducing the rights of God; a priest, mediating God's grace in the sacrament (Vacandard 1908).

In addition and beyond this path to perfection was another avenue for the more ardent wayfarers, those who thirsted to experience true illumination and mystical union with God. Only these were expected to engage in the *systematic* practice of meditation (Phillips 1955). Under ordinary circumstances, the deliberate quest of illumination was pursued exclusively by the so-called "religious"—monks and nuns—who strove to attain the status of perfection. The rich tradition of mystical itineraries is chiefly a monastic one until the Fourteenth century, when pious men and women sought to achieve the status of perfection without wholly abandoning the world (Egenter 1928). The significance of this desire of laymen, especially those of the Low Countries and the Rhineland (Seesholtz 1934; Clark 1949) to practice innerworldly asceticism, albeit under priestly direction, will not be lost on readers of Weber (1958a).

As must be apparent, the ruling perspectives of the two institutions of conscience were quite different. The outlook of the Forum of Conscience was predominantly legal or forensic; the sovereign end of the practice of meditation was the shedding of the old Adam, the total rebirth of the soul. In this endeavor the purgation of conscience was only the first step on the ladder. The consummation devoutly hoped for was the mystical embrace of Christ with the illumined spark of the soul.

Let us now deal with each of these institutions in turn.

Interestingly, it is the Twelfth and Thirteenth Centuries—the era of the Crusades, of Western recovery of the Mediterranean, of expanded urban liberties and mass social heresies, of vernacular literature and the new Universities (Haskins 1939; Herr 1961)—which witnessed the extraordinary advance of the new logic of conscience and the emergence of a new system of administration of the cure of souls (Lottin 1942–60). This fact alone should suffice to warn us against the naive assumption that the idea of conscience could not appear until the Reformation because of the oppressions of the Medieval Church. It was Abelard (d. 1142) who revolutionized the dialectic of moral agency and decision. The titles of his major works—*Sic et Non (Yes and No), Ethica seu Scito te Ipsum (Ethics or Know Thyself)*—powerfully dramatize his dual effort: to develop the implications of the new moral sense, to apply reason in harmonizing the ambiguities of tradition (Abelard 1935; Sikes 1932). Like Luther after him—the contrast is as compelling as the comparison—Abelard was strongly attracted to St. Paul. The dictum *Quod non ex fide peccatum est* (Rom. XIV) suddenly seemed to require a complete reinvigoration of the human will and, therefore, an exhaustive analysis of the shades and grades of evidence, opinion, knowledge, commitment. (Luther, by contrast to Abelard, drew the opposite lesson from the *Epistle to the Romans*. He became the champion of the "serf will" and the foremost enemy of a casuistry of intention.)

Long before Aquinas, authoritative medieval theologians and jurists were construing conscience as the *proximate* (not the *remote*) rule of right reason. Specialized treatises tracing the obligations of conscience in the here and now, spelling out how individuals were obligated to act in every case they encountered in the conduct of their lives, began to appear. In these works, conscience extended into every sphere of action, ranging over the whole moral life of man from the making of contracts to the making of war (Kirk 1927). After 1215, when annual confession became the obligation of all Christians, these treatises became the guides to Christian souls everywhere. The influence of handbooks on conscience survives wherever Catholic religious life is practiced.

Only one sphere, strictly speaking, was beyond conscience in the Middle Ages—the sphere of Revealed Faith. Two positions, which seem contradictory to the illumined conscience of later days, were vigorously affirmed by all the scholastic moralists:

1. everyone was under the strictest obligation to act in accordance with the findings of his convinced conscience;

2. a convinced conscience *(conscientia certa)* was not necessarily a right conscience *(conscientia recta)*. Not the individual conscience but Eternal Revelation, the natural law, the canon law and other binding rules were the ultimate imperatives of the individual conscience.

The transvaluation of the value of conscience, its detachment from the practical life of man and its expansion into and confinement to the sphere of Faith did not occur until the Reformation. The story of the storms which culminated in this situation will be discussed in our next section on Protestantism. Here we will look more closely at the development of the meditative tradition. It was within this institution that there emerged the notion of an illumined transmoral conscience (Tillich 1945) which was to prove the undoing of the Forum of Conscience and the source of the ideas of Inner Light (Burrage 1912; Woodhouse 1951) and of the Enlightenment Concept of Reason (Crocker 1959; Hazard 1953; Sampson 1956).

There is a vast literature reporting early Christian and medieval efforts to experience the vision of God and enjoy Him in mystical union. Thanks to Dean Inge (1899), Father Pourrat (1922), Evelyn Underhill (1933), Henri Brémond (1928), Bishop Kenneth Kirk, and others, we are now able to trace the development of the philosophies and techniques of meditation in the successive works of such celebrated masters of the contemplative life as the pseudo-Dionysius the Areopagite (*ca.* 500), Johannes Climacus (d. 649), Richard (d. 1173) and Hugo (d. 1141) of St. Victor, St. Bernard of Clairvaux (d. 1153), St. Bonaventura (d. 1274), Meister Eckhart (d. 1327), Thomas à Kempis (d. 1471), the anonymous author of the *Theologia deutsch* (*ca.* 1350), which left its mark on Luther.

With endless variation of images, these authors explore the spiritual ills of men, the arduous pilgrimages which need to be undergone if peace is to be won for the soul, the indispensable role of masters in the achievement of what we would call cures and insight. The psychological and religious dimensions of these works are suggested by their eloquent titles, for example: *The Celestial Hierarchy, The Mind's Itinerary into God, The Cloud of Unknowing, The God of Love, The Imitation of Christ, The Spiritual Exercises.*

Each phase of the spiritual pilgrimage is minutely examined in the light of the individual author's experience and conviction. Thus Johannes Climacus, the Byzantine author of the *Ladder of Divine Ascent,* is singlularly revealing on the subjects of gluttony, shameful fantasies, and the value of subordinating one's will to the master under all circumstances. St. Bernard (d. 1153) is supremely eloquent on the mystical love of Christ.

It is St. Bonaventura, the noted 13th Century Franciscan thinker, who provides the systematic psychological and theological analysis which helped to codify the distinctive convictions and procedures of the meditative tradition (Phillips 1955). The titles of two of St. Bonaventura's works express his central perspectives:

1. *The Threefold Way*—In this extraordinarily influential work, Bonaventura systematically sets forth the triple way of the contemplative life: the *purgative* way, the *illuminative* way, and the *perfective* or *unitive* way.

2. *The Mind's Itinerary into God*—This work, explains a recent editor,

> . . . is addressed to those who are ready to answer the divine call to live the mystical life and to taste of God's sweetness in ecstatic union (Bonaventura 1956, p. 19).

The purgative act is practiced in meditation, prayer, and contemplation. Purgative meditation has as its main object, self-examination; its main purpose is to bring to bitter consciousness the soul's moral disorder and the grave danger which it entails, thus achieving a complete detachment from all sinful inclination. Purgative prayer transforms meditation into weeping and deploring sin and into asking for mercy; its main affections are pain, shame, and fear. Purgative contemplation, finally, leads the soul from shame to fear, from fear to pain, then to imploring prayer, to rigor and severity, and finally to ardor which culminates in the desire for martyrdom, the ultimate purification of love, and makes the soul rest and fall asleep in mystical peace under the shadow of Christ.

On the illuminative way, the soul is mainly concerned with a penetration into truth. The illuminative act is likewise practiced in meditation, in prayer (which is less clearly expressed by Saint Bonaventura), and in contemplation. Illuminative meditation turns the ray of intelligence to the multitude of sins forgiven by God's mercy, broadens it then to show all the benefits of God, natural and supernatural, and finally turns it back to the Giver of all of them, Who has still greater rewards awaiting the soul in heaven. Illuminative prayer, according to Saint Bonaventura, has, it seems, as its main task to ask for mercy and help in union with the Holy Spirit, groaning in us by an ardent desire, in union with Christ by trusting hope, and in union with the Saints by their intercession. Illuminative contemplation finally leads to the splendor of truth by imitating Christ, or to be more exact, by an impregnation of our mind with the passion of Christ, and that again in seven steps: first there is a humble submission of reason to a God who was crucified, followed by deep compassion, admiration, grateful devotion, the putting on the form of the suffering of Christ, and finally, the ardent embrace of the Cross, in which and through which the splendor of truth will dawn.

On the perfective or unitive way, the soul is mainly concerned with charity. It is the perfective act that is now practiced in meditation, in prayer, and in contemplation. In meditation the spark of wisdom must be kept aloof from all attachment to creatures, must be enkindled by turning to the love of the Bridegroom, and must be elevated beyond the senses, the imagination, and the understanding into a blaze of desire for the Bridegroom who is absolutely desirable. In perfective prayer the soul is prostrated in adoration and deep reverence, in benevolence and complacence, becoming one with God in the fire of love. Here Saint Bonaventura adds the six degrees of the love of God. In perfective contemplation the soul again reaches the sweetness of love in seven degrees: vigilance for the coming of the Bridegroom is first; then confidence in Him; third, a deep longing for Him; fourth, a rising beyond oneself to the height of the Bridegroom; fifth, complacence that dwells on the comeliness of the Bridegroom; sixth, joy in the abundance of the Bridegroom; seventh, a union of the soul with the Bridegroom in the sweetness of love. (Bonaventura 1956, pp. 20–22)

We must not allow Bonaventura's theological language and homoerotic symbolism to drive us into minimizing his psychological insights and philosophical ingenuity. His exhaustive investigating of the "threefold way" can be examined with profit by contemporary psychiatrists. His depiction of the *Mind's Itinerary* and the "reduction" of the arts to theology are important steps on the way to modern thought. It is in a way accidental that Meister Eckhart rather than the Seraphic Doctor, as Bonaventura was called, had so profound an influence on the backgrounds of the Protestant Reformation. Both men elaborated the image of the soul's rebirth as a result of mystical union with God, which eventually undermined the forensic institutionalizations of thought and sentiment. Medieval illuminism provided inspiration to Luther, the revolutionary sectarians, the English dissenters, the American Quakers, and the myriad Continental Romantics whose voices sound in the philosophy and literature of the last two centuries.

The illuminist philosophies which had bloomed in the meditative tradition were to attain their fullest flowering in radical Protestant cultures and their transcendental offshoots: inner light mysticism, Enlightenment rationalism, the sundry romaticisms of the 19th and 20th centuries, including Existentialism (Nelson 1962d). Each step along the way was a blow on behalf of the liberation of conscience until at the end, the wholly unencumbered conscience was itself called into question by the union of utilitarianism, Darwinism, historicism, and Freudianism. The religion of the inner light had eventuated in the dwarfing of the moral conscience. The 20th century has been marching under the banner of "Beyond Conscience" (T. V. Smith 1934).[3]

The evolution contrasts markedly with Post-Tridentine patterns of development in the Catholic culture-areas. Surprising as it may seem to many, the bounds of "liberty of conscience" underwent audacious expansion among the probabilist theologians and jurists of the 16th and 17th centuries (Pascal [1656–57] 1920; Doellinger-Reusch 1889; Lea 1896, II, chap. 21). Paradoxically, here, it was the eventual success of the innovating Jesuit program for the "liberty of opinion" in the face of the ardent opposition of Port-Royal, Pascal, the conservative Dominicans and the Popes themselves which preserved the unique medieval orchestration of conscience, casuistry and the cure of souls.

Overlooking critical structural contrasts, a recent writer, Professor George Mosse (1957) of the University of Wisconsin, has insisted upon the identity of outlook across the barriers of the Reformation. The Puritans of New and Old England, alike, he contends—such men as William Perkins (d. 1602), William Ames (d. 1633), John Winthrop (d. 1649)—were hardly different from the Jesuits in their blend of piety, worldly prudence, policy and probabilist casuistry. Indeed, he insists, the great Puritan divines and statesmen were committed in principle, if not in name, to the program of *raison d'état* of Machiavelli (d. 1527) and Botero (1540–1617).

I reserve for the following section (and other writings in progress) the proofs of my view that these equations of Catholic and Protestant outlooks rest on faulty observation of surface resemblances. As I see it, the distinctively Roman Catholic and Protestant culture-areas have generally related to the moral and social ambiguities of the modern and contemporary eras in notably different ways (Groethuysen 1927–30).

These differences can—I am persuaded—be explained in significant measure by the different courses followed by the Roman Catholic and Protestant communities in the struggles over freedom of conscience and opinion in the 16th and 17th centuries.

Again we have run ahead of our story. We turn now to Luther and the Reformation.

VI. THE TRANSMORAL CONSCIENCE: FROM LUTHER TO FREUD

The Protestant Reformation begins a vast new experiment in the culture of the self and the systems of self-direction. The important details of the early history of these developments are not yet even now agreed upon by impartial scholars, and there is the sharpest difference of opinion as to the original associations and contemporary outcomes of the teachings of Luther and Calvin. Our own day has witnessed marked changes in the style of interpreting the meanings of the Reformation (Bainton 1956; Holl 1961). It is no longer possible to say, as used to be claimed, that thanks to Luther's renewal of

Christian liberty, the free man possessed of his own free conscience was now released from the fetters of medieval priestcraft and the superstitious doctrine of the efficacy of works. Everyone now knows that Luther was not an enlightener in the manner of Diderot and Voltaire, even of Kant (Bainton 1951). Yet, it does not seem sensible to regard the Reformation as simply a reactionary throwback to the Middle Ages. To call it an "escape from freedom," the seed bed of Nazi totalitarianism, as Erich Fromm (1941) did, is to regard the cultural circumstances from a hopelessly alien perspective.

More recent characterizations of the era by David Riesman (1950) and Erik Erikson (1958) represent an improvement in the social psychological studies of the relevance of Protestantism for the culture of character. Riesman and his collaborators relate the Reformation to the changeover from the traditional to the inner-directed social character, touching hardly at all on the unfolding of historical circumstance and teachings. Erikson interprets Luther's triumph over his agonies of conscience as a decisive episode in the forging of a new cultural identity. In truth, we are hardly beyond the infancy of our understanding of the inner history of conscience, character and culture in the modern world.

The exact influence of Luther on the notions of self and spiritual direction is no easy matter to state. The following must be counted among the decisive facts:

1. In his early years as a reformer, especially in his pre-Reformation treaties of 1520 and his appearance at the Diet of Worms, 1521, Luther assumed the posture of the Liberator of Conscience (Bainton 1950; Boehmer 1924). His condemnation of the medieval religion of works culminated in the burning of the Corpus of the canon law *(Corpus juris canonici)* and the so-called angelic *Summa* on the cases of conscience by Angelus Carletus de Clavasio. Luther thus publicly signalized his aversion to the medieval organization of the moral and religious life, above all to the triune integrations of conscience, casuistry and the cure of souls.

2. Once Luther had proclaimed the Gospel meanings of justification "by faith alone" (and true—lifelong—repentance), mandatory annual confession, and the fourfold role of the priest in the administration of the sacrament of penance were without foundation (Hardeland 1898; McNeill 1951).

3. Gone too was the basis of the medieval concept of the moral conscience and the moral effort through casuistry to make conscience operative in the world. Luther's strongly anti-Pelagian theology ruled out the concept of the attainment of Christian perfection (Nygren 1953) through the imitation of Christ, the ultimate pradigm (Nelson

1949a, Appendix). The Reformation from the time of Luther was set against the medieval system of spiritual direction.

As against these stresses of Luther, we have to recall others which present Luther in a very different guise:

a. Luther recoiled in horror from the conclusions drawn from his teachings on conscience by the left-wing supporters of his movement. In his withdrawal, he relapsed into the medieval truism that *conscientia* (conscience) was meaningless without *scientia* (knowledge), *Gewissen* was folly without *Wissen* (Castellio 1935). This endorsement of medieval intellectualism was a blow against the unrestricted emancipation of the conscience from superpersonal norms. It also allowed the continuance of persecutions for conscience, although now under the charge of blasphemy.

b. Luther's attack on casuistry was coupled with a proclamation of unqualified temporal authority in the political sphere. Conscience was now increasingly confined to the religious realm. Inner freedom and outer bondage tended to occupy entirely separate domains in Luther's stark dualism (Trinkaus 1955). It was this less familiar side of St. Paul's influence which gave lay rulers a control over men they had not had in medieval civilization (Troeltsch 1911).

Again and again, efforts have been made in the long history of Protestantism to restore analogues of the medieval framework free of the alleged medieval excesses or corruptions. These results have always been unavailing. Confession, casuistry, moral and religious counsel, organized spiritual direction have lacked for fundamental support within the framework of Protestantism (McNeill 1951; Nelson 1963a; Troeltsch 1912). The individual in Protestant cultures has the choice and the obligation of doing God's will without the aid or regulation of learned casuists, counsellors, and confessors.

According to Weber (1958a), Groethuysen (1927–30), and others, a fundamental reorientation of the social and cultural patterns of the Western world could not occur until the medieval administration of self and spiritual direction fell before the onslaughts of Luther, Calvin and their followers. So long as a distinction was made between the special calling of monks who lived "outside the world," systematically observing a rule in their pursuit of the status of perfection and everyone else *in* the world, who lived irregularly, without benefit of a rule, in the midst of continued temptation; so long was there a brake on the incentive of ordinary men and women to forge integrated characters with a full sense of responsibility. The Protestant notion of a disciplined character nourished by a resolute conscience replaced the medieval sense of life as a round of sin and penance.

This aspect of the influence of the Reformation has been the subject of continuous debate by sociologists, historians, and culturally inclined psychologists. Recently there has been a shift in the focus of discussion. Interest has lately been centering on the alleged disappearance under our very eyes of the inner directed Protestant ethic in favor of a so-called "social ethic" as a result of the spread of the power of the large-scale organization in the institutional structure of the United States. The limitations of this paper do not permit a full-scale review of the evidence in these pages. One observation may be allowed however: current discussions both of the origin and demise of the Protestant Ethic generally neglect to distinguish between the many dimensions of the problem. At no time has Protestantism been lacking for a collective church ethic. The connections between the contemporary organizational ethic and the normative patterns and life style of the Protestant church have been overlooked in the recent characterizations of Riesman, W. H. Whyte (1957), and others.

Many of the most impressive institutional consequences of the Protestant variants of conscience, character, and culture have yet to be appreciated in their full implication. A selected number of episodes will be briefly mentioned in the following paragraphs:

Perhaps the most important development in the Protestant era recalls the struggles and wanderings of the dissenting groups that came from England, Holland and Germany to the United States where the notion of the inner light was to have its foremost influence. The nonconformist illuminist sectarians effected a transvaluation of the value of conscience by subordinating the moral conscience in the medieval sense to the inner light (Burrage 1912; Cohn 1961; Woodhouse 1951). Neo-Platonic illuminism which had been so significant in the medieval practice of meditation decisively triumphed over medieval rationalism which nourished the medieval administration of the moral self and the forum of conscience. The important link between late medieval mysticism and Protestant illuminism was the concept of the spark or witness of God (Hof 1952) in the soul *(scintilla animae, syneidesis).* Modern rationalism of the Enlightenment variety owes more than most men know to Platonic and neo-Platonic mysticism. The shift from the religious to a more secular orientation occurs as early as the seventeenth century (Cragg 1950; Tawney 1926). Once the move had been made to the new notion of inner light by the dissenting groups of the seventeenth century (Solt 1959; Woodhouse 1951), the medieval orchestration of conscience, casuistry and the cure of souls was undone.

Over the centuries, Protestantism has seemed to vacillate between rationalism and fundamentalism (Troeltsch 1911; H. R. Niebuhr 1957). Although these two have been at swords point a thousand times since the onset of the Protestant Reformation, they do not seem entirely dissimilar from the point of view of the medieval concept of conscience. Rationalism is illuminism detached from its mystical source and symbolism. Forgotten is the image of the rebirth of Christ in the soul, leaving the sober afterglow of a reason freed of irrational constraints and declaring the truth by the sole authority of its inner light. Fundamentalism is biblicism, the desperate effort to maintain a fixed point of authority against the threat of the dissolution of landmarks by the work of reason. Both rationalism and biblicism have little need for the learned doctors of theology and canon laws, the learned directors of the soul who crowd the medieval scene.

Romanticism is illuminism in a new guise. It is the deification of the individual (and collective) ego (Santayana 1916) and the apotheosis of the unconscious forces which have been discovered to be the ego's foundation and underside. Romanticism joins all other variants of illuminism by rejecting the contextual integration of conscience, and casuistry, and the cure of souls. Romanticism makes the emancipated feelings the sovereign legislator for each man and for all mankind; directly applies these feelings to the complicated circumstances of the daily life resolving the riddles of tangled interest by reference to the command of love and the dictate of will; relieves itself of the need for spiritual counsel by treating explosive impulse as ultimate norm. The most impressive and fashionable expression of contemporary Romanticism is Existentialism. Reducing the matter to a simplified equation for our present purposes, we may say that Contemporary Existentialism seems to be Romanticism triply armed by three of the most forbidding constructions of modern thought: Husserl's constitutive phenomenology, Heidegger's neo-Gnostic fundamental ontology, and Kierkegaard's neo-orthodox theology of crisis (Nelson 1962d).

The nineteenth century witnessed the near demise of the older arrangement of conscience, casuistry and the cure of souls, and the surge forward of a series of surrogate religions. One of the most powerful among these new religions may be described as the *religion of the transcendental self, the transmoral self beyond conscience* (Tillich 1945). Its myriad expressions are elaborated in all the masterpieces of art, literature, philosophy, and even science. Perhaps the most revealing expressions are the intense and stark journeys into the interior which began with Rousseau, Goethe, and Fichte. Every last corner of the phenomenology of existence and spirit is probed in the pages of such philosophers and literary explorers as Hegel, Kierkegaard, Nietzsche, Baudelaire, Rimbaud, Conrad, Mann, and Joyce. We are in great need of a thoughtful full-length study of the spiritual itineraries (Hopper 1947, 1957) during the last two centuries from the point of view suggested in this paper. Without this, we can hardly hope to understand our own times (Balakian 1947; Heller 1957; Praz 1951; Raymond

1950; Sypher 1962). If we would appreciate what Freud and psychoanalysis mean in our present era we need to know *to what degree Freud is the heir of the religion of the transcendental self, to what degree he is its undertaker.* I have elsewhere suggested that a key to this riddle lies locked in Freud's "Divine Comedy" which he called by the name, *The Interpretation of Dreams.* The following lines suggest half of the answer:

> [When the *Interpretation of Dreams* was ended], there was little life left in the gallery of guises—Byronism, Promethianism, Parnassianism, dandyism, diabolism, pietism, scientism, moralism and so many others—assumed by the philosophies and substitute religions of the modern era. (1957b, p. 8)

It seems appropriate to close this section of our paper with some reflections on two historical paradoxes marking the relations of Protestantism, the major source of the religion of the self, and psychoanalysis, whose cultural implications are even at this moment being violently disputed.

1. Psychoanalysis did not originate in Protestant settings. It emerged in Catholic Vienna and its pioneer was a Jew.
2. The highest development of psychoanalysis to date has occurred in Protestant America.

So far as the present writer knows, nobody has yet thought to wonder about the first paradox. I take the liberty of putting my suspicions in the form of questions. Is it possible that the limitations upon the confessional in Protestant lands were too great to admit the growth of organized spiritual direction? Protestant culture tended to produce individuals who understood their responsibility and wills in ways that inhibited recourse to others. Jews and Catholics have never shared the Protestant religion of self-reliance.

How then explain the second paradox? Our most interesting hypothesis on this score has been provided by a French publicist, Raoul de Roussy de Sales (1938). It was precisely, claims this author, because of what we are calling "instrumental activism" in American Puritanism that once it was decided to organize the overcoming of neurosis, no cultural limits were placed upon the achievement of a liberation from sin and guilt in relation to the superego. It was precisely sectarian Protestantism which encouraged the conviction that world and self could be permanently purged of imperfection and confusion. Nowhere else has there been so much conviction in the positive power of unashamed love and self-expression. The social constraints upon the triumph of any such notions on the Continent have always been very notable. America is a country in which, in Max Weber's language, the psyche was to receive its most comprehensive rationalization. If time allowed it would be interesting to trace out the ways in which the social democratic outlooks so clearly noted by de Tocqueville in the 1830's (Lipset and Lowenthal 1961), contributed to the permeation of American culture by psychiatric and psychoanalytic ideas.

If in one critical respect Protestant antipathy to spiritual direction set up barriers to the promotion of psychoanalysis, in another it provided the patterns for the relation of therapist and patient. The weakening of the separate priestly class implied that the relationship of spiritual direction lasted only so long as the client was unable to act on his own with responsibility. This emphasis is a Protestant element within psychoanalysis. From the beginning Freud emphasized that the goal of treatment was the achievement of autonomy on the part of the patient, the ability to regulate his own life by norms of his own devising. Freud was in many ways closer to Kant than to Nietzsche.

It is now time to consider the relations of psychoanalysis to 20th century culture.

VII. PSYCHOANALYSIS AND 20TH CENTURY CULTURE

A new era in the history of spiritual direction begins with Freud. Our knowledge of the formative phases in this development has now been vastly extended thanks to the recovery during the last 15 years of important documents and unpublished manuscripts. The following sources are especially revealing: Freud's intimate letters to his friend and mentor, Dr. Wilhelm Fliess (Freud 1887–1902); the surprising *Project for a Scientific Psycholoy* (1895); the candid letter of Breuer on his collaboration with Freud (Cranefield 1958). These new materials carry us far beyond the indications published in Freud's Clark University Lectures of 1909 (Freud 1910), *On the History of the Psychoanalytic Movement* (1914), *An Autobiographical Study* (1925), and occasional biographical papers.

The turning points in the crystallization of psychoanalytic methods of treatment may be sketched as follows:

1. While still a relatively young man, Freud had the good fortune to be associated with the experienced Viennese internist, Dr. Joseph Breuer, who, as we now know, had a highly developed theory of aetiology and cure of mental disorders. In Breuer's view, hysterical disturbances resulted from undergoing of traumatic experiences which left painful memory traces in a state of hypnoidal suspension. Through the application of hypnotism, the patient achieved the

ability to recall the traumatic episodes. The recall was accompanied by the fresh experience and cathartic abreaction of suppressed affects and noxious ideas (Breuer and Freud 1895).

Too many contemporary writers have spoken of the break between the two men purely in terms of their different estimates of the role of sexual factors in the aetiology of neurosis (Sullivan 1959; Nelson 1959a). As Freud tells the story, his first decisive technical departure from Breuer was the abandonment of hypnosis in favor of unrestricted and undirected free association.

2. Another critical moment in Freud's development is associated with his departure from the teachings of Charcot and Janet on the subject of aetiology of hysteria. Whereas the French school spoke in terms of congential failures in the capacity for psychosynthesis, Freud (1910) insisted on stressing the role of unconscious repression of conflicted affect. The implications of this shift for the concepts of therapy can hardly be exaggerated. Freud was thus launched on the road to construing analysis as a relationship of antagonistic cooperation between therapist and patient. Analysis was, above all, the struggle against the resistances which crystallized in the transference relationship.

(It may be remarked parenthetically here that Freud failed to give due weight in this period of his development to the valuable contribution embodied in the Charcot-Janet position. How *are* psychosynthesis and executive integration of the ego effected? How, indeed, does the individual under the stress of massive doses of heterogeneous and inconsistent stimuli manage to achieve a stable identity? Fortunately, this way of conceptualizing the problem was not to disappear into the mists. Owing to the influence of Janet and Emile Durkheim on a number of notable writers, notably Elton Mayo, an important social-psychological and psychiatric theory of malintegraton developed in the United States.)[4]

3. Freud reports that the decisive steps in the separation of psychoanalysis from previous therapies occurred as a result of his efforts to understand his own feelings in the course of (a) his friendship and correspondence with Wilhelm Fliess, and (b) his intensified professional relations with private patients. We are indebted to Kris, Erikson (1957), Jones (1953–57), James Strachey (Freud 1900), and others for important studies of these years.

It will not hurt to repeat here:

It was the need to understand his own feelings, above all, which led Freud to the decisive findings of early psychoanalysis. Psychoanalysis in all its senses—an approach to the general theory of human behavior, a method of clinical research, a technique of treatment (Erikson 1957)—came to fruition in the course of his own self-analysis (1895–99).

Thanks to Strachey's variorum edition of *The Interpretation of Dreams,* we can now trace the steps in Freud's momentous journey within. Freud was by no means the first person to undertake this painful pilgrimage. In addition to the philosophers and theologians noted earlier in this paper, one would need to mention such intrepid searchers as Goethe, Kierkegaard, Amiel, Rimbaud, Dostoyevski, Strindberg, Nietzche (Heller 1957; Hopper 1957). Freud's distinction consists in the fact that he devised a prosaically scientific way of charting the depths he had explored. Neither villifying nor deifying the inner demons he had uncovered, Freud doggedly sought to map and explain the workings of the unconscious in man's passage through life.

Contented that he had penetrated the riddle of dreams, Freud spent the next six years (1900–1905) chiefly in surveying two other domains still shrouded in darkness; the action of the unconscious in the psychopathology of everyday life (Freud 1901), the itinerary of the "libido" in the child's psycho-sexual maturation in the setting of the family culture (Freud 1905).

The principal cornerstones, as Freud conceived them—*The Interpretation of Dreams* (1900) and his *Three Essays on the Theory of Sexuality* (1905)—had now been set in the edifice.

4. Freud explains elsewhere that he was spurred on to develop psychoanalysis as a distinctive set of procedures in the hope of improving upon available methods of therapy, notably the electrotherapy of Erb, the relaxation therapy of Weir Mitchell and the rational therapy of Dubois (Freud 1914). His own approach, psychoanalysis, was in use many years before he began to set down his thoughts about therapy. Though the word psychoanalysis was used for the first time in 1896 (Freud 1951), his major papers on technique did not begin to appear until the years 1910–15 (Freud 1953–74, vol. 12). The ensuing discussion of psychoanalytic technique takes its point of departure from these papers.

The exact connections of psychoanalysis with our theme will not be grasped unless we look with a fresh eye at the distinguishing features of the so-called "classical psychoanalytic treatment." This is more easily said than done. Familiar professional manuals rarely explain with sufficient detail and discrimination what implications are to be drawn from the collections of stipulations and procedures presented as "basic psychoanalytic technique" (Menninger 1958). For the present purpose, I shall place the distinguishing features of classical psychoanalytic treatment under two headings, comprising *ten articles:* Part I, *The Analytic Contract,* entered into by patient and doctor, comprising *five* sets of conditions which the patient agrees to observe and *two* conditions which the doctor accepts; Part II, *Analytic*

Techniques, three articles naming critical aspects of the procedures agreed upon by qualified practitioners.

In my view, the seven articles in Part I are not in themselves *techniques* of treatment, as they are so often said to be, but rather the conditions precedent or the *mise-en-scène* of treatment. Part II represents minimal agreements as to technique among persons engaged in the practice of psychoanalysis. The least well understood, and, indeed, the most controversial article is no. 6, on "interpretation" which is hardly ever defined even in the specialized papers:

Part I contains the following seven articles:

1. The patient agrees to come for treatment at stated times, a fixed number of times per week (six, five, or fewer).
2. The patient agrees to pay a fixed fee in an agreed upon manner at stated times.
3. The patient agrees to adopt a reclining position with the analyst out of sight behind the couch.
4. The patient agrees to desist from "acting out" in the course of treatment.
5. The patient agrees to report his thoughts and feelings without restriction or censorship. This is the so-called "cardinal rule" of psychoanalysis.
6. The analyst agrees to analyze the patient, that is to communicate to the patient "interpretations" by which the patient gains "insight" into his problems.
7. The analyst (implicitly) agrees to terminate the treatment when the patient's condition has been sufficiently improved or been removed.

Part II involves the following three understandings agreed upon by analysts:

8. To the greatest extent possible, in the manner of a surgeon, the analyst is to maintain an attitude of strict neutrality and impenetrability to the patient.
9. Properly speaking, psychoanalysis is analysis of the resistance to cure.
10. In the course of treatment, the resistances concentrate in the transference resistance or neurosis. Therefore, psychoanalysis is the analysis of the transference neurosis.

It is not possible here to deal with more than a few of the issues relevant to the ten articles. I shall especially stress the *cultural* implications of the cardinal rule on free association (art. 5); the limitations of the analyst's role to "interpretation"; the significance of the emphasis on the analysis of the transference resistance; and the value commitments underlying all 10 articles. I shall also ask what important considerations of a cultural nature are "bracketed" (temporarily treated as out of bounds) by classical psychoanalysis. What cultural consequences follow from this methodic suspension?

As Freud tells the story, the advantages of "free associations" were first brought home to him by evidence that Breuer's patient, Anna O., appeared to improve notably when she was freely allowed to practice what she called her "chimney sweeping" (Freud 1910). Freud's dislike of hypnotism and his awareness that hypnotherapy had restricted usefulness led him to prefer "free associations" as a device for gathering information (Freud 1910).

We have only to look closely at the analytic interview against the background of its cultural and social contexts to perceive many functions of free association not emphasized by Freud. The patient encouraged to associate freely is in effect being advised that the analytic session may be regarded as an opportunity to try his wings on the ocean of his unconscious with full assurance that the analyst will buoy him up if he threatens to sink or drift. Since it is uncontrolled fantasy the patient fears, the supervised practice of free association is a way of developing greater ease in the management of one's own inner demons. In this way, the patient acquires enhanced ability to fight off the frightening feelings and thoughts—the shame, guilt, disbelief, anxiety, panic—occurring in the wake of the stream of associations.

As analysts well know, the ability to be relatively uninhibited in associating increases in the course of the analysis. It is both the effect and the proof of the patient's expanded power to tolerate his wildest fantasies. Supervised association in the analyst's office performs the function of trial exercises of "regression in the service of the ego" (Kris 1952).

One may observe parenthetically that great creative artists have long understood the necessity of enlarging the horizons of awareness by deliberate regressions. An extraordinary anthology of passages on the avenues to expanded consciousness could be gathered from the writings of such notable figures as Goethe, Byron, Stendhal, Rimbaud, Kierkegaard, Joseph Conrad, Joyce, Nietzsche, Gide, D. H. Lawrence, Mann, and the playwrites currently associated with the so-called Theatre of the Absurd (Esslin 1961; Nelson 1962b, 1963a and c).

An especially memorable passage will be found in an autobiographical statement by Conrad:

> Remember that death is not the most pathetic—the most poignant thing—and you must treat events only as illustrative of human sensation—as the outward sign of inward feelings—of live feelings—which alone are truly pathetic and interesting . . . That much is clear to me. Well, that imagination (I wish I had it) should be used to create human souls: to disclose human

hearts—and not to create events that are properly speaking *accidents* only. To accomplish it you must cultivate your poetic faculty—you must give yourself every sensation, every thought, every image—mercilessly, without reserve and without remorse: you must search the darkest corners of your heart, the most remote recesses of your brain—you must search them for the image, for the glamour, for the right expression. And you must do it sincerely, at any cost: you must do it so that at the end of your day's work you should feel exhausted, emptied of every sensation and every thought, with a blank mind and an aching heart, with the notion that there is nothing—nothing left in you. To me it seems that it is the only way to achieve true distinction—even to go some way towards it. (Letter of October 28, 1895 to Edward Noble in Conrad 1954, pp. 731-33)

In the consistent opinion of artists and their publics, the greatest figures in the history of literature have been those who have had the courage to plunge into the whirlpools of the unconscious in order to discover truths that have been repressed and denied.

The psychoanalytic conception of the relations of master-client is markedly different from the earlier conceptions of this relationship. Innumerable writers before Freud recognized the importance of the emotional connection which he was to call transference. None thought to say that in the treatment proper all other symptoms tended to collapse into the transference neurosis; that as helpful as transference was in promoting therapy the transference was the foremost *resistance* to cure; that cure was not effected unless the transference, *negative* as well as positive, was "worked through."

Psychoanalysis is the first schema of direction of souls in the West which conceives transference in this many-sided way. All earlier methods emphasized the religious duty to strive for a permanently positive transference and allowed no place for the possibility of a negative transference. The new conception of transference implies a new conception of society and self. Freud himself hedged a bit on the matter of the negative transference. His reluctance to become involved in the treatment of the narcissistic neuroses and the psychoses may be explained in part by his unwillingness to become involved in the stormiest tempests of the negative transference.

The exaggerated emphasis on the loving relation of therapist and patient in recent days may well be a mark of the discomforts associated with the challenges of Freud's universalistic consensualism. In this respect, interpersonal psychiatry and existential psychoanalysis allow for regressions from the Freudian position. This is especially evident in the existentialist's neglect of negative transference. Considerable insight into the importance of provoking the negative transference may be found in several schools of Zen Buddhism (Senzaki and McCandless 1961).

Universalistic consensualism is the central value on which Freud built his system of psychoanalysis. Anything which restrained the equal freedom of both parties militated against a therapeutic analysis. For this reason he insisted on the equal observance by therapist and patient, alike, of the formal stipulations of the analytic contract. Freud was the first great director of souls who recognized the threats to liberty built into the strongly emotional connections of master and disciple. His exceptional care about the establishment of fees, hours of appointment, etc., subserved this central function (Menninger 1958).

Easily the best way to discover the strains and gaps which developed in the Freudian system is to study the succession of crises in the psychoanalytic movement (Munroe 1955). Ernest Jones and Erich Fromm (1959) notwithstanding, all error was not on one side, nor all truth on the other. Nor is it sensible to argue that every act of dissent was a blow for liberty and every defense of the Establishment a justification of intellectual obscurantism. In fact, there is much to be learned about all the principles and doctrinal questions concerned by viewing the history from the perspective of the changing polemical contexts.

Each decade since the origin of psychoanalysis produced its own crises. We may confine ourselves here to referring to the central points at issue in the differences with Adler, Jung, Ferenczi, and Rank in the second and third decades of the present century. To speak of this in turn:

Underlying the separation from Adler was the fact that Freud had allowed his distinctive stress on the dominion of the unconscious to becloud the role of the ego. Adler pressed so hard on this front that he soon found himself in a camp of his own (Freud 1914; Colby 1951). For the abandoned unconscious, Adler (1939) eventually substituted the notion of conscious identification with the "social interest" (*Gemeinschaftsgefuehl*). Adler's emphasis on the ego was not without influence on Freud's later restoration of an ego psychology *within psychoanalysis* (Ansbacher 1956).

Jung pointed to a more serous gap in the Freudian schema. He was the first psychiatrist to recognize the magnitude of the spiritual crisis which marked the dissolution of traditional religion in the 19th and 20th centuries (Jung 1931). His insistence on the symbolic archetypes in the collective unconscious (Jung [1912] 1962) grew out of a desire to provide guidance to lost spirits who were unable to find meaning in existence without the aid of a metaphysical involvement. The wild exaggerations which came to characterize Jung's work do not vitiate his frequently profound insights. Freud's theories of man and culture did require enrichment.

The psychoanalytic crises of the '20s are not so well remembered today as are the crises of 1912-13. In many ways, however, the crisis of the '20s left a deeper mark on the subsequent development of psychoanalysis as a therapeutic system than did the more widely publicized deviations of Adler and Jung. Here the issues developed out of dissatisfaction with the inhibiting biases of ultra-Freudian theories of aetiology, dynamics, and technique. Ferenczi and Rank both sought to move psychoanalysis toward a greater emphasis on the pre-Oedipal sources of neurosis. Both were deeply stirred by the influence of object-relations in mental development. Each in his own way was convinced that Freud's approved techniques of treatment were unsuitable for the sorts of cases—so-called "borderlines"—increasingly coming into treatment in their day (Bromberg 1959; Munroe 1955; Jones 1957).

Regrettably, Freud's polemics against the excesses of Rank and Ferenczi drove object-relations theory and treatment techniques into a corner from which, surely in orthodox circles, they do not truly dare to emerge even today.

Robert Waelder's recent codification of what he calls basic psychoanalytic theory provides an especially revealing illustration of the severity of the orthodox reaction against the alleged insurgence of a new therapeutic irrationalism (Waelder 1961). He refuses to mention any deviations from the so-called model techniques which have developed in the last three decades on the grounds that all of them are merely restatements of the excesses of Rank and Ferenczi. Instead he calls upon his colleagues to pursue the study of psychoanalysis as a basic science. Only by such efforts can psychoanalysis hope to discover a way of ending emotional and mental disorder. In the name of the pure science of psychoanalysis and, persumably, the ultimate drug, Waelder turns his back on the possibility of responsible innovation in psychotheraphy.

The situation at present may be described as follows:

1. Freudian psychoanalysis has made great inroads into clinical psychiatry at the medical schools and at many major psychiatric faculties and mental hospitals (Lewin and Ross 1960).

2. To the informed, these external successes do not obscure the fact that orthodox analysis has been undergoing a loss of vitality at its own core (Alexander 1961). The efforts of Hartmann (1958) and others (Rapaport 1960) to build a bridge between psychoanalysis and general psychology by developing a general theory of ego development and function have been resulting in notable shifts in Freud's fundamental orientation. The loss of vitality is particularly evident in the unreadiness of classical theory to keep pace with progress in the sphere of technique. Eissler (1953) has sought to make room for

deviant procedures by permitting what he calls "parameters of deviation from the model technique" on the understanding that these deviations will be liquidated at the earliest possible opportunity in favor of the pure psychoanalysis. The outcome of this scholastic compromise recalls the circumstances of astronomy before Copernicus and Galileo. The multiplication of "epicycles" (*read* "parameters of deviation") is allowed to proceed indefinitely in order to preserve the model theory (Ptolemaic theory, "pure psychoanalysis"), in the face of the "irregularities" of the current data and the "premature" demands for a more comprehensive codification of established findings (Coleman and Nelson 1957a).

Experimental psychiatry has been making new headway in the development of chemotherapy. As usual, utopian claims (Dubos 1961) are made by sanguine publicists.

Among the most perceptive practitioners there is increasing sensitivity to the new challenges presented to the therapist by the deepening crisis in our cultural situation (Alexander 1961). Classical psychoanalysis has yet to adjust to the fact that great numbers of men and women are unable to discover meaning in their lives and times (Wheelis 1958).

The slowness of orthodox analysis to relate to the deepened crisis has provided an opportunity for dubious faiths newly borrowed from the Continent. The various forms of existential analysis (*Daseinsanalyse*, American existential and onto-analysis) are not really new systems of relieving individual souls so much as surrogate religious in the guise of clinical psychiatries and philosophical anthropologies (Nelson 1962d). The willingness of all but a few leaders of these movements to continue obsolescent techniques of treatment is a mark of the insignificance of the role of psychotherapy in existential analysis (Boss 1957; May and others 1958; Rogers 1961). The bitter indictments of Western science and the striking congeniality to neo-Gnostic, theosophic, and Eastern mystical strains of thought in the newer existential psychiatrics represent a high point in the advance of the "transmoral conscience" at the expense of the remaining hopes for a fresh integration—purged of the traditional excesses—of conscience, casuistry and the cure of souls.

It remains to be seen how well distinctively Western orientations will fare in the stressful years ahead.

SOME CLOSING REMARKS

Nature and nurture, alike, equip individuals variously to promote their ends in the worlds they inhabit. If, then, we wish to estimate, perhaps even to predict, their powers to respond to the challenges of their several environments, we must include reference to the many disparate factors likely to be relevant in their varied life situations. The physical coordinates are the easiest to study: How well do different organisms tolerate changes of different sorts in the atmosphere they are asked to inhabit? In what patterns are physiological and neural mechanisms linked in the processes of adaptation to bear stress (Selye 1956)?

Soon we find ourselves spurred to introduce reference to "non physical"—social, cultural, interpersonal, and intrapsychic—aspects of the field. How differently matters appear when we trouble to ask: when does a perceptual stimulus, indifferent or innocent to one, become a catastrophic psychic stress to another? How do different cultural schemes define stresses and responses? How do individuals variously estimate the relative shares of stress meted out to them (Merton and Kitt 1950; Merton 1959; Merton and others 1959)? In what ways are subjects affected by the situational and interpersonal contexts of the stresses they are bound to undergo—the times at which, the places in which, the persons in whose company, the authorities at whose instance, they undergo stress?

We will not go far in discovering how well an individual will probably do, given a battery of stress challenges of different sources—a psychometrician might be disposed to call this his Adjusted Multiple Stress Potential (AMSP) or Quotient (AMSQ)—by cleaving to a purely physicalistic conception of response to stress (Selye 1956). All measurements of human capacity must reflect rather than ignore the distinctively symbolic character of human existence. The so-called behavioral sciences have no way of escaping the cultural dimensions of human behavior.

The time has now come to say some parting words about the workings of the allocative institutions we have referred to above under the head of Cultures as Symbol Economies:

Every social system necessarily engages in the production and distribution of coveted symbols. Available resources are assigned to competing uses with a view to maximizing desired value outcomes, which, inevitably, remain in scarce supply relative to effective demand. Within this framework, systems of spiritual direction emerge which acquire great influence in determining the abilities of people to bear the passions and infirmities societies and systems define as stressful. The capacity of an individual to perform constructively in the midst of stress is a function of the society's success in maintaining a favorable balance of the supply of symbolic (and other) resources at its disposal.

Mental healers perform critically important functions in the symbol economies of societies. They invent, distribute, and consume significant symbols. Along with other cadres engaged in whatever measure in spiritual direction, they play strategically ambivalent roles in framing definitions of the social, cultural, and personal states of affairs. At different times, as I have elsewhere sought to show, mental healers have acted, in the language of W. Robertson Smith (1889), as prophets and priests; in the language of Arthur Koestler (1945), as yogis and commissars. We are, in truth, in dire need of fresh research and insight on these fateful matters.

The future of self is extremely obscure in the present historical interim. Depending on their political and philosophical commitments, groups and individuals are describing the self as culture's foremost achievement, mind's vilest metaphysical illusion or society's most noxious disease.

It is still too early to tell how well the Western sense of self will fare in the galactic era ahead. There are powerful forces working at cross purposes in this regard. Whatever the outcome, spiritual directors and systems of direction will continue to play strategic roles in defining stressful situations and aiding men to cope with them. Original nature is too fitful in expression and incoherent in aim to serve Everyman as a trusty guide. So long as each of us is required to be symbolically endorsed by others; so long as all aspire to taste vindication in however vague a sense, we search for our meaning in a design not of our own devising.[5]

The emotional illnesses of men will change their shapes; the techniques of psychological cure will adopt new strategies; mental healers and spiritual directors will claim and be accorded new roles. Selves may light up the skies or they may disappear behind the clouds. But of one thing we may be sure:

Conscience, Casuistry, and Cure of Souls—Madnesses, Methods, and Masters—these and other of man's puzzling cultural inventions serving now as solvents, now as sources of stress—will be with us to the end, so long as human society endures. Better to face this prospect squarely than to waste ourselves in vain attempts to "escape from the gloomy actual" by entering upon "the quest of oblivion, sought in the delusion of ideal harmony" (Huizinga 1924; Nelson 1954)!

NOTES

1. Thus to mention one prominent example: the Italian Renaissance receives little mention in the following pages. The chief reason for this neglect is that we are still largely in the dark about the ruling self-concepts and

schemas of spiritual regulation in this vastly over-interpreted but under-researched era (see Cassirer and others 1948; Kristeller 1961; Trinkaus 1940; Nelson and Trinkaus 1958d.) The same cautions clearly apply to the state of our current knowledge of Eastern systems of spiritual direction. An earnest attempt to go beyond traditional stereotypes is made by Herbert Fingarette (1963).

2. The terms "Occident," "Orient," "East," and "West," as used in the present pages, are largely metaphorical in character. They are not meant to imply archetypal unities of differences, but simply clusters of stresses in as yet too ill-defined culture-areas, in which marked diversities become more evident the more carefully we study the data. New results along these lines are currently being promised by an active University of Illinois group (Miron, Archer, and others) applying Osgood's "semantic differential" tests to selected societies.
3. The absence of any historical examination of Riesman's evidences in the recent full-length review of his work by S. M. Lipset, Leo Lowenthal, and others (1961), is only one symptom of the present state of the problem. Even more revealing is the lack of any history of the idea of conscience since the time of Abelard. The present writer hopes to improve this situation before the present decade is ended.
4. Strangely, this thread of influence emerges in only an intermittent and vague way in Mayo's booklet on Janet (Mayo 1952, pp. v., 120–21). See, however, Mayo (1945). Additional clues will be found in the author's (1963c and 1965c).
5. A particularly revealing orchestration of these themes in library form is to be found in Genet's classic, *The Balcony*; see Nelson (1963a).

FOUR

Conscience and the Making of Early Modern Cultures: Beyond Max Weber

I. THE SETTING OF THE PROBLEM

Sociologists and historians have too long been ill-served by partisan wrangles over Weber's views on the relations between the "Protestant Ethic" and the "Spirit of Capitalism." Weber is himself not entirely free of blame for the confusion. The evidence indicates that from the start of his work he allowed himself to be trapped in an ambiguity from which he never quite extricated himself. Although he makes it clear in one of his little-known letters that he was interested, above all, in unravelling the spiritual foundations of the vocational asceticism of contemporary occidental civilization, a number of passages—especially the sensational first and last paragraphs of his celebrated essays of 1904 and 1905—did not exclude the possibility that he was arguing that Calvinism, capitalism and Mammonism were cut of the same cloth. Surely, this perspective does not appear as a noteworthy element in Weber's work during the last fifteen years of his life when he was doing his path-breaking studies of the course of "rationalization" in the historic civilizations and religions of the world.

Sociologists and historians who have been touched by Weber's influence continue, therefore, to be at a loss as to how they are to organize investigations into the development of European sociocultural patterns and institutions from roughly the end of the Middle Ages to our own day.

We will center our attention here on the sixteenth and seventeenth centuries, the era of the Protestant and Catholic Reformation and Counter-Reformation, the "Scientific Revolution," and the emergence of new nations and polities. Whether or not we name them here, the persons of our drama include Luther, Galileo, Des-

Originally published in *Social Research* 36, 4 (1969): 4–21. Reprinted with permission.

cartes, Pascal, Cromwell, Locke and many contemporaries including American colonists, and, as our last points of reference, Bernard de Mandeville and Pierre Bayle, two of the foremost polemicists in the remaking of traditional concepts of conscience and character.

This essay suggests that we need to start from a new point of departure. Our new sightings have to fuse certain of the realizations of Weber's last years, and neglected aspects of work done by others in a related spirit, notably that of Durkheim, Simmel, Veblen, and Weber's renowned but too little understood predecessor, Sir Henry Sumner Maine. Evidence from recent historians and philosophers of mathematics and physics—such as Whitehead, the late N. R. Hanson, Thomas Kuhn and others—is drawn upon in the hope of improving Weber's account of the history and present bearings of "rationalization" in contemporary society and culture.

The Weber to whom we have to attend in the newer analysis of the makeup of the cultures of the modern era is the Weber who during the last decade has taught us to watch the various ways in which behavior and belief change under the impact of rationalization processes of different sorts, of which the changes described in *The Protestant Ethic* represent only one instance. His mature analysis is expressed with particular force and scope in the Introduction he drafted in the last year of his life for his *Collected Essays in the Sociology of Religion,* published posthumously. It is this essay which challenges us to go beyond him in analysis of the network of notions and processes expressed in the idea of "rationalization."

Seminal as Weber's studies have been, advances in the analysis of the rationalization phenomenon have been slowed by two of his own preconceptions:

(a) Weber persisted in envisioning the histories of the great civilizations and world religions as a phasic sequence of tendencies to rationalization. Each civilization—the ancient Hindu, Chinese, Ancient Hebrew, Medieval Christendom, as well as the Protestant Ethic itself—illustrated one phase of a major movement along the rationalization axis. In the main, other-worldly mysticism gave way to inner-worldly asceticism; irrespressible impulses gave way to rationalization of conduct; magic to science, and so on.

(b) In the same spirit, he apparently could hardly help conceiving the processes of rationalization in terms of the models of the second law of thermodynamics with its familiar implications for a dissipation of the original heat energy, ending in what Talcott Parsons has called the "nightmare of entropy." Few will forget the frightening predictions of the closing pages of his *Protestant Ethic.*

These stresses push Weber into slurring over several critical distinctions, well enough known to him, on which I will need to insist in the present essay. I refer to the shades and grades of *formalisms* and *formalizations, rationalisms* and *rationalizations,* and the non-material as well as the material technologies. Continuities and discontinuities between these ideas need to be carefully noted.

If Weber does not truly anticipate our story—and indeed if he at times throws us off the scent—it may be perhaps because he was in a number of respects too much a son of his own age. The immense force of industrial-political rationalization which Weber saw in his own country and during his visit to the United States in 1904, led him to run together the various levels and layers of economic, social and cultural change. It is no wonder that Weber's views on rationalization incline again and again in the direction of the views of "spirit" and "mechanism" associated with the names of Feuerbach, Marx, Simmel, and others.

The last decade and a half of Weber's life witnessed ever mounting pressures towards programs of total coordination of all functions—political, industrial, cultural. Accelerated thrusts in the direction of "scientific management," "efficiency engineering," "time-and-motion" analysis, and other extensions of the so-called system of Taylorism had their issue in ever more forbidding prophecies of the total mechanization of spirit such as those of Walter Rathenau. The possibility that "spirit" might one day again gestate new forms and that "monstrous" mechanisms might in some respects and in certain circumstances subserve the purposes of "spirit" understandably became more faint as the ravages of war and its aftermath spread.

My own perspectives in these pages—and in a number of related current studies—will stress the distinctions between the series of progressive thrusts coming to a head in the sequence of industrial revolutions since the eighteenth century and the series of intellectual revolutions culminating in the twentieth century scientific-techno-cultural revolution through which we are now passing. Before the end of the eighteenth century these two revolutions—the economic-technological revolution and the intellectual revolution—had few and infrequent connections. The growing interpretation of the lines of force and the fateful intensification of the links between them in our own lifetime constitute both the threat and promise of the future.

To gain fresh momentum in the analysis of the rapidly expanding symbolic universes and to achieve fresh awareness of the increasingly influential careers in our day of non-material technologies, we must be ready to turn for help to a number of thinkers—philosophers and jurists, no less than sociologists—who are not so well remembered in discussions of "rationalization" as Weber is. In the present connection, I would first mention one of Weber's neglected predecessors, Sir Henry Sumner Maine. It is to the underlying insights of Maine's

Ancient Law (1861) and to several of his neglected lectures that I refer here rather than to mooted details of his social-historical discussions.

I would especially recall his stress on the role, in the development from what he calls "stationary" to "progressive" orientations of patterns, images and ideas, of cultural logics and other symbolic structures and technologies, notably the juridical mechanisms which channeled action. Maine's brilliance as a sociologist of culture and cultural institutions has yet to be fully acknowledged. Among the most important instances of such logics and technologies undergoing development, in my view, are *rationales,* structures of reasons, explanations, procedures establishing requirements in respect to truth, legality, virtue, fittingness and other directive norms (cf. Nelson 1968a, pp. 162–163).

Basically, in complex societies of the Western world there have always been bodies of protocols which correlate all notions and evidential canons, associated with the proof or disproof of arguments for or against any given declaration or claim, whether the declaration be about what is or was or ought to be. Social action and accounting without such rationales is unthinkable. Where the established rationales go out of phase or lack a compelling and vital center, social and cultural regressions to the so-called "state of nature" regularly occur. First, as our vernacular has it, "everything is up for grabs." Later, in the memorable phrase of Hobbes, "clubs are trumps."

Neither societies nor individuals can long remain in a situation so anomic. Rationale systems are the air we must breathe in order to survive. This becomes especially true as societies become more complex and collective representations differentiate.

The everyday men of societies are hardly in a position to discover for themselves what deserves credence or how men ought to act in any given case. As societies grow more complex, increasing amounts of the activities of numbers of specialized intellectual workers come to be devoted to the organization and communication of rationales and directive cultural technologies.

Every group requires to have at its disposal some symbols of its faith, some acceptable and accepted canons of interpretation. Furthermore, in many crucial respects the purpose of formulae is quite the same whether the formula be theological, scientific or mathematical. A formula is a formalism, a major turn on the royal road of formalization which carries us to even higher plateaus of abstractive power and to symbolic technologies of even greater professional authority. Speaking socioculturally, formalisms serve the following notable purposes, among others: as mechanisms of thought condensation, they constitute an order-creating social memory, correlating impressions and perceptions, fixing the cues to the required or desired behaviors and beliefs, setting the frame for further questions and the process of inquiry. Today more than ever before in the history of our civilization it is possible for ordinary men as well as professional scholars to see how essential a part formalisms and formalizations play as prerequisites in the processes of codification, axiomatization, mechanization, computerization, and so on.

Historians and philosophers of science have become extraordinarily aware of the extent to which scientific developments are functions of antecedent and innovating paradigms. This notion has been very powerfully put by Kuhn in his book, *The Structure of Scientific Revolutions* (1962). And there have been others who have studied "world-hypotheses," "nuclear metaphors," and other notions which are correlated with the idea of the central scientific paradigms (Pepper 1957; Lovejoy 1936).

I repeat: *to understand the making of early modern cultures we need to understand the making of early modern minds, and, therefore, we need to have a proper sense of the changes in the central paradigms as well as the restructuring of axial institutions in society.*

Contemporary research confirms—and, indeed, Weber's view implies—that complex societies require a set of coordinating symbolic systems of cues composing, in my view, no fewer than five sets of needs and "programs" of which I have elsewhere described four under the heads of dramatic designs, directive programs, defensive systems, and symbol economies (chapter 2 above).

For our purpose, it suffices to speak of two logics in two decision spheres of science and moral action, in the broadest sense of these notions. In respect to decisions in the areas of scientific proof (in the widest understanding of that phrase) we may speak simply: claims to truth in whatever measure any one cares to make—propositions of any sort one may wish to state—all these need to be confirmed by reference to some set of validated evidence, procedures, or proofs. The second center of rationales is what we are here calling the *logic of moral decision*—in the sense of all human action and opinion having to do with *praxis* (practice). Clearly, within the Western world, it has always been necessary for anyone who claims to act in the light of one or another maxim, or who adopts one or another moral opinion, to offer proof that his intentions or the consequences of his acts are congruent with acceptable values and rationales.

From the point of view of the sociologist of knowledge (so-called), it is sometimes interesting to note that the rationale—protocols offered or, indeed, the entire rationale-system—may smack of rationalization, not so much in the Weberian sense, as in that associated with Mannheim and Freud; that is, the covering over as by an ideology, or a noble notion or defense of some other interest, whether known or unknown. All this is, however, not so decisive an aspect of the matter from the point of view I am adopting here. I would stress that all behavior gets to be norm-dependent in one way or another, in the sense that all behavior, whether or not it originates in a value, has nonetheless to be referred to a value or defended by a value—i.e., made congruent with a rationale.

II. THE REVOLUTION IN THE RATIONALES: LUTHER, GALILEO, DESCARTES, PASCAL, AND OTHERS

In this section I shall place special stress on the different ways in which various societies and nations north and south of the Alps responded to strategic changes during the sixteenth and seventeenth centuries. These changes occurred in the rationale-systems, the cultural logics and non-material cultural technologies. The triangulated structure of ideas and procedures which had its chief institutional embodiment in the Forum of Penance and the Court of Conscience will be my point of special concentration.

During the entire period under discussion it will be recalled that the logics were interdependent, in fact they were woven together in a single fabric of propositions centering around the notion of conscience. As the continuing usage of a number of modern European languages should serve to recall, the Latin *conscientia* had imbedded within it a dual reference: the moral conscience of "the proximate rule of right reason in the moral sphere," *and* scientific knowledge. It is, therefore, no wonder that all important cultural and social innovations in our period had to involve attack upon, or reconstruction of, the logics of decision in the spheres of *action and thought,* in the scientific and moral domain alike.

Our story can be put very simply. The sixteenth and seventeenth centuries saw two major revolutions within the frames described here. One was the successful revolution called the Reformation, which was a breakdown and reconstruction of the Court Christian and the received logics of moral decision. All fundamental images and ideals of self, spiritual direction, group life, were reshaped in the ensuing correlations of newer elements and reconstituted older elements released by the dissolution of the older complex. The other revolution was the scientific revolution, as developed in the work of Copernicus, Galileo, Descartes and others. This, also, was directed against the received logics of decision, only in this case it was the logics of scientific proof and knowledge. Here, it will be noted, Copernicus, Galileo and Descartes aimed shafts against the Establishment's accommodating casuistries of explanation and proof, namely, fictionalism and probabilism. They claimed to have true knowledge, objective certainty and subjective certitude.

Different and even contradictory as they were in many of their views and motivations, there is one critical sense in which the pioneers of the Reformation and the pioneers of the scientific revolution were at one. All—Luther, Calvin, Galileo, Descartes, Pascal—attacked the late medieval casuistry of conscience and probabilism at their very roots. Moreover, their attacks against every shade and grade of conjecturalism, fictionalism, and probabilism were put forward in the name of subjective certitude and objective certainty. From the point of view of this essay, it hardly matters that Luther thought ill of the Copernican hypothesis or that Pascal was a devout Catholic and not a Calvinist (see chapter 7 below).

Under the first heading one might need to deal with the views of Luther and Calvin on conscience, the old law, usury, conflicts in Great Britain over the common law, equity, the liberty of prophecy, inner light, social compact, the Court Christian. Under the second heading, we might consider two celebrated controversies in which the rationale-systems were in the balance: Galileo's encounters with the Inquisition, and Pascal's *Provincial Letters* (1656) and the bitter controversies over the system of moral probabilism and the direction of the cure of souls in Italy, Spain, Portugal and elsewhere.

Let us now attempt to mediate some of the controversies related to Weber, Tawney, Troeltsch, and more recent scholars, notably Herbert Luethy, H. R. Trevor-Roper, and S. N. Eisenstadt. The brilliant but neglected work of Bernard Groethuysen (1968) on the origins of the bourgeois world-view in France will be brought into focus in the analysis of post-Reformation Catholic developments in the Sub-Alpine lands. A thumbnail sketch of the documentary evidence on these issues comprises the next section of our paper. We reserve for the closing pages several suggestions about the meanings for ourselves and the future of the afterlife of the elements generated in the two-front struggle to recast the logics of conscience and knowledge. It is proposed that a key to our own cultural situation may be found in the fusion of the eventual outcomes of those two great movements, the Protestant Reformation and the revolution in science and philosophy.

III. HISTORICAL EVIDENCE

(A) The Middle Ages

The axial directive programs of medieval society, as institutionalized in the Court of Conscience, rested on a few clear assumptions. All human action was answerable to "right reason," of which conscience was the proximate rule and the revealed law was the ultimate norm. The bearing of this belief and thought upon particular situations was mediated to believers by experts possessed of acknowledged professional skills and credentials. The "presiding officer" over the execution of the law at the local level was the judge in the "Inner Forum" or Court of Conscience, the priest in the confessional. Indeed, this judge exercised a complex fourfold function I have analyzed elsewhere. Wherever this sense of existence prevailed, everyone supposed that the dictates and principles of conscience could become—and needed to become—meaningful at every turn in everyday life.

Solutions for the conduct and regulation of man's life and all his relations in the market place, the battlefield, the court the home and elsewhere, were, therefore, developed in innumerable treatises on the cases of conscience. All the urgencies of life and the aims of men as they moved about in the daily round were indeed grist for the busy mills of the casuists.

Unimaginative scholars have fancied that the numerous exceptions in these treatises prove that the Court of Conscience was a mockery and that Luther's criticisms were directed against a dead shell. Nothing could be further from the truth. It is to the credit of Weber that he understood better than any of his critics that Luther was not a libertine casuist but a spokesman of a *new conscience,* a conscience beyond casuistry, indeed, as Paul Tillich once put it, a *transmoral* conscience.

(B) Luther and the Early German Reformation

The critical points at issue in the struggles over the Court Christian and the rationales of existence and thought were not everywhere the same during the longer period which encompassed the fifteenth to the eighteenth centuries. It is this which has caused so much confusion among sociologists of culture and historians of ideas and institutions.

In Germany the attack against the triangulated structure of conscience, casuistry and the cure of souls was immediate and total, and the solution was drastic. The authority of conscience in the religious field was expanded so as to constitute the area of total freedom, but the religious sphere was wholly separated from the sphere of moral regulation and of political control.

In Germany, therefore, there occurred the dichotomization between the realms of inner freedom and outer boundage. These realms, which contained respectively religion and the world, were hermetically sealed from each other as though constituting separate universes. Where this dichotomization persisted, the oppositions of conflicts between religion and the world could, and sometimes did, become titantic conflicts. Luther's Pauline solution finds echoes, for that matter, in Weber and others. Luther insisted on the absoluteness of the freedom of the individual conscience in the deepest recesses of the now distinctively religious sphere, but this liberation was purchased at the sacrifice of a fabric or casuistry in the moral or the political sphere. Conscience and casuistry were divorced from each other; no new structure of the cure of souls was available to make up for the loss of the discredited Catholic Forum of Penance.

(C) Lutheranism in the Seventeenth and Eighteenth Centuries

It would be an error to suppose that this drastic solution prevailed without any change throughout the centuries of Lutheran history. The culture of every complex society or national establishment will necessarily undergo certain changes as the social processes are worked out. Thus, as one might expect, there was a new Protestant neo-scholasticism, there was a new casuistry in the seventeenth century and there was, in the course of the eighteenth century, an effort to renovate a system of spiritual direction that was, however, freed from the restrictions felt by Luther in the Catholic structures.

(D) "True" Orders and Reformation Consciences

We may now draw certain conclusions about the Reformation which may be concisely stated as follows:

1. The Reformation claimed to replace a corrupt modern order by the true primitive order. The true primitive order was not everywhere conceived or developed in the same way, but wherever such an order was elaborated, it proved not only to have many new elements, different in structure not only from those which had prevailed in the Middle Ages, but also from those which had characterized the apostolic community of the early church.

2. During the Reformation, as during other eras of accelerated change, the disestablishment (de-regulation) of the old order became a necessary first step to the development of a new order. Older elements left temporarily suspended in a free-floating state were correlated with newer emerging patterns: the essential sociocultural functions were served in dramatic ways in the newer settings.

3. The new order developed new integration of life, both personal and political, through a rearrangement of existing boundaries. Some important established institutions were scrapped, others were given expanded functions, and in the end the older maps were redrawn,

fixing new coordinates for all focal points of existence and faith: religion-world, sacred-profane, civil-ecclesiastical, liberty-law, private-public.

4. The reformers denied that there were true scriptural bases for such institutions or norms as a Court of Conscience, a law of conscience, cases of conscience, human judges of conscience, obligations to confess to human judges of conscience, obligations to do penance imposed by human judges, powers of human judges or courts to absolve from guilt toward God, a Papal Canon law or an Apostolic penitentiary dispensing a treasury of Merits, distinction between so-called counsels and precepts of Christ, distinction between mortal and venial sins, distinction between purely religious and purely secular callings and ways of life.

5. The reformers did not reject the idea that conscience could be obligated by the Divine Law, and in certain instances, by the imperial mandate and royal decrees.

6. In general, the contrasts between the Protestant and Catholic patterns may be expressed as follows: The Court of Conscience and the Court Christian disappeared in the Protestant lands, and with them the historical notions of a law of conscience, human judges of conscience, a casuistry. The decisive change in the culture and administration of "conscience-casuistry-cure of souls" implied profoundly radical changes in the images of self, the cultures of spiritual self-realization, and the systems of spiritual direction. The way was open for the progress of newer philosophies of the Inner Light, the light of Pure Reason, along with newer sanctifications of positive civil law, a pantheon of religions of the transmoral conscience whose flowering became a prominent feature of Protestant culture-areas in the nineteenth and twentieth centuries.

7. Protestant lands also underwent many other institutional innovations in the course of the partition of the Court-of-Conscience complex. New scope and authority were given to the Inner Light, sparked by the Holy Spirit. This was the Holy Spirit within each individual and within groups. This inspiration came to serve as a basis for vastly expanded involvement of new participants in a variety of different relations to self and world: charismatic activism, quietist mysticism, convenanted corporate consensualism, natural rights individualism, a religion of Pure Reason. In addition, a host of new concepts and attitudes governed men's thought and actions: private judgment, personal responsibility for self-regulation, the rule of law (that is, the rule of common law as opposed to prerogative decree, equity, etc.), the rule of general or public utility and a newer calculus of welfare and happiness, public ethic and logic of consequences as

opposed to a private ethic and logic of conscience, Political Arithmetic as well as Moral Arithmetic.

(E) Sub-Alpine Catholicism

It is astonishing how slow historians have been to deal seriously with the structures of thought and rule which were elaborated in various Catholic lands south of the Alps in the period of the so-called Counter-Reformation. Only recently a number of newer historians, notably H. R. Trevor-Roper and Herbert Luethy, have been insisting that the real key to the development of capitalism and industrialism north of the Alps was not the congeniality of Northern Europe to newer economic values, but total hostility and even reaction of the South to any modernizing influence. I consider that this approach also is overly concerned with a single aspect of Weber's views and fails to account for, or provide some clue to, the complex civilizations that were elaborated in Catholic culture-areas in the sixteenth and eighteenth centuries in Italy, France, Spain, Portugal, Holland and elsewhere. Neither of these authors accords enough weight to the critical controversies relative to the sociocultural frameworks of definition and control described here.

Adaptations occurred in Roman Catholic lands; indeed, some of the adaptations were of an extraordinary libertarian cast, but their spirit and substance were essentially different from those of the changes undergone by Protestantism. In Catholic culture-areas "liberty of conscience" in the moral sphere was expanded, the casuistry of opinion and action incredibly ramified, and the cure of souls and priestly direction retained. The over-all process was one of flexible accommodation by the Church and its main organs to the changing structures of society and to changing social interests.

The triangulated structure of representation and regulation compressed in the Forum of Penance and the Court of Conscience underwent adjustments of various sorts, but the essential structures, institutional and spiritual alike, persisted. Indeed, immense new authority, scope, and suppleness were given to the Forum of Penance by the reform of the cure of souls by Loyola and the reform of the casuistry and the direction of conscience by the Jesuit spokesmen of the new system of the lesser probability.

In Italy, Spain, France, Portugal, Holland and elsewhere great struggles raged from the sixteenth through the eighteenth centuries over the niceties of the system of the lesser probability. Galileo, Descartes, Pascal, the Jansenists and Jesuits cannot be understood without referring to this background, whose importance for the student of the history of culture cannot be exaggerated. The emergence in the midst of all this of a new middle sort of man, the "self-made" Catholic burgher brilliantly described by Bernard Groethuysen (1968), has been generally overlooked. This new modern sort of Catholic layman felt at home in the "secular city" but saw little need either to flaunt or to flout the Church's doctrines.

Even before the bitter conflicts had subsided, a new international army had formed about the Pope. Outer and inner missions were underway in all areas of the earth. The older Church of the middle ages had given way to a new Baroque style of "benevolent autocracy" in which fictionalist philosophies of science and lesser probabilist theories of opinion were attempted as flexible accommodations to the new needs. Although the world remembers best the dramatic crises in the Church over Galileo and Pascal, the main outcome so far as the organization of the Church was concerned was the preservation of the triangulated structure in a world in which self-made men formed an important new heaven.

IV. CONCLUSION

Enough has been said above to suggest why I find myself in disagreement with those who would put Weber's purposes in *The Protestant Ethic* in terms of a local historical proposition about the relations of economics and religion in sixteenth century Europe. In the end, the proving grounds for Weber's views are not Prussia or even England, but the Soviet Union, the Far East, the Near East, Africa, in short, the world.

The making of early modern cultures has been deliberately studied here from the special point of view of the revolutions in rationale systems in the sphere of conscience in its dual bearings in order to illustrate the limitations we impose on ourselves when we restrict our focus to purely political and social-economic perspectives. Weber has deserved very much better understanding from scholars than he has so far received except from a few like Talcott Parsons, James L. Adams, and a number of younger scholars they have variously influenced, especially Robert Bellah, David Little, and Jan Loubser.

Weber seems to me more accurate than Tawney and those historians who have followed in his footsteps. Troeltsch, Tawney and others go to extremes in insisting that the Reformations of Luther and Calvin were essentially medieval in cast. The weaknesses of these views appear especially prominent in the highly condensed summary of the position by S. N. Eisenstadt in his MacIver lecture (Eisenstadt 1965a, especially p. 671). The heart of the difficulty seems to be that

Eisenstadt, like Tawney and Troeltsch, adopts the contemporary assumption that social-economic and political institutions are primary and fails to sense the decisive and fateful character of the revolutions in the regulative rationale-systems, especially the logics of decision as I have described them above.

My own earlier discussions of the views of Luther and Calvin on the exegesis of Deuteronomic commandments on brotherhood and usury, the old law, friendship and equity (Nelson 1949a) were not intended to show that the two reformers had new capitalistic economic views, but that they had opened the door to decisive changes in assumptions and rationales which had controlled men's minds for over a thousand years. The notion of the new sort of "brotherhood" I have called "Universal Otherhood" (where all become brothers by becoming equally rather than differentially others), was born in the era of the Reformation and it has remained one of the hallmarks of Protestantism. Today, as I have suggested elsewhere (Nelson 1965j, p. 596 [and chapter 5 below, pp. 81f]), this ethic is being exported to every corner of the earth along with the newest energy packages, automation, the computer, the bank, the department store, the central office and the Board room.

The "Protestant Ethic" so often and so prematurely pronounced dead continues to move among us in every shifting forms, and in few places do these motley shapes promise to have such active spectral lives as in the very lands where this "ethic" is today being opposed—and will tomorrow be opposed—with all the faith, resources, and technique embattled men can command.

FIVE

Civilizational Complexes and Intercivilizational Encounters

A PREFATORY NOTE

The main stress of the present essay is on the importance of variant "structures of consciousness" and the contacts of historically complex societies which have been the cradles and homes of world religions. My chief reason for adopting this emphasis is that it seems to provide exceptionally apt ways of relating to the questions we are obliged to ask ourselves in the comparative study of both civilizational complexes and intercivilizational encounters.

The present draft allows much to remain implicit which will need to be made explicit at a later date. Thus, little detail is offered here on many of the so-called "real (in the sense of material) factors" influencing the shapings of cultural patterns, civilizational complexes and the impacts and outcomes of intercivilizational encounters. Cultural as well as social facts of great importance in these connections are treated lightly; e.g., the forms of language and script, the degree of formalization of the classificatory systems.

The first and last sections of the paper reflect some of the results of case studies which I and others have conducted in this area.

I. SETTINGS AND PERSPECTIVES

My efforts grow out of a number of deep convictions and commitments: Sociology is at a turning point in respect to the horizons it is obliged to confront and the perspectives and methods it is obliged to

---

Originally published in *Sociological Analysis* 34, 2 (1973): 79–105. The original draft of this paper was subtitled "Structures of Consciousness from the Perspective of the Comparative Historical Sociology of Sociocultural Process." A portion of the statement was presented at a special Seminar on "Civilizations" at the Annual Meeting of the American Sociological Association which met on August 30, 1971, in Denver, Colorado. Reprinted with permission.

adopt in order to make sense of the perplexing and tumultuous sociocultural processes of our time. We dare no longer suppose that these processes can effectively be gotten at by confining ourselves to settings which are local, parochial, or instantial in terms of the level of interaction (for example, small groups); the social-system setting (the family); the geographic setting (the local community); the temporal horizon (the current moment, the present).

We are obliged to see that many of the most important phenomena of processes and productions of our time are occurring everywhere across the world and they are occurring most intensely in those levels and in those settings which have been least systematically studied by sociologists or anthropologists. I refer to the *societal level,* the *civilizational level* and the *intercivilizational settings* and *encounters.* In truth, these last-mentioned settings have always been of enormous importance and it would be a mistake to suggest that there have been eras when great complex polities have escaped being in the throes of processes at these levels.

To understand and explain the perplexing sociocultural processes of our time it becomes necessary to acknowledge that we are in the throes of no fewer than a half-dozen revolutions which originate in different contexts, express multiple interests, proceed at varied paces and are hardly likely to be contained by any set of societal remedies or interventions. Among these revolutions one might mention:

1. *The Scientific-Technological-Perspectival Revolution* which has been proceeding with ever intensified pace, scope, and impact since the beginning of the 20th century. This set of developments embraces the knowledge revolutions in theoretical physics, the formal sciences, the biological, social and behavioral sciences; the revolutions in extra-scientific, artistic and symbolic mappings; the revolutions in transportation, communications, systems of control, productions of new forms and uses of energy.

2. *The Spatio-Temporal Revolution* related to the above and most evidently illustrated in the reshaping of our environment and ecology under the impact of the global technologies of transport communications, atomic energy—all resulting in the precipitous shrinking of a world now forced into anguished conjunctions in the midst of abrasive contacts.

3. *The Rationalization Revolution* involved in the spread across the world of systems of instrumental and functionalized rationalization —multi-national as well as national structures of management and labor organization, business enterprise, political and military operation.

4. *The Socio-Morphological Revolutions* resulting from explosions and implosions of populations and dissonant ethnic subcultures.

5. *The Equal Freedom, Equal Share, "Freedom Now!" Revolutions* carried on in the name of social justice and equal participation: "Freedom Now!" for all strata and groups previously deprived of the same—youth, women, ethnic minorities, racial minorities.

6. *World-Revolutions in the Structures of Consciousness and Conscience* include struggles within all strata and groups across the world in relation to each other over the by-products and issues of modernization, industrialization, rationalization and Westernization.

The extravagant promises and the erratic workings of the touted "cybernetic" political and economic structures yield perplexing results. The simultaneous presence of bounties and deprivations mingle hope and despair in a game appearing to operate according to rigged actuarial tables with unfavorable odds. In these straits, millions are goaded to demand full and immediate fulfillment of their *ultimate* hope for joy, love, peace, plenty. A crucial form of this oppositional trend, recently at peak intensity after intermittent appearances over the past two centuries, assumes the character of a comprehensive antinomianism, better described as *A-nomism*. In this view, the norms of scientific rationality and so-called "objective consciousness" which have been distinctive features of European civilization, are identified as the malignant blights that have brought the West to its present disintegration. The cry is general for a new beginning which carries us far beyond—and away from—Solon and Pericles, Abelard and Roger Bacon, Galileo and Descartes, Locke and Jefferson, Max Weber, Freud, Einstein, and others.

All six sets of these newer revolutions were foreshadowed in developments of earlier centuries, especially the 19th century; all gained great momentum in the first decades of the present century; all have been proceeding at vastly accelerated paces since the First World War.

The interference effects which flow from one of these to another now immensely exacerbate, complicate, and intensify the stresses and strains; all are undergoing change at the very same time.

Surely one of the most critical components in the newest phases of the situation is the incommensurability between our technical means and the political and social divisions of our day. As a result of supersonic transport, space travel, and other extraordinary expansions in the scope of action, civilizational structures and complexes which were once in infrequent contact are now in one another's backyard.

There seems little doubt that complex societies are almost always likely to undergo heightened feelings of threat, ecstasy, even virtigo when conclusive contacts occur with other great societal complexes at the level of their respective rationales. Many of the most abrasive aspects of contemporary politics result from the equal inability of all powers to create iron walls which would insulate them against the unwelcome intrusion of alien modes of thought and perception. In centuries past, relations among great powers at this level were relatively rare and ordinarily not fraught with grave perils.

II. FIRST REMARKS ON TERMS AND CONCEPTS

These newer developments place us all under many demanding obligations. An obligation which falls with special weight upon scholars is the mandate to forge a language adequate to express the differences and dimensions central to the establishment of a comparative historical sociology of civilizational complexes and intercivilizational relations. So far our main barrier in this sphere has been our excess readiness to accept ambiguous and inflated versions of commonsense concepts. We ought to contrain ourselves to use our terms precisely so that we can identify their bearings and their components as accurately as possible.

Wherever possible in the following pages, I shall use the key terms strictly rather than loosely. Thus I reject the widespread habit of using the word *civilization* to refer to the sociocultural processes of selected great nations. Instead, I prefer to reserve the term "civilizational complex" for a segment of the paradigmatic cultural patterns in the *sphere of the expressive* and *instrumental productions of societies* or *societal complexes.* In speaking of complex societies in this way, I refer to *political societies* which contain numerous internal differentiations, including nations, classes, institutions, and varieties of cultural experience.

I want to reserve the word *culture* to refer to those structures of expression and commitment which may be said to embody the most distinctive designs developed within and by the people of these societies. In general, for reasons discussed in detail elsewhere, I place strong stress on those structures of expression and commitment which have been elaborated into symbolic configurations of one sort or another.

I wish to speak of many of the habits and the mores of people as *shared behavior* which need to be differentiated from higher level cultural structures.

To speak once again of the central notion of "civilization," I freely admit to sharing a preference for the style of analysis initiated by Durkheim and Mauss.

By the *civilizations* of peoples I wish to refer to the governing cultural heritages that constitute the accepted milieus of 2 + n societies, territories, or areas which generally enjoy or have enjoyed a certain proximity. These strongly-based acceptances will normally be discovered to constitute configurations of the following elements: identities of language, the highest level of the technology of the group, which I would call the "prime material facilitations and skills" of the group; the central patterns of reciprocities including juridical rules; the fundamental canons governing the decision-matrices in the spheres of opinion and act (see chapter 4); the taken-for granted structures of consciousness, comprising cultural world-views, logics, images of experience, self, time, the beginning and the end, the extraterrestrial powers. These schemas have been described earlier (in chapters 2 and 3) under the headings of dramatic designs, directive systems, systems of spiritual direction, and symbolic economies.

III. HIGHWAYS AND BYWAYS IN THEORY AND RESEARCH: THE PIONEERS

The time has clearly come to take some bold steps in going beyond the conceptual frames now available for the comparative study of civilizational complexes and intercivilizational relations.

Actually, a vast amount of material lies at hand in the writings of comparativists in the study of religion, science, and societies. These materials have never been closely sifted and systematically codified. Indeed, in this respect, the views of such investigators as Maine, Durkheim, Mauss, Max Weber, Needham—to speak first only of the pioneers—have never been brought into close contact with one another. The rich literature which has been made available by area specialists in history, anthropology, and political science during the present century has not been gathered together. The result has been that the frames of reference needed for such a codification have been lacking.

My point of maximum concentration here will be on the concepts needed to differentiate among the ruling structures of consciousness in relation to the critical matrices of experience and regulative order, or, if one prefers, the systems of social and cultural action and function. If some of the phrases I use seem unduly related to recent popular writings in this field, it is a matter of chance. As every reader knows, the foremost specialists and theoreticians have long been working toward the development of frames of this sort.

The structures I propose can readily be shown to rest upon the fundamental work of the classical sociologists, anthropologists and comparative historians—notably, as already indicated, the work of Maine, Durkheim, Mauss, and other members of the Durkheimian school, Weber and a great number of contemporary authors dealing with cultural structures, intercultural relations and international relations.

I begin with some issues which hold center stage today.

A. Needham and Weber

In the last two decades Joseph Needham has placed before the world the most important series of challenges made by any of the major comparative historians, sociologists, and philosophers of science since the days of Max Weber.

Needham and Weber are only two of the men of the present century whose work has revealed the need for and promoted the possibility of a *systematic* study of the *comparative sociology of civilizational complexes and intercivilizational relations.* Conceptual advances, some of equal weight, have been made by a number of men, chiefly Durkheim, Mauss, and Sir Henry Maine. A wider list would include such names as: A. Bozeman, J. Huizinga, A. Kroeber, R. Redfield, R. Bellah, C. Geertz, R. Lowenthal and the too-soon departed pathfinders, M. Hodgson, J. Levenson and G. von Grünebaum.

We have no time to lose in striving to collate—to compare and contrast—the views of Needham and Weber and other contributors in respect to the distinctive characteristics of major civilizational complexes. We must, however, avoid falling into the trap of seeing all of Weber's and Needham's arguments and evidences for the differences of patterns of East and West from the polarized points of view they may—mistakenly—seem to represent.

B. Weber

In both of the perspectives central to this paper—the varieties of civilizational complexes and the test cases of intercivilizational encounters—there is no turning away from the findings of foremost historians, archeologists, comparative geographers, anthropologists and so on. Again Max Weber offers the handiest point of departure for the renewal of our discussion. Doubtless realizing the immense complexity of establishing casual change in the cultural sphere, Weber began with the supposition that different societal complexes could be found to have adopted one or another sort of emphasis within the frame of possible orientations to world and religion. By so doing, he assumed the answer instead of working out intermediate steps.

In the main, he did not face several critical questions connected with the developments of phases of the structures of consciousness and conscience, nor did he undertake to test out his hypotheses by examining the opportunities for analysis of intercivilizational contacts. Thus one will look vainly in his pages for a discussion of the immensely revealing Jesuit mission in China in the 16th and 17th centuries (See Needham 1954, esp. vol. 3; Rowbotham 1942; Franke 1967). Similarly his self-denying ordinances prevented him from offering us light on a whole series of revolutionary transformations from the 18th century onward (Franke 1971; also Crozier 1970).

Despite Weber's massive, even incomparable, contributions to the exploration of our themes, it must be admitted that he failed to offer sufficiently discriminated articulations of the theoretical frameworks needed for coping with the challenges presented by his own data. Except in a very few writings to be discussed again below (I refer to the "Author's Introduction" to his posthumously published *Collected Essays in the Sociology of Religion*, and to his monograph on *The Rational and Social Foundations of Music)* his analysis, especially his analysis of rationalization and rationalism, slips into errors of emphasis of the following sort: Driven by complex scholarly preoccupations and methodological commitments, Weber regularly shied away from open investigations of sociocultural process. His key concepts, such as rationalization, social action, social relations, were elaborated again and again in a manner strongly savoring of social psychology. In this spirit he failed to give due emphasis to the distinctive power and influence of intertemporal decision-matrices in all the spheres of human action and thought—what I have been calling the operative structures of rationales in respect to the definitions and organizations of all of men's intertwined worlds: their environments, their experiences, their options.

Despite Weber's great sensitivity to the myriad social effects of historically-institutionalized varieties in the orientations of meaning, despite—perhaps because of—his extraordinary sophistication in methodological matters, Weber shied away from dynamic analyses of the processes of change and the implications of these processes in the structures of consciousness. In the same way, Weber all too rarely exploited the chance to follow up the great insights of his later years which are most clearly represented in the "Author's Introduction" and the monograph referred to above.

Again and again Weber assimilates too quickly all the forms of rationality and shaping of rationales to the model of means-ends instrumental rationalization. His overriding desire to establish his theses in respect to the controversies over the spirit of capitalism led him to give short shrift to many perspectives especially critical for the comparative, in-depth, historical study of sociocultural process in civilizational perspective.

I refer mainly to the following: the changing shapes and workings of the different historical crystallizations of the structures of consciousness within the culture-areas deeply influenced by great world religions; the variabilities in the terms of reference of the opposed cultural logics; the phases in the elaboration and the mechanisms in the workings of the symbolic technologies.[1]

Despite these reservations, one cannot be too grateful to Weber for opening up the problems of the rationalization process and the phases in the progress of the Spirit of rationalization. Who knows what Weber might have gone on to do had he lived on after 1920? I like to think that he would have gone forward to give due weight to the problem of the importance for social development of the spreads of universes of discourse, generalizable terms and concepts, formal logics or codes with generalized terms accessible to wider universes of discourse. In the end it was his failure to connect these with the extensions of wider communities, the openings out of new accesses of participation, which may have blinded him to the wider horizons of the comparative history and sociology of socio-cultural process in civilizational perspective.

C. *Needham*

As odd as this may sound to many fellow admirers of Max Weber, I must plainly say that in my point of view Needham represents an advance over Weber in respect to a clarification of the dimensions of the questions we have just discussed. Indeed, I would argue that Needham goes far in offering answers to the questions of universal import he has been putting to himself and others. The difficulty is that he puts these questions in different ways in different places and so gives the appearance of inconclusiveness both in the forms of the question and in the forms of the answers.

The two main weaknesses in Needham's analysis are: (a) his very incomplete knowledge of the patterns of development of the West and the levels of development attained from the 12th century forward; (b) his almost total neglect of the work and insights of the theoretical and empirical advances of sociological forerunners in the study of civilizational complexes.

Notwithstanding these limitations, Needham proves correct on many scores: (1) He senses that the Chinese found it very difficult to get to a notion of a law of nature comparable to that which was rooted in Greek thought. The context-overdetermined stress of the Confucian ethics on the Five Relationships and the aesthetical-magical Taoist images of Yin and Yang tended to preclude the emergence of a general law of nature. (2) Needham correctly perceives the essential importance in the new mathematical physics of structures of abstract rationality that were universalistic in character. His stress on universality represents an advance over Weber's discussion of rationalization which often thwarts itself by over-concentration on the means-ends maximization model of instrumental reason. (3) Needham has a certain perception of the inhibiting effects in China of the magical base of the imperial structure and the Mandarin aristocracy.

Regrettably, Needham is unable to free himself sufficiently from his polemical stains to recognize that the fundamental structures of abstract rationality in the West received very full elaboration well before the time of Galileo in fields other than experimental science. He misses altogether the critical importance of the breakthroughs in the structures of consciousness and the rationales of conscience, opinion and act, which occur in the new universities, and the new forensic, associational, and political settings starting with the 12th and 13th centuries.

D. Durkheim and Mauss

Oddly enough, as important as Weber's works are, we have to turn to Durkheim and Mauss for fuller clues in respect to the shapes of the contents of the civilizational complexes. Durkheim and Mauss were especially sensitive to the extent to which cultural structures of a certain degree of generality could come to serve multiple purposes as vehicles of different ideas and interests. The neglected *"Note on the Notion of Civilization,"* done by these pioneers in 1912, is a model of intensive reexamination of conceptual frames which had hitherto proved a barrier to the development of a proper comparative sociology of civilizational complexes and inter-civilizational relations.

Durkheim and Mauss arrived at the very notion of civilization by courageously facing up to the anomalies of the Durkheimian school's foundation theorem. The original insistence on linking every social process to a social body broke down in the course of systematic reflection on the spread of world religions and universal sciences far from their point of origin and original audiences across boundaries of every sort. It took great courage on the part of Durkheim and Mauss to acknowledge that they needed to modify their basic frames by introducing a fresh term to embrace those phenomena which were intersocietal and intercultural; and so they went to work on the analysis of "civilization." They referred to this level of process as the civilizational level and they included within this level languages, religious ideas of a high degree of abstraction, scientific notions, and rules holding between nations (the so-called *ius gentium* or the *ius inter gentes*).

E. Maine

Sir Henry Sumner Maine will one day be recognized as having performed great feats in expanding our powers to relate to the two main varieties of challenges we have emphasized in this paper. The criticisms which have been directed against details of his work are of little account when compared to the extraordinary suggestions of his distinctions among the sorts of complex societies and the possible grounds of the differentiations in their patterns. Here I restrict myself to speaking only of the following distinctions:

1. As everyone must recall, Maine was passionately concerned with explaining the differences in both the patterns and pathways between what he calls *stationary societies* and what he calls *progressive societies.* By "progressive societies" he meant those groups ready to attempt the amelioration of their own plight by deliberative innovation and legislation. In "stationary societies," fear of change results in an extreme reluctance to go beyond settled habit and prescriptive rule. Maine's analysis of traditionalist casuistries of improvement such as "legal fiction" and "equity" allows us to recognize these as very ingenious answers to the problem of changing without seeming to change.

2. At the base of Maine's thought is his awareness of the critical importance thoughout history of the extension of the boundaries of the moral and political communities. He fully perceives that the passage to the territorial society represents a very great move in the direction of civilizational complexity. In the language I have applied in my previous studies, we see a passage "From Tribal Brotherhood to Universal Otherhood." Associational ties reduce the authority of familial control and piety. Increasingly the unit of which the civil law tends to take account is the individual and not the family.

What may well be the single most important contribution of Maine to the discrimination of complex societal and civilizational structures has been noticed by hardly a handful of historians. Maine paid immense attention to the stages in the development of the directive systems and symbolic designs of great societies. He came to believe that key variations occurred in the following settings, among others: (a) the original shapes of the directive imperatives; (b) the extent to which the original judgements were open to review; (c) the linkages between the strongly-sanctioned imperatives and theological ritual designs; (d) the influence of the date of the codification of the structure; (e) the changes effected through passages to abstract universals of philosophy.

On all these scores Maine came to see that it was possible to establish strong differences in the relations of components of sociocultural process in India, Greece, Rome, and the Hebrew nation. In India a long delay in the codification of the directive imperatives led to the total sacralization of bodies of custom, habit, ritual, and theology, resulting in a deep inability to modify the structure through supply casuistry or fresh legislation. The immense variety of languages, social groups and prescriptive rituals in India stood in the way of a full ventilation of the principal structures and a full rationalization of intelligence. The moves to neutralization, generalization, universalization, rationalization were checked at every turn.

F. Once again: Weber, Needham, China and the West

Having begun this section with Weber and Needham, I return to their names and work in ending it.

Undoubtedly the most exciting of all intercivilizational encounters is that which involves the relations of China and the West in the 16th and 17th centuries and China and the West in the 20th century. I will speak now only of the first of these sets of encounters: Clearly Matteo Ricci and those Jesuits who were brought to China came to play a considerable role in the development of Chinese technology and science. It was the need to maintain the sacro-magical system in respect to astronomy, chronomancy, and geomancy which allowed Ricci, Verbiest, Schall, and others to come to such high positions in the framework of the imperial departments. The Jesuits sought to accommodate themselves in every respect to the Mandarinate and to the Confucian style; so much so that they were prepared to make notable concessions in linking Confucius to Christ and in making room for elements of Catholic ritual. In the end the Jesuits were obliged to leave China because of nativist reaction against the influence they were asserting on the emperors. Moreover, the exceptional liberties the Jesuits took in respect to the theological and ritual accommodations occasioned an international controversy which was settled only at the topmost level through a decision of the Papacy and the Papal Congregation on Rites.

From our point of view, one aspect of the contact of the Jesuits and the Chinese cannot be too strongly emphasized. Full breakthrough to generality in the *terms of reference and universalities in the communities of discourse* never occurred during this era of critical contact. The words used by Ricci to describe the motives behind the translation into Chinese of Euclid's *Elements* are helpful in this regard:

> It was during this time that the Fathers undertook a work which at first sight might not seem to be wholly in keeping with the purpose of their mission, but once put into practice proved to be quite beneficial. Doctor Hsu Paul had this one idea in mind: since volumes on faith and morals had already been printed, they should now print something on European sciences, as an introduction to further study, in which novelty should vie with proof. And so, this was done, but nothing pleased the Chinese as much as the volume on the Elements of Euclid. This perhaps was due to the fact that no people esteem mathematics as highly as the Chinese, despite their method of teaching, in which they propose all kinds of propositions but without demonstrations. The result of such a system is that anyone is free to exercise his wildest imagination relative to mathematics, without offering a definite proof of anything. In Euclid, on the contrary, they recognized something different, namely, propositions presented in order and so definitely proven that even the most obstinate could not deny them. (Ricci 1953, p. 476)

Nor should it be assumed that the use of the telescope for the purpose of improving predictions of eclipses in any way involved a basic breakthrough in the structure or generative role of the sciences in respect to the production of knowledge.

The power of the sacro-magical structures is not yet ended in China, but there is substantial evidence that the faith-structures and the rationalized structures of conscience are gaining ground in this era of "Cultural Revolution" and "The Thoughts of Comrade Mao." Older particularisms are giving ground to the newer universalities implicit in the notions of nation, people, and pure and strict science. Few documents on cultural themes are so revealing of how far the Chinese have gone toward universality in their policy of "Westernization with honor" as Chou-en-Lai's statement on language reform. Chou's explanations of the decision for the Latin over the other alphabets (including the Cyrillic!) need to be read by every student of the civilizational politics of the time of transition to the 21st century and the third millennium:

> . . . the Latin alphabet is a set of symbols of international usage, and no country can claim it as its own. We cannot say that Frenchmen use the English alphabet any more than we can say that Englishmen use the French alphabet. We can say only that Frenchmen use the French alphabet and the Englishmen use the English alphabet. Likewise, when we adopt the Latin alphabet, in which we make necessary adjustments to suit the needs of the Chinese language, it becomes the phonetic alphabet of our language and is no longer the alphabet of ancient Latin, still less the alphabet of any foreign country. An alphabet is a means of transcribing pronunciation. We make it serve us just as we make trains, steamships, automobiles, and airplanes serve us. (And, from the point of origin, all these are imported.) It is also like using Arabic numerals for counting and calculating, the Gregorian calendar

and the Christian era for recording the year, kilometers for measuring distance, and kilograms for measuring weight. The adoption of the Latin alphabet will, therefore, not harm the patriotism of our people. (In Crozier 1970, pp. 128–131)[2]

IV. CIVILIZATIONAL COMPLEXES, ENCOUNTERS AND STRUCTURES OF CONSCIOUSNESS

The comparative historical sociology of civilizational complexes and intercivilizational encounters can only be approached a step at a time. All becomes a jumble if we plunge forward with undue haste to account for the immense variety in the shapes of these complexes. Many long-established clichés readily prove to rest on slender foundations of fact, to conceal as much as they reveal. A number of other cautions will help. We would err if we supposed that the critical civilizational patterns have acted over great distances with constant force. Whenever adaptation occurs it occurs selectively.

The evidences are equally plain on a number of related points. To speak of a single civilizational atmosphere in the history of any large societal environment is to miss the fact that conflict over civilizational ascendencies is endemic in large societies. From our point of view there is no warrant for confining "civilization" to one set of constructs, to one sort of fabric.

Civilization can, indeed, be based on diverse geometries. However, insofar as civilizations come to serve as cultural prerequisites for the relatively enduring organization of different sorts, they will generally be found to involve patterning of related elements. These patternings are found to comprise arrangements of coordinates defining cultural ontologies, epistemologies, and logics; directive systems, dramatic designs, and sociopolitical frameworks; and technologies of different sorts—symbolic as well as material.

I turn now to a series of differentiations in the *structures of consciousness* as they are revealed by close studies of the histories of those great civilizational complexes which have been associated with great world religions. We need to discriminate a series of patterns in the structures of consciousness and the degree of collectivity or individualization in the forms of representation. I make no pretence of offering a complete inventory of these structures or of setting forth an evolutional sequence which is irreversible. On the latter, I specifically reject the idea of "irreversibility" as being unhistorical. I simply refer to three types which have been historically prominent across the centuries and which can easily be discovered to be useful in defining the shapes of civilizational complexes and the impacts and adaptations which arise whenever there have been conclusive contacts of the major societal complexes.

A. Consciousness-Type 1

The first pattern of structures of consciousness is characterized by the predominance of collective acceptances of responsibility to make amends for collective wrongs or falls from a state of undividedness. The ruling supposition is that all lapses from unity in both macrocosm and microcosm have to be atoned for by collective acknowledgements of crime and wrong, assumptions of liability, propitiations, sacrifices and commemorations.

From a standpoint of comparative sociology of cultures the leading clues have been provided by Max Weber. What I have in effect described is what Weber would call the sacro-magical structure of consciousness. The ruling instances of these sacro-magical forms may be found in all societal and civilizational complexes dominated by prescriptive etiquettes and rituals, all oriented to the total fulfillment of laws believed to be ontological in character and to have their sanction in the cosmic orders, the commands of the ancestors and the primordial traditions that require that they be preserved through literal performances of fixed obligations. A number of variations are possible in the social structural guarantees of these ritual performances. These differences, however, do not fundamentally alter the central supposition that the total society is under total obligation collectively to fulfill the ontological demands and to offer totalistic propitiations, placations or commemorations, if the harmonies of the cosmos are to endure. Weber illustrates the authority of such systems in ancient India, China, and Judaism, and sees the break from those systems as constituting one of the decisive passages in the direction of rationalization. It needs to be perceived that structures of this sort lie below the level of the social religious formations he so powerfully analyzes in his *Collected Essays in the Sociology of Religion,* especially in those segments having to do with India, China, and Judaism.

Weber's frames need to be supplemented by reference to the work of Durkheim, Mauss, and Maine, especially the essay by Durkheim and Mauss on *Primitive Classification* and Durkheim's introduction and conclusion to his *Elementary Forms of the Religious Life. Consciousness-Type 1,* as we are here describing it, may be spoken of as a structure where, for whatever reason, collective representations prevail and affirm the absolute authority of magical-prescriptive structures which are fulfilled by all groups without notable deviation. Within this structure of consciousness developments may occur in the direction of a wider rationalization, but if these developments are essentially elaborated as reinforcement of the magical-prescriptive systems, they need still to be spoken of as instances of *Consciousness-Type 1*.

From the point of view here adopted, *Consciousness-Type 1* may be ascribed no less to the Chinese than to the Hindus; no less to the Hindus than to the Jews, even after the appearance in China of the Confucian ethic and the Mandarinate, or in Judaism even after the prophets. The central factor is the dominance of a sacro-magical structure which binds the entire community in propitiations, commemorations and expressions of the collective conscience.

B. Consciousness-Type 2

From my point of view, a decisive breakthrough occurs in the emergence of another structure of consciousness, *Consciousness-Type 2*, in which all are enjoined to enter into meditation, producing a faith-structure of consciousness. The faith-structure requires that all individuals and groups, however differentiated, are under the obligation to engage in continuous purgation and catharsis of evil thoughts and feelings and to discover a way by which they can appropriately conform by being *informed* by the faith-consciousness. The faith-consciousness already implies a kind of *logos* or world soul in which participation is accessible through psychic conformities. The passage to the faith-structure of consciousness is illustrated most clearly in the passage from Judaism to Christianity and has been very powerfully described in the work of Edwin Hatch, *The Influence of Greek Ideas and Usages in the Christian Church*. Faith-structures of consciousness are already a premonition of a next phase in the development of consciousness, the move toward the rationalization of the contents of faith; that is, the systematic analysis of the contents of and evidences for faith, the appearance of a science called theology. In *Consciousness-Type 2*, where faith-consciousness prevails, the form of participation, whether collective or individual, is existential and is manifested in practice, activities, imitations, exemplifications.

The faith-structures allow for many complex differentiations. One critical differentiation is the appearance of specialist virtuosi in meditation and in the achievement of mystical union whose skills in the attainment of "essentiality" are believed to provide vicarious expiation for others who have not been able to attain the level of expertise necessary to qualify in their own right. I would observe that here I am departing somewhat from the ways in which Weber talks of the cross-cuttings of his critical types in respect to the modes of orientation to religion and world. The faith-structure of consciousness may, indeed, issue in a new legitimation of the inherited patterns in the sacro-magical prescriptions or they may lead to a radical devaluation of the activities of the world and produce the patterns that Weber ascribes to other-worldly asceticism. They may, indeed, issue in the intermittent treatment of the sacro-magical and even moral values as second-order values.

Faith-structures of consciousness represent a phase of many different religions. The above-mentioned book by Edwin Hatch permits us to see the transition from the prescriptive structures of forensic Judaism to the faith-structures of consciousness resulting from the permeation of Judeo-Christian beliefs by Greek and Roman influences during the Hellenistic and Roman Imperial eras.

The soul's commitment to faith does not guarantee that a unity will be achieved between the moral and the theological. Actually, the faith-structure of consciousness is one which treats the moral order in different ways at different times, sometimes as the direct revelation of the theological, sometimes in a way which is relatively indifferent.

Where faith-structures of consciousness prevail, one often discovers the presence of groups and sects (claiming to be exempt from natural causality and natural law) which are supposed to be in possession of special revelation to commit acts not ordinarily allowed. Faith-structures of consciousness are manifested strongly in mystical acosmic, mystical millenarian, mystical militaristic sects, and inner-worldly mystical groups. Among these groups marked by faith-structures of consciousness were the Gnostics, the Sufis of Islam, the *Perfecti* among the Cathari, the Families of Love and other groups of enthusiasts who felt themselves possessed by God's grace and power.

Not infrequently faith-structures of consciousness are characterized by a *transmoral consciousness,* a conscience beyond good and evil. Those who are perfect and enjoy direct illumination feel themselves free from the obligation to live in accordance with the ordinary moral or civil law.

The key to the faith-structures of consciousness is that individuals committed to faith feel themselves to be part of the truth, a manifestation of the divine in expression of the universal will or sovereign design. Existence in the faith is truth.

C. Consciousness-Type 3

Consciousness-Type 3 comes into operation when the collective structures of representation have weakened and large segments of a population are not agreed as to the contents of the faith, the evidence of the faith and the implications of the structures of faith and action, belief and opinion. The most decisive illustrations of this breakthrough to the *"rationalized structures of consciousness"* are to be found in the 12th and 13th centuries in Europe when the new logic and the new dialectic bring forth a new theology.

The principle architect of this new theology is Abelard, who, in a series of extraordinary writings, clearly set forth the method by which the contents of faith are to be differentiated and sifted so as to produce a logically provable structure of *rationales* which undergird belief in all spheres of opinion and act. The most critical encounter in the transition between *Consciousness 2 and 3* is the conflict of St. Bernard of Clairvaux and Peter Abelard. (See chapter 12, pp. 218f)

It may be admitted that the terms of my reference may recall the thought of many writers from whom I would separate myself, the principal one being Auguste Comte. My position clearly implies that a universalistic theology represents a decisive breakthrough, the emergence of a prime science. The distinctions between what I would call the theological, metaphysical, and positive structures of consciousness are in my view distinctions within the *Consciousness 3*, rationalizėd structures of consciousness. The differentiations within *Consciousness 3* are not nearly so difficult to attain as are the passages from *Consciousness 1 to 2* and from *Consciousness 2 to 3*. Indeed, I would contend that the passage from *Consciousness 1 to 2* probably represents the most decisive move in the direction of the functionalization of intelligence, the generalization of the terms of reference, and the universalizing of the community discourse. It hardly matters that the passage to a faith-structure of consciousness allows the intermittent dominance within communities of faith of men especially skilled in the achievement of states of contemplation and undividedness. The key fact is that so long as the fundamental frames are faith-consciousness, all who have some access to the faith-frames may lay claim to being messengers of the faith, exemplars of the faith, or incarnations of the faith. In such a situation social class and caste may readily be broken through by manifestations of power or charismatic authority in respect to the annunciation.

The differentiations which have developed within *Consciousness 3* were already implicit in this pattern from the very outset. No sooner is a rationalized theology under development than we have separations of the parts of the analysis with the help of Aristotle, Plato, and others. There is immediate opportunity to distinguish between moral philosophy and natural philosophy. The terms of discourse are absolutely general and increasingly available through formalization into abstract languages which are tehnical (artificial in the strict sense), and open to mathematical formulation at one or another level of abstraction. Moreover, the forms of proof appropriate for these different spheres are readily discriminated. The passages to formal sciences, theoretical sciences, and experimental sciences, are much easier than have hitherto been alleged.

I call attention here to only one of the many inferences which may be derived from the immediately preceding discussion. Those who wish to engage systematically in the comparative study of civilizational complexes must always attend closely to the fundamental focus of differentiations *within as well as among* the structures of consciousness. Idiomatic mixes and balances of elements of all the respective structures above may be certain to appear on close inspection of the civilizational areas and the workings of large historically-embedded societies. Scholarly studies in this field have to be as deeply concerned with stratigraphy as are archeology and geo-history.

It is only by adopting these wider points of view that we can escape falling into traps which await the unwary. Thus, for example, due attention to cultural stratigraphy helps us to make sense of the wide-spread recurrences in our own day of aspects of sacro-magical structures of consciousness mistakenly believed to be extinct. By the same token we avoid too quickly identifying as a Western society one whose seeming differentiations within the rationalized sector only serve to call attention to the insignificance of rationalized intelligence or universalistic intent at the focal point of the society's logics and its agencies of decision and function. Comparative historians and sociologists are familiar with many highly bureaucratized societies unable to break through traditionalism at critical levels with respect to the structures of consciousness, despite public commitments to Westernization.

V. BALANCES AND ASCENDANCIES OF INTERCIVILIZATIONAL ELEMENTS: HISTORICAL SKETCHES

Rationalist Elements in the West

Without saying so in so many words, Max Weber clearly identified Greek rationalism and Roman rationalism as twin pillars of the development of Western civilization. He would almost certainly have agreed that Greek rationalism may be said to represent the highest classical development of the logics of opinion, inquiry and demonstration; and similarly Roman rationalism, the highest classical development of the logics of organization, of action and of administration. We dare not stop here to ask why Weber did not deal more directly and at greater length with these recurring thrusts and interweaving influences (Nelson 1949a, 2nd ed. 1969).

From the point of view adopted in this essay, it is necessary to note that different eras and movements in the history of the West have been marked by different emphases on one or another of the crystallized structures of classical rationalism. (I say nothing at the moment about the analogous issues of the degree to which these variations in the ascendency of one or another expression of given structures of consciousness prevail in one or another Eastern society.) Surely so far as Europe is concerned it may be said that where Greek rationalism has been predominant it has often been the case that the critical decisions as to action and function have had to meet the tests of theoretical science and philosophy. Where Roman rationalism has been prominent a strong tendency has regularly asserted itself to declare or derive the ontologies from the requirements or imperatives of *praxis* or the logics of control. Our own time has offered myriad illustrations of the hegemony of the instrumentalized logics of rationalized control.

Classical Greece, especially Athens, offers the paradigmatic case of a society and culture in which we see the most conclusive passage from the tribal society to the polis, from the family to the public institutions, from particularism to universalism in spheres of science and philosophy. Nowhere so greatly as among the Hellenes did there occur a movement to achieve structures of the highest degree of universality. It is therefore no wonder that Western civilization rests upon a Greek foundation: Plato and Aristotle, Euclid and Archimedes, Ptolemy and Hero of Alexandria. And it rests as we have just seen, upon a Roman foundation—Roman law and administration.

The subsequent development of civilizational complexes and intercivilizational contacts proves to be a study of the viability of integrating the classical inheritance. This is the meaning of the processes of translation and adaptation of Greek metaphysics, Greek dialectic, and Greek science. If we are to understand the different degrees of adaptation of the Greek inheritance we have to understand the measure to which the Muslims, medieval Hebrews, and medieval Christians—and for that matter, the early modern Chinese and Japanese—allowed their own structures to be permeated by the underlying universality of Greek logic and dialectic.

A great blunder is often made here by scholars of narrowly construed behaviorist-positivist outlook, that is, those who identify the progress of knowledge with "positive progress" within *positive sciences*. From the *civilizational* point of view adopted in this essay, it is not nearly so important whether in any given science a given people did or did not actually make an advance upon the Greeks in respect to one or another discipline—for example, chemistry, optics, or mathematics. The fundamental issue is whether there did occur a comprehensive breakthrough in the moralities of thought and in the logics of decision which open out the possibility of creative advance in the direction of wider universalities of discourse and participation in the confirmation of improved *rationales*.

AN AXIAL SHIFT – THE 12TH AND 13TH CENTURIES

The great break in the balances of intercivilizational ascendencies comes in the 12th and 13th centuries. It is the era in which we have the Crusades marked by highly abrasive contacts at every level—military, political, cultural, and theological. In these conflicts we have encounters of Western Christianity and Islam, Western Christianity and Byzantine Christianity, Western Christianity and the Mongols, Western Christianity and China, Western Christianity and Africa, Western Christianity and the Jews.

I will confine myself here to speaking only of selected aspects of these contacts. My main stress will be on the complexes and contacts of Western Christendom, the Muslims and the Jews.

For reasons we do not yet fully understand, the 12th and 13th centuries saw an extraordinary resurgence of Western power on the battlefield, in commerce, in science, in theology, natural philosophy, and mathematics. Translators and scholars did extensive work in recapturing the Greek heritage through the medium of Arab and Hebrew texts. The great advances made by Arabs came to be incorporated into Western science. This is especially true in the fields of optics, algebra, and medicine. Also involved in this diffusion of culture were the more critical components of the Chinese and Hindu inheritance. From the Chinese there came the compass, forms of sailing, and the approach to printing with the use of paper. From the Hindus came the so-called Hindu-Arabic numeral system which reached the West in 1202 through the translation of the so-called *Book of the Abacus* by Leonardo Fibonacci of Pisa.

The most critical borrowings came from the Muslims. It was they who placed at the disposal of this period much of Aristotle and Plato that had been lost, together with critical commentaries by Averroes and others.

It was this new body of writings, constituting the new logic and the *Libri naturales*,[3] which were to be the point of departure for the great searching of the Western spirit. The immense changes which follow in the wake of the new logic have already been referred to in connection with the development of the rationalized structures of consciousness in the spheres of opinion and action identified with Abelard and his successors. From the year 1210 to the year 1325 there occurred a complete overhauling of the structures of legitimation and theoretical *rationales* of Christian theology and natural philosophy. Natural theology was a half-way house; it contained the promise of natural science. The distance between Adelard and Abelard in the 12th century and Galileo in the 17th is not so great as one may suppose. Bernard of Clairvaux, Walter of St. Victor, the Cistercians, and the Victorines sought desperately to restrain the progress of Aristotle's natural philosophy. In 1210, at a synod of Paris under the direction of Robert de Curzon (the extraordinary Cardinal Legate of Pope Innocent III), Aristotle's *Libri naturales* were declared to be illegitimate material for study until corrected. In 1255 Dominican students were summoned to familiarize themselves with Aristotle's natural philosophy. In 1277 Bishop Stephen Tempier of Paris condemned 219 propositions of which a good number were believed to be derived from the commentaries of St. Thomas Aquinas. In the same year, propositions developed in the schools were placed under condemnation by Robert Kilwardby, the Archbishop of Canterbury. (See below, pp. 127ff.)

Despite this, despite the censures intended against Thomas and the censures against William of Occam, the scholastic theologians moved ahead claiming the liberty of philosophizing, the right to entertain as speculatively probable any and all propositions which were not utterly known to be theologically false.

In the year 1325 Thomas Aquinas became a saint, the Angelic doctor of the church, a required master for all Dominicans and for many others. This story is a shorthand of the extraordinary advances in legitimation of borrowed *rationales.*

Therefore, no mystery is attached to the fact that despite the very evident superiority of the Muslims and Hebrews in the special sciences in the period from the 9th through the 12th centuries, it was in the end the West which broke through the controlling power of the faith-structures of consciousness and moved toward universalities in the terms of reference and the communities of discourse.

In Islam the *'ulamā'* prevailed: the highly developed sciences never had the power to overcome the ruling forensic theological frames.[4] The Islamic structure was powerfully undergirded by the ritual-prescriptive and faith-structures of consciousness, and between the two structures it was extremely difficult for the terms of reference to acquire generality and the communities of discourse to acquire the universality necessary for the passage to the wider functionalization of intelligence and the wider formalization of language. The great struggle in Islam occurred in the course of the battle over the *kalam.* The *'ulamā'* never achieved the degree of rationalization which was achieved in Romano-Canonical jurisprudence; the various schools of Muslim law continued to be based upon [the Qur'ān and] *hadith* ("tradition").

It must also be noted that a very considerable set of moves occurred within the framework of the Hebrew community across the entire world. Special sciences, especially medicine, did undergo development. Moreover, as in the case of Maimonides and then later Crescas, their work illustrated a passage toward theologies deeply philosophical in character. However, the dominant tradition is the *halachic* tradition which maintains the prescriptive structures and the juridical frameworks of the Mishnah and the Talmud. The result is that there was no escaping the immense power of traditionalists in defining the structures of experience and the terms of discourse for any investigation. Wherever Judaism remains conservative or orthodox, the same essential realities prevail and the breakthrough to *Consciousness 3* is restricted. In Israel today, the appearance of a full breakthrough rests upon a certain separation of spheres which is predicated upon *ad hoc* understandings.

The clue to the differences among Muslim, Jewish, and Western Christian adaptations of the Greek heritage may be found precisely here. Muslim and Jewish structures were stubbornly communalistic, resting upon the articulation of the Holy Law in every aspect of existence. The enormous authority of the *'ulamā'* in Islam and the *halacha* in the medieval Hebrew community stood in the way of full rationalization of both of those structures.

It was only in Christianity, where the breakthrough to natural theology had occurred early, where the accelerated passage to universalistic structures came with the new logic and the new science of the 12th and 13th centuries, that the way was prepared for the distinctive patternings of Western civilization, the accelerated thrust toward modernization, rationalization, and universalization.

VI. A LOOK BACK — A LOOK AHEAD

Universalism and Communalism

The history of intercivilizational relations over the past two millennia has repeatedly involved encounters of complex associational polities marked by advanced bureaucratic structures with nations in earlier phases of communalistic identity. From the time of Alexander the Great, there has been a series of struggles between complex structures claiming universalistic missions and community-societies committed to the preservation of communalistic identities and ritual magic or faith-structures of consciousness. The patterns of response which have developed in the wake of these conflicts have been very similar since the days of Alexander the Great. For the most part, universalistic notions have spread from an imperial center to various peripheries in all the societies involved.

New mediatorial elites, sometimes freshly constituted as a new aristocracy, have participated in syncretistic religions and the elaboration of patterns congruent with the rationalized consciousness and the universalist *rationales* of the bureaucratic society. In all such countries there have arisen "prophetic" groups which have sought to preserve rooted solidarities, pieties and identities in movements of *recollectivizations* of consciousness. The struggles of the Jews against their Hellenistic and Roman rulers from 168 B.C. to 135 A.D.—from the time of Antiochus Epiphanes to the Destruction of the Temple and the Bar-Kochba Revolt—are outstanding illustrations of this.

So long as the *oikumene* is able to maintain itself with a reasonable measure of stability there is a preservation of the forms of rationalized consciousness which are characteristic of civilizational complexes of the West.

The demise of the Late Roman Empire manifested itself in the spread of competing mystery religions and the new forms of apathy and indifference. The prevailing structures of consciousness which evolved from the breakup of the empire were mainly faith-structures, the intensity of the conflicts among the competing theological movements testifying to a great deal more than political factionalism or, indeed, even to economic hardship. They testify to conflicts in the structures of consciousness being etched out of the welter of new experience.

Revolutions and Rationales

All countries of the world, since at least the French Revolution, have been and are the scenes of bitter struggles over the relative values of rooted collective consciousness and schemes of rationalized intelligence. Throughout the 19th century, the struggle over the structures of consciousness took a variety of forms within individual nations. Strong pressures to modernization and Westernization marked the history of lands such as Russia, Japan and India. But these pressures were matched by recurring demands for rejection of Westernization. Slavophilism, Japanese nationalism, Pan-Islamism, and latterly, African nativism, have been the results of religious and ethnic, as well as political and economic interests.

Today we are again at a turning point in the successive civil wars and revolutions of the cultures of conscience and consciousness. This time it is the

Western rationalized consciousness which is on trial in all countries of the world. The frightful events of our century have convinced great numbers that the civilizations reared on the so-called "myth of the *objective consciousness*" are beyond repair.

So far as Europe is concerned, Italian Fascism and German Nazism marked the mass inception of a critical reversal in the itinerary of Western civilization. The fearsome views and acts associated with the Fascists and Nazis were expressly declared to have the purpose of preserving the purity of the collective racial and national unconscious from contamination by alledgedly soulless deracinated *Westernized* intelligence.

The "cultural revolutions" which are occurring everywhere across the world today are marked by intensive thrusts of collective process. Vast efforts are in progress to forge new collective identities and new forms of collective experience and expression. These collective forms are now seen most clearly in youth movements, but they are to be found wherever separatist groups pursue the right to collective identity.

These movements in a collectivist direction suggest that the distinctly Western efforts to develop the individual personality and the individual conscience may have reached their zenith. Thrusts in the directions of personhood and rationality have been dominant elements in Western civilization from the 11th century. Respect for developed structures of *rationales* in relation to what I have called both the *moralities* and *logics of thought and action* have been steadily declining in the present century. It now appears that these thrusts may have reached their zenith in the years between 1890 and 1920.

From a current perspective, it seems that the half-century since 1920 has been one of especially intense passages towards collective conscience and collective consciousness. Those who call us to the collective reappropriation of redeeming symbols through the collective unconscious may indeed prove to be more popular prophets than the defenders of rationality. The tides are now turning strongly against the very idea of *rationale,* and, indeed, every image of *nomos* and *nomism.* On every side we are seeing efforts to achieve new forms of corporate embodiment, new fusions of identity, and new patterns of non-contractual consensus.

The shifts we have witnessed and are now experiencing in the spheres of collective sensibilities and in the structures, *rationales,* and streams of conscience and consciousness, are no less critical for the future of culture and civilization than the changes in the economic and political spheres which have followed in the wake of heightened industrialization.

VII. SOME CLOSING PREDICTIONS

There is a high probability that the latest forms of the objective structures of consciousness and conscience now being deplored by the Western critics of civilization will move forward in the Orient. Wherever external forms are accompanied by historic residues—for example, of Greek thought, Roman law, medieval theology, Renaissance humanism, 17th century science and political philosophy, and so on—we may expect them to undergo the sorts of restrictions and compartmentalization which have already been manifesting themselves in the Far East. Recent developments in Japan offer an important paradigm of this process, and I believe that India will undergo similar phases. And the experience of China is likely to be the same.

It may safely be predicted that Marxism will play a strategic part in these processes during the next phases. The reason is, as I have hinted elsewhere, that Marxism appears to offer an immensely potent and adaptable mechanism for encouraging modernity and Western rationalized modes, while preserving flexible frames for the pragmatic maintenance of certain collective, non-reflective patterns. Precisely because it has proved its ability to perform this feat in a number of lands in West and East alike, Marxism promises to be the most critical form in which a Western *national* form of socialism apparently free from the blemishes of German National Socialism or Italian National Socialism will spread across the globe.

Marxist structures in variant forms are proving compatible with storm and stress phases of breakthroughs in India, China, and Japan, just as they were in Russia. Under the mantle of Marxism in the variants that are arising, there can occur the adoption and adaptation of Western science and technology. This is true despite the restraints which are inspired by different cadres from time to time; however painful they prove to many in the interim, the retardations prove momentary and transient. In the same vein, there can also occur the adoption and adaptations of certain philosophies of the West and even certain logics and structures of the West, but not exactly what I have been calling "moralities and logics of action." Wherever Marxism takes hold, we see the expansion of populistic collectivism and new balances of social elements, workers, students, and political party candidates—balances which will always appear to be heading toward an open future rather than to a maintenance of undesirable privileges and prerogatives of the past.

As this statement draws to a close, I will admit to a private doubt that many "developing" countries will not find it possible or desirable to preserve older or create newer civilizational cultural structures. I do not mainly refer in this connection to advanced facilitations or high science. I refer, rather, to the patterns of reciprocities, mutual answerabilities, and relations of trust; above all, I refer to the *rationales* and canons of opinion and act; respect for and protection of rights and interests of persons, including their *abstract* rights and interests. As I see matters, the new nations are unlikely to have structures defending the juridical rights of individuals, whoever they may be, against the powers of government, or to allow for the development of structures that will flourish without impairment in neutralized places; nor will it be possible for there to be neutralized cultural institutions which will carry on their own functions within this totalistic environment. Constitutions, congregations, churches, universities, independent judiciaries are losing their attractions for embattled people.

Given the dynamic totalism of the newer societies which are in the making, we dare not expect the full or even partial institutionalization of those structures which seem to me to be integral to inherited notions of civilization. It seems highly unlikely that the remaining years of the present century will witness an encouragement of other prerequisites of civilization, above all, widely dispersed rights to join together voluntarily for mutual benefit in cultural and associational purposes; freedom of conscience and consciousness; acknowledgement of answerability to universal rationales.

NOTES

1. In this latter sphere, Weber shifts his stress markedly from work to work. He is seen to best advantage in his little-known work on music (Weber 1958d); he is too condensed on arithmetic and algebra in his opening discussion on Hinduism; he misses the critical importance of the neglect of geometry by the Chinese (see Weber 1958a, pp. 13ff).
2. Chou continues:

> One remaining question with which we are all much concerned is the future of Chinese characters. We all agree that as a written record they have made an immortal contribution to history. As to whether or not they will remain permanently unchanged, whether they will change on the basis of their original forms, or whether they will be replaced by a phonetic language—Latin letters or other phonetic scripts—we need not draw a hasty conclusion. Any language is, however, subject to change, as evidenced by the changes of the characters in the past. There will be changes in the future. We can also say that there will be a day when the languages—written and spoken—of the different peoples of the world will gradually become one and the same. The trend in the development of the languages of mankind is that they gradually approach one another until, at long last, there will be no wide difference. This is a good and not a bad anticipation. As to what scheme will be adopted, it is too early to hazard [a guess]. On the question of the future of the Chinese language, there may be various views. We can bring them out for discussion and debate. I shall not dwell upon it any further since it does not fall within the scope of the task of the language reform. (In Crozier 1970, pp. 128–31)

3. This dispute had many very fateful implications which have been studied by scholars in specialized works. Here I would only observe that one eventual outcome was to promote the notion that there were two Books both attesting God's Power and Will, one the *Book of Revelation* and the other what came to be called the *Book of Nature* or the *Book of Creatures.* During the 15th century this idea gained new momentum and eventuated as the notion that the Book of Nature, being written in numbers, is a "universal manuscript" which is accessible to all (who know numbers) to investigate and understand and is uniform in its meaning, whereas the *Book of Revelation* requires interpretation by special-interest elites (theologians) who rarely seem to agree on the meaning of any passage (see chapter 9 below, pp. 156ff).
4. It would be slightly more accurate to say that the *fuquhā'* (the jurists, those learned in *fiqh,* "law") prevailed in official circles; and thus it is quite true that *kalam* ("theology") never obtained the significance or autonomy in Islamic thought that it obtained in Christianity. It was the study of *fiqh* which was the dominating intellectual discipline in Islam. The term *'ulamā'* refers broadly to all those who are learned in religious matters which include both *fiqh* and *kalam.*—Ed.

PART III

Comparative Studies of Sciences in Historical and Civilizational Perspectives

SIX

On the Shoulders of Giants: The *Comparative* Historical Sociology of "Science"

Evidence mounts that we need to review our points of departure in the sociology of science. The social-psychological, micro-sociological perspectives which have long governed the research done on scientific institutions and communities have now largely spent themselves.[1] The great issues which were the original setting of these inquiries are only rarely recalled today.[2]

The wider sociology of science, in the sense intended here, currently needs to be more clearly distinguished than it is from the more segmental issues with which it is too often identified to the detriment of the systematic progress of broader-gauged research. I refer to such familiar topics as: the interpersonal social relations among the social rankings of scientists in university departments, or extra-academic research institutes or agencies, or the fashions in the forms of citation in professional papers.

An indispensable step in attaining the objective here stressed is to renew the spirit and recollect the substance of the work done by the classical and contemporary pioneers of the comparative historical sociology of social and cultural process. I refer to such men as Maine (1861, 1871), Durkheim and Mauss (1963, 1971), Max Weber (1946, 1958a), the young Robert Merton (1970), Joseph Needham (1954), and others who have offered new horizons in special area studies. Although only a few of the aforementioned—notably Merton and Needham—place primary emphasis on the sociology of "science," they are all acutely sensitive to the need to find wider clues to the changes in the images and definition of realities. Extraordinary acute insights remain to be recovered from their work.

Originally published as "On the Shoulders of the Giants of the *Comparative* Historical Sociology of 'Science'—in Civilizational Perspective," in *Social Processes of Scientific Development*, edited by Richard Whitley (London: Routledge, 1974), pp. 13–20. Reprinted with permission.

If we are to have sociology of science in the spirit required for our times of "axial" conflicts in our horizons and sensibilities, we shall have to give renewed stress to the fact that "science" refers preeminently to *interepochal* structures of consciousness, languages, modes of awareness, cognitive mappings, structures of rationales, agreed-upon procedures, symbolic technologies and so on (cf. Nelson 1968a). In culture-areas marked by very comprehensive spreads (and equally comprehensive rejections) of the rationalization of consciousness, we find sciences constituted by languages of an extremely high degree of abstraction; their decision-matrices generally consist of highly formalized mathematical or logical sets—so-called nomological nets. Any effort to define the social factors and outcomes of science must do so in relation to the sum total of all the elements in the societal and civilizational settings in which one or another science locates. This means that the sociology of science must be historical and comparative. To link Durkheim and Maitland, we may say that the sociology of science is comparative and historical—and systematic—"or it is nothing."

Purely retrospective assumptions invariably bar access to the ways in which *sciencing* occurs and "sciences" are arranged in different places at different times. If our sights are only set to locate so-called precursors of present-day "truths," we fall into the trap of writing separate histories of separate sciences as though these sciences were individuated from the start, consecutive in development, and unconnected with wider and deeper changes in the structures of existence, experience, and expression (see Agassi 1963, for an approach to this emphasis).

A large part of the confusion in this field results from the too often unexamined assumption that the present level and form of sciencing in the most advanced Western lands needs to constitute the criterion for the life and form of science in all other times and climes.

The contemporary situations in respect to the mixes in both the so-called "West" and "East" of the active elements and components of cultural complexes—the mixes of rational-non-rational, abstract-empirical, technology, magic, superstition—demand a wider understanding of the meanings and forms which sciences have taken in different settings than we possess today. Robin Horton (1971) and a number of others have recently raised crucial questions regarding the ethnocentric character of our available definitions of science. Even though I cannot accept the peak assumptions and conclusion of Horton and others who adopt his view, I would agree that we need to look closely at the way in which sciencing occurs in settings of the widest variety. My reasons for stopping short of agreement with the full relativist position on "sciencing" will have to be left for anothei occasion. Here it need only be noted that the link between these arguments and the largest cultural political questions of the present cannot be too strongly emphasized. . . .

In a series of papers I have written over the past decade I have been urging the need to stand on the shoulders of those giants who have applied wider comparative historical perspectives to the understanding of what I have called "civilizational complexes and intercivilizational encounters" (chapter 5 above). Throughout this work I have sought to recover the rich ores mined by a number of writers now often forgotten, men who pioneered in the comparative historical study of religions, jurisprudence and forms of sensibility in civilizational perspective.

The most important spur which now beckons us forward today is the challenge of Joseph Needham. The issues which Needham has posed for the development of the comparative historical sociology of science are no less important than those implicit in the work of Max Weber. Indeed, our next major step in this field of studies will be the collation, comparison and contrast of the perspectives, data and hypotheses put forward by Needham and Weber in respect to the distinctive patterns, paces, and outcomes of scientific development in the so-called "east" and "West" (see Needham 1954, 1963, 1969a; [and chapter 10 below which takes up this task]).

It is the view of the present writer that all efforts to achieve this end will quickly lead us to acknowledge that the so-far largely untapped resources left us by Maine, Durkheim, Mauss, the Durkeimians, the young Merton, and Max Weber need to be exploited if we are to meet Needham's challenges—his implicit as well as explicit challenges.

An indispensable reference point of this analysis will be the study of factors working to promote and those working to retard the forging of the new types of universalities and universalizations necessary for the institutionalization[3] of innovation in the "advancement of science."

In this connection there would be need to be as resourceful and precise as possible in spelling out the reciprocal—what some would call "dialectical"—interanimations of the multiple universalizations occurring at the different levels of social structure and cultural process. If the highest levels of universalization are to be attained, there have to be concurrent movements in the following directions:

(a) Bars to freedoms of entry and exit from the communities of learners and participants in the communities of discourse (and interpretors thereof) have to be lowered; inherited invidious dualisms have to be transcended.[4]

(b) Incentives to produce and distribute warranted knowledge, including new mappings and innovating procedures, have to be multiplied by deliberate policies.

(c) Blocks to the achievement of ever-higher levels of generality in the language structures—written and spoken—have to be surmounted. In this spirit we must attend to the thrusts to higher levels of abstraction in the forming of special languages meant to serve as maps of terrain explored and remaining to be explored.[5]

There is no sociological issue which does not come under scrutiny when we become intent on tracing out the dialectical crossings of universalization processes and effects prerequisite to the spread of rationalized science. We no sooner confront Needham's challenges directly than we find ourselves tapping the resources of the classical pioneers who sought to explore the complex issues related to passages to wider universalities created by new forms of symbolic reference; the spread of universalization processes in the widening of communities of discourse; the expansions of opportunities for full participation in the community of learners, interpreters and teachers; the overcoming of particularistic inhibitions to the spread of communication and community; the encouragement of the "individuals" intent upon advancing the exploration of every facet of their experience; the availability at every level of social and cultural differentiation of accepted tests of warranted assertibility.

Questions of this sort involve us in every aspect of the fortunes of social institutions and cultural structures of "science."

Other notable points and corollaries which arise in this expanded view might be stated as follows:

(1) Comparative historical sociology of science must begin with the supposition that histories are multiple, and that all sciencings occur in the course of histories and are themselves histories. There is nothing that requires us to suppose that any one place will serve as the permanent abode of any or all sciences. In the same way, we dare not suppose that different sciences will forever have the same relation to one another or that groups or cadres linked to the sciences will share similar orientations at all points.

(2) We accomplish little when we treat actual histories as a reservoir of instances which go to prove one or another abstract proposition of structural-functional analysis or general systems theory. In some of their manifestations these approaches are less the roads of an advancing sociology of "science" than surrogate faiths.

(3) To see "sciencing" and sciences *in all the contexts of socio-cultural process* and interaction, we have to adopt *civilizational perspectives* in the study of civilizational complexes and intercivilizational encounters.

Under this last rubric, I would call attention to the available research on the intercivilizational encounters of the twelfth and thirteenth centuries—involving societal settings, mutual influences, and unexpected outcomes due to the encounters of Chinese, Muslim, Hebrew, Byzantine, and Western Christian peoples; the encounters of China and the West in the sixteenth century and from the time of the Matteo Ricci mission to the expulsion of the Jesuits; the struggles over the new science in nineteenth century India; the conflicts over Western Science and civilization in Tsarist Russia and the Soviet Union; the Nazi challenges to Western science.

We dare not close without noting that the stress throughout this paper [—and in the studies to follow—] is on the *comparative historical sociology of science* IN CIVILIZATIONAL PERSPECTIVE. In recent years there has been a resurgence of interest among sociologically-oriented philosophers of science in the historical sociology of science in the modern European nations and the USA—especially in the nineteenth and twentieth centuries. Some of the work done in this field seems more easily tied to the "empirical" referents than the work we have been describing. Unhappily, the range of sociological issues which come into view in this more limited perspective are likely to be narrowed by the bracketing of civilizational and inter-civilizational settings. Principal contributors to this latter field include: P. Duhem (1954), J. Ben-David (1971), L. P. Williams (1962), A. Thackray (1970), R. S. Turner (1971). A recent piece by Stanley Coben (1971) on "The scientific establishment and the transmission of quantum mechanics to the United States, 1919–32," represents this form of work at its best. Coben's essay offers the strongest contrast to the micro-social-psychological studies which constitute a large part of the work which describes itself as the sociology of science.

NOTES

1. There is a large literature on this sort of sociology of science with continuing installments in various journals, especially *Social Studies of Science.* For a survey of this work, see Cole and Cole (1967, 1973) and Ben-David (1971).
2. See, however, Merton's important new Preface dated 1970 to the recent reissue of his doctoral dissertation (Merton 1970, esp. pp. xvii–xviii), and this writer's review (1972).
3. It will be noted that the horizons of our stress on "universalizations" range beyond the more restricted images of "universalism" now accepted as a component of the ethos of the "scientific community." Most recently some of the larger aspects of the problem have received renewed emphasis by Merton. See his strong statement, "Insiders and Outsiders" (1972).

4. Critical clues to this issue may be found in Durkheim (1915), Durkheim and Mauss (1963, 1971), and Weber (1946, 1958a).
5. Only a few sociologists have addressed themselves to this question which has been generally left to philosophers and mathematicians such as A. N. Whitehead, B. Russell, M. Klein, and T. Dantzig. We are in need of additional "sociolinguistic" studies of higher level abstract languages. An exciting awareness of the thrust of these problems will be found in Chou En-Lai's statement on the reform of languages (see Crozier 1970, pp. 128–31; and above pp. 91f).

SEVEN

Sources of "Probabilism" and "Anti-Probabilism" in 16th and 17th Century Science

The time has come to bid adieu to a series of hoary stereotypes which are dying hard in familiar textbooks. Far from being noncomformist sceptics intent upon destroying the superstitions of the unlettered, the pioneers of modern science and philosophy were ardent searchers after "truth," questing for *subjective* certitude grounded in *objective certainties*. This quest irresistibly led them into a struggle against the two main fronts of the accommodating conjecturalism of the learned.[1] The two highly sophisticated—and in the opinion of some of the innovators, diabolically sophistical—traditions, which especially appeared to bar the way to their establishment of the imperative new foundations for exact knowledge and compelling belief were:

1. the *fictionalist* theory of "hypothesis" and "saving the phenomena";[2]
2. the *probabilist* casuistry of "conscience" and "opinion" (see Kirk 1927; Deman 1936; Nelson 1963b).

These two traditions—and principal targets of the innovators—have yet to be adequately discriminated by scholars.

Thanks to a number of lively recent discussions we can be brief in our remarks about the conflict over the meanings of "hypothesis" and "saving the phenomena." Among theologians seeking a viable compromise, this attractive tradition deriving from Plato, Aristotle, and Simplicius, was taken to imply the idea of "saving the phenomena" by mathematical suppositions for which physical truth is not claimed and to which physical truth is not ascribed. It was this approach which Osiander sought to attribute to Copernicus and it was this

Originally published as " 'Probabilists,' 'Anti-Probabilists,' and the Quest for Certitude in the 16th and 17th Centuries." *Actes du Xe congrès international d'histoire des sciences*, Ithaca. *Proceedings of the Xth International Congress for the History of Science* I (1965): 269–273. An earlier draft was read to the Congress at its Ithaca meeting of August 28, 1962. Reprinted with permission.

same approach that Cardinal Bellarmine later sought to urge upon Foscarini and Galileo. The resistance to this fictionalist concept of hypothesis by Copernicus, Galileo, Kepler, and Newton is now well-known.[3]

The second tradition, of which mention was made at the outset, originates and develops independently of the status of mathematical schemas in cosmology: I refer now to the neglected tradition of the probabilist logic and epistomology, loosely called probabilism. Probabilist theses arise less among mathematicians or mathematically-oriented physicists or philosophers than among speculative natural philosophers, logicians and, most prominently, for reasons I have indicated elsewhere (Nelson 1962a), among moral theologians and philosophers. The logic of probable opinions begins with Aristotle and the Stoics, and undergoes a notable development in the 12th and 13th centuries, where it can be clearly seen in the *Summulae logicales* of Peter of Spain (Pope John XXI) in the year 1271. The precise details of the history of the probabilist logic and epistomology from the late 13th century onwards have not yet been studied with sufficient care, but it is apparent that strong probabilist motifs will be found in William of Occam, Nicholas of Autrecourt, Pierre d'Ailly, as well as among a host of lesser moral theologians and philosophers.

The central theses of Pre-Tridentine probabilism are these: (1) it is folly to anticipate a greater certitude than a given subject matter allows; (2) demonstrative knowledge is to be had in mathematics but not in physics; (3) wherever certitude is not available, philosophers may freely entertain probable opinions; (4) in the moral and practical (by contrast to the speculative) sphere, an individual confronted with opinions of varying probability, must prefer and act upon the *safer* or more *probable* opinion. The two latter stresses later come to be known as "tutiorism" and "probabiliorism."

Probabilism did not undergo great elaboration or become a comprehensive system until the promulgation in 1577 of the revolutionary thesis of *minus probabilismus* by the Dominican, Bartholomeus de Medina in his commentary on the *Prima Secundae* of St. Thomas, published at Salamanca in 1577. The theory of the lesser probability declared that a moral agent confronting alternatives was not obligated to act upon the more probable. He acted prudently enough if he acted upon the less probable in the face of the more probable, even when the less probable was less safe.

It is impossible to exaggerate the revolutionary consequences of this thesis. Within the next decades the Dominicans had become frightened at the outcomes of the new approach and the thesis became identified with the Jesuits, who elaborated its implications with extraordinary subtlety and dialectical boldness. Before very long, the theory of the lesser probability was being applied to the spheres of faith and the sacraments as well as to the sphere of everyday conduct. Conservative forces in the Church were frightened and sought to moderate the latitudinarian direction and antinomian consequences of the new outlook. Indeed, the issue was the occasion of violent struggles among orders; between Church and State, in Portugal, France, Italy, and the Low Countries, wherever, in truth, Catholicism prevailed. The Popes were themselves compelled to enter into the controversies and eventually, in the case of Popes Alexander VII and Innocent XI, to condemn lists of excessively relaxed propositions. Despite all this, a moderate position, *minus probabilismus* in speculative and moral affairs—not in faith and sacraments—gained the ascendancy among the professional moral theologians (see Doellinger and Reusch 1889; Lea 1896, esp. vol. 2).

Can echoes of the various phases, medieval and post-Tridentine, of the developments of probabilist logic and epistemology be found among the pioneers of early modern science and philosophy?

We have only to look once more at the major documents with care to find our illustrations:

1. Throughout Galileo's *Dialogues Concerning the Two Chief World Systems,* he is at pains to maintain probabilist frames of reference. Repeated references are made to the relative probabilities of different opinions.

2. The Holy Office was obviously aware of Galileo's strategies of argument. Extensive and pointed references to the dialectic of probability are found in the statements prepared by the theological qualifiers charged with preparing materials for the promulgation of the definitive declaration.

3. At three critical points, as only a few commentators have observed, the sentence against Galileo, issued on June 22, 1633, refers to the vexing problems of probability. The most decisive allusion, occurring at the very heart of the sentence, firmly links the "vehement suspicion of heresy" to Galileo's apparent ascription of probability to the condemned thesis.

4. Perhaps the most revealing insights into the role of the probabilist contexts in Galileo's trial will be found in the now forgotten *Theologia fundamentalis* by the Cistercian Cardinal and moral theologian, Johannes Caramuel y Lobkowitz. Caramuel puts the question plainly: Was it indeed the case that the Copernican hypothesis was divested of all probability by the action of the Congregation of the Index? Strictly speaking replies Caramuel, No! Why then was Galileo treated as he was? Caramuel's explanation hinges on the use of subtle arguments about the grades and shades of probability.

In a formal sense, Galileo was, in truth, not a transgressor inasmuch as, speculatively speaking, he was entitled to accord a measure of probability to a philosophical view condemned by a Congregation. By choosing, however, to defend the *speculative* probability of a thesis which had been rendered *pratically* improbable by interdiction, Galileo had revealed himself as disobedient and practically temerarious. Moreover, insofar as he had proved himself in contempt, he could rightfully be punished as a heretic. In refusing as he had, to obey the Church, notably the Congregation of Cardinals, he had rendered himself suspect in regard to faith.

The primarily political character of Caramuel's outlook is indicated at one point in his extraordinary explanation. The probability of the theory of Aristarchus, he observed, was never in dispute during the two previous centuries, when the issue was confined to circles of philosophers. The problem changed its complexion altogether, however, when Scriptural passages were injected into the controversy by the philosophers. Here, ultimate authority for interpretation was reserved to the Pope. To permit the philosopher-exegetes to go unchecked would be to furnish alms to the enemies of Catholic Faith. The spark had, therefore, to be put out before it became a conflagration like that of the Protestant heresy.[4]

5. Turning now to Descartes: A polemic against probable reasonings, conjectures, and probabilists runs through all the work of Descartes. Hardly any of his commentators have thought to connect this with the probabilistic logic of his Jesuit teachers and of great numbers of the Masters of the universities. Instead it has been mistakenly supposed by more than a few modern authorities that Descartes is inveighing against mathematical theories of probability (as evidenced in the work by Grant). The late R. P. Blake has helped us to see that Descartes had a more highly inflected position on the subject of experience and inductive inference than is generally attributed to him. Yet the fact remains that Descartes wished to liberate science and philosophy from probabilist conjecturalism.

6. It may seem odd to include a discussion of Pascal's *Provincial Letters* (1656) within this setting as the *Letters* are clearly not a treatise in mathematics or physics, but a deeply committed polemic against the new trends in moral theology. The *Provincial Letters* are easily the most celebrated, though not the most impressive philosophical, attack upon the probabilists to see print in the 17th century. Pascal's motives are not an infinity removed from those of Descartes and Galileo. Pascal, too, quested for certitude and certainty. He wished to establish human conduct upon unshakable foundations. His certitudes were moral principles grounded in Scriptural revelations and the dictate of the uncontaminated conscience. Pascal directs his barbs against a host of latitudinarian probabilists whose applications of the

new principles of the lesser probability appeared to put all true religion, ethics, and moral action in doubt.[5]

Indeed, Pascal's attack upon the probabilists was only one of a great number of criticisms of probabilism. Opposition to the probabilists had been mounting steadily since the second decade of the 17th century. Eventually the wave of criticism against the extremists among the innovators led to the overhauling of the post-Tridentine system of the lesser probability *(minus probabilismus)*.[6]

In sum, as I have elsewhere written:

> Whatever their hopes—whether they wished to advance pure science, logic, secular progress, true (primitive Christian or "natural") religion—the pioneers of early modern science and philosophy sought to establish new foundations for exact and compelling knowledge and belief. The prevailing probabilist dialectic that inhibited the quest for new data, the recourse to well-designed experimental research, strict logical proof, and mathematical formalization inevitably became a major target of the 17th century "enlighteners." (Nelson 1962a, p. 615)

NOTES

1. A preliminary statement of these views will be found in the present writer's (1962a).
2. The sources and current discussions are now conveniently summarized in two papers by Professor Grant (1962a, b).
3. Students of the intellectual developments of the later medieval and early modern eras will not fail to draw one inference from the above: the tide is, finally, beginning to turn against Pierre Duhem. Leading philosophers and historians of science are currently calling into question his philosophical fictionalism, his theological fideism, and his handling of the documentary sources. Few, if any, of his celebrated theses seem as obvious today as they did to many scholars a decade or more ago. This is decidedly true for his claims that Copernicus and Galileo were philosophical tyros when compared to Osiander, Bellarmine and Pope Urban VIII (Cardinal Maffeo Barberini). The lively discussions by R. P. Blake, A. Crombie, A. Koyré, T. Kuhn, K. Popper, E. Rosen and others cited in Grant's papers may now be supplemented by A. Grünbaum (1960) and E. Rosen (1961, especially pp. 81–98).
4. The statement above is not offered as a complete analysis of Caramuel's discussion. I trust to review critical aspects of Caramuel's life and thought in a series of studies now in progress. (See chapter 8 below.)
5. We need not here decide whether his attacks upon Escobar, Sanchez, Diana, and others fairly represent their work. It may, however, be remarked that none of the so-called refutations of his *Lettres* is free from blemish. See H. F. Stewart's helpful introduction to his edition of the *Letters* (Stewart in Pascal 1919/20).

6. Presently there were to appear a number of new approaches designed to avoid the admittedly defective outcomes of the initial system of the lesser probability. Aspects of the story are recorded in the writings of Deman (1936), Doellinger and Reusch (1889), and Lea (1896).

EIGHT

The Early Modern Revolution in Science and Philosophy

The early modern revolution in science and philosophy has been arduously studied and discussed for at least two centuries now.[1] Contemporary scholars and publicists (especially Koestler 1959) are now everywhere continuing, with undiminished vigor, to explore archives (P. D'Elia 1960), plumb sources, and to dispute one another's interpretations (see Koyré 1949, 1957b; and Rosen 1964 on Duhem). Nonetheless, few would claim that the roots and outcomes of the 17th century "breakthrough" had been reliably established (see D'Abro 1950; Hanson 1958; Stallo 1960; Einstein and Infeld 1950; and the essays in Crombie 1963). Indeed many of the central issues are hardly more clear to us than are the circumstances and ideas which explain the breathless "take-off" of our own century (Rostow 1962).[2]

It is important—but it no longer suffices—to point to the influences of movements of thought associated with such philosopher-scientists of antiquity as Pythagoras, Plato, and Archimedes, or their disciples of the era of the Renaissance (see Burtt 1925; and Clagett 1959a). Knowledge about the work of the later medieval thinkers and experimenters of Padua, Paris, and Oxford is much more widespread today than it was in the time of Renan (1852) or Duhem (as seen in Rosen 1964b; and Moody 1966). The present generation of American scholars has made very great strides in its investigations of these areas (see Clagett 1959a, 1959b; Murdoch and others in I. B. Cohen and Taton 1964, I). One unexpected outcome of this work had been a

Originally published as "The Early Modern Revolution in Science and Philosophy. Fictionalism, Probabilism, Fideism, and Catholic 'Propheticism'," in Robert S. Cohen and Marx W. Wartofsky, eds., *Boston Studies in the Philosophy of the Philosophy of Science*, volume 3 (Dordrecht, Holland: D. Reidel, 1967), pp. 1–40. An earlier version of this paper was read at the Berkeley joint meeting of the American Association for Advancement of Science and the Society for the Scientific Study of Religion on December 27, 1965, in the program on "Religion and Science." Reprinted by permission of D. Reidel Publishing Company.

profound review of the evidences offered as proof of medieval priorities. At this very moment historians of medieval and Renaissance science are questioning whether Duhem's supporters are justified in ascribing the new astronomy to Nicholas of Oresme (Rosen 1964a), the new mechanics to Albert of Saxony and Thomas Bradwardine (Moody 1966), the new principles of method to the Averroists of Padua (Gilbert 1963).

Even more inconclusive than the situation presented by historical research is that found among sociologists and cultural psychologists (Hirsch 1962; Hagstrom 1965). In a word: Marxist or other deterministic explanations which have emphasized the force exerted by accelerated changes in technology and division of labor related to explosive "capitalist" trends during the so-called commercial revolution remain problematical (Hessen 1931; Borkenau 1934; Clark 1937; Grossman 1935; Hall 1963b; Koyré 1957b, pp. 148–150, 1965; Olschki 1927; and Zilsel 1941–42). Notwithstanding the "impressions" derived from a hasty reading of outstanding American sociologists, the "Protestant Ethic" cannot be identified as the indispensable factor inspiring the *emergence* of modern science or fostering an atmosphere favorable to the pursuit of disciplined experimental investigation.[3] Nor, finally is it possible to agree with the conclusions of a recent study of the scientific intellectuals which pointedly questions the claim that Protestantism or any other religious outlook has ever exerted a positive influence on the progress of science (Feuer 1963, especially p. 7).[4]

The social and cultural outcomes of the 17th-century revolution are now being investigated with less heat and more light (by Hazard 1964; and Baker 1948, for example) than was generally the mode two decades ago (see Maritain 1928; Eliot 1950). One reason for this may be growth in the appreciation of science on the part of historians and literary men who have strong interests in the changes in worldview. To compose the equations as they seem to stand today, we need to add to the familiar names of Burtt, Lovejoy, Cassirer, Whitehead, Willey and Marjorie Nicolson, such newer or older neglected authors as Dijksterhuis, Koyré, Rossi (1962), Spektorsky (1910, 1917), and even Husserl (1954).

Granting the unsettled state of learned opinion, it is no wonder that popular accounts of these developments so frequently distort the historical record. What is disturbing, however, and what will lie at the heart of the present paper, is the fact that exceedingly well-informed scientists and scholars again and again go wide of the mark out of mistaken confidence in untested presuppositions about the prerequisites of scientific and philosophical development. The chief source of the problem is that distinguished representatives of widely opposed outlooks persist in reading back into the past a set of perspectives and commitments belonging to our time, above all the assurance of the superiority of fictionalist and instrumentalist views of scientific and philosophical theory (see Popper 1965; and Einstein 1953). Surprisingly, the consensus includes a host of world-renowned spokesmen of every shade: liberal-humanist pragmatists (Dewey 1929); leading associates of the Vienna school of logical positivists; outstanding physicists, including, for a time, Einstein himself; Catholic and non-Catholic historical scholars under the influence of Duhem, celebrated equally as a physicist and historian of science (see Lowinger 1941, and the works cited there), as well as being a devout son of the church. (On Duhem's religious views, see Picard 1922; and Duhem 1954, Appendix).

It is not our plan here to study the ground for the unexpected convergence noted above. Odd associations occur more frequently in the history of thought than popularizers and ideologists know or say. The present paper cannot pretend to study this issue in the depth it deserves; yet we dare not postpone clearing up one delicate matter which easily lends itself to misconstruction. The aforementioned spokesmen *do converge* at important points, but they are very far, indeed, from constituting an organized ideological party or a political action group. They differ in any number of respects relating to outlook, interest, policy, sophistication in their arguments, command of the materials—whether historical, philosophic, or scientific—and explanations of the social-historical situations (Joravsky 1961; Mueller-Markus 1964). They are surely not one in their motives. The failure of controversialists to state this situation clearly has had distressing results, which need no illustration here (see Levin 1908; and Nevsky's epilogue; as well as the critical remarks about the work of Lyubov Alexandrov in J. Edie 1965, vol. 3, pp. 462–463).

II. "PRIESTS" AND "PROPHETS" IN CATHOLIC CULTURAL LIFE

The ensuing discussion of our theme announced above has its source in a half-dozen or so points which are yet to be accorded the recognition they deserve by scholars and non-professionals alike:

(1) Copernicus, Galileo, Descartes, Pascal and many others in the front rank of innovators of the early modern revolution of science and philosophy received their intellectual formation in Catholic "culture areas" (as noted above in chapters 4 and 7).

(2) The second point, a corollary of the first, figures only negatively

in the initial phase of our argument: Major innovators in philosophical method, experimental design, mathematical formalization, did not require the intervention of Protestantism or the ethic of *inner-worldly asceticism* (stressed earlier in Nelson 1962a, pp. 615–616).

(3) To destroy the pieties of the unlearned was no part of the plan of the pioneers, nor were they crypto-libertines or skeptics.

(4) Rather, their animus was directed against a complex of views which had the endorsement of their own teachers and which prevailed among the leaders of the religious establishments of their own day, including Cardinal Bellarmine and Pope Urban VIII, the prime antagonists of Galileo.

(5) The entrenched network of ideas which they hoped to shatter rested on two sets of perspectives which are only now beginning to be accurately discriminated: (a) The *fictionalist* views of "hypotheses saving the phenomena" which had long prevailed in astronomy; (b) the *probabilist* teachings on opinion which enjoyed almost uncontested sway among scholastic professors of logic and moral philosophers in the 16th and 17th centuries (as set out above, p. 000; for some introductory literature and texts, see de Blic 1930; Deman 1936).

(6) In opposition to these outlooks, the pioneers opposed their conviction that *objective certainty and inner certitude* were the indispensable signs of science, true philosphy, and just belief. Copernicus and Kepler directed challenges mainly against *fictionalism;* Descartes and Pascal launched their shafts mainly against *probabilism.* Galileo aimed his arrows at both traditions.

To understand the roots and the intensity of their commitment to certitude, we will need to be more attentive than we have thus far been to the influence of their continuing involvement with the religiously correlated doctrines of conscience. No corner of human belief and action was exempt from the sphere of conscience (Kirk 1927; Lea 1896, vol. 2; [and chapter 3 above, pp. 43ff]). It is, indeed, in the now forgotten treatises of conscience that we will find the most comprehensive discussions produced in the 16th and 17th centuries of the shades and grades of opinion, evidence, doubt, ignorance, probability, certitude, and certainty.[5]

There is at least one more neglected perspective which gives promise of throwing light on the scientific and philosophical crises in which the pioneers were involved. I refer to the hypothesis put forward by the pioneering Scottish historian of religion, W. Robertson Smith (1846–1894) and recently extended by other sociocultural scientists (Smith 1889; Lods 1937; Weber 1958a).

On this view, the history of the various religious communities is marked by recurrent struggles between those who adopt or are assigned the perspectives and roles of *priests* and those who assume the perspectives and roles of *prophets.* Those playing the role of "priests" in *either* the religious *or* secular spheres often develop a tendency to shield the *arcana,* that is, to protect the holy deposit of truth, the sacred rites, and precious utensils from the threat of profanation by the uninitiated. As time passes, however, they are driven to make complex adjustments to changing situations, but these are felt to be "correctly" appreciated only by those who have received proper training and accreditation in authorized schools and facilities. Within such an atmosphere, demonstrable certainties concerning the questions of the greatest importance are rarely to be had. (This must not be obscured by the fact that ordinary believers are expected to have faith in the doctrines which are proposed to them by their superiors.)

At critical intervals in the histories of societies, however, there appear those who insist on proclaiming a mighty (subjective) certitude of their own or in pointing to an (objective) certainty which appears to be clearly established by a new found revelation provided by conscience or some other compelling sign. Such men feel called upon to serve as *prophets,* to make the truth *manifest* in word and deed.[6]

Within this setting, it may be held that Copernicus, Galileo, Descartes, Pascal and a number of their contemporaries in their several ways are expressions of "prophetism" arising in Catholic settings as responses to conformist accommodations favored by the leading officials and intellectual sympathizers of the Church hierarchy on both theological and political grounds. All of the pioneers might have spared themselves great pains if they had been willing to adopt the compromising alternatives of "fictionalist" and "probabilist" theories of natural knowledge proposed to them by tradition-oriented guardians of faith such as Andreas Osiander, who wrote the posthumous preface to Copernicus, Cardinal Bellarmine and Pope Urban VIII (see Rosen 1959, especially pp. 24–25; Bornkamm 1943; Dillenberger 1960; and Broderick 1928, for Bellarmine's fictionalist commitments).

Our introductory reflections seem to point up the need for further exploration of three questions in the following pages:

(1) The pioneers *had to fight* in the name of certitude and truth. They did not really have the option of remaining ensconced inside their disciplines by the acknowledgement of the merely "hypothetical" character of their experimental results and theoretical views.[7]

(2) To be an innovating physicist or philosopher in bygone days meant to risk becoming embroiled in dangerous conflicts with

theological authorities, perhaps to place one's life in the balance for the sake of an idea. Had the pioneers not risked everything in struggles *against* fictionalism and probabilism, today's physicists would not have been as free as they now are to champion fictionalist and probabilist positions.

(3) Twentieth-century scientific culture is likewise involved in another result of this research. I refer to the indication that "fictionalist," "conventionalist," "probabilist," and even "empiricist" theories of science (which have been so popular among the world's leading theoretical physicists and philosophers of science since the beginning of the century) do not necessarily lead to secularist, liberal or experimental democratic outlooks, notably in morals or politics. The historical record suggests that on more than one occasion in the history of civilization—non-Western as well as Western—fallibilist or fictionalist theories of *science* have been a foundation for *infallibilist fideism,* that is, unqualified accord to the authorized articles of faith, in the spheres of theology and religion (Vignaux 1959; and see Planck's 1937 writings on religion and science, especially, pp. 151–187). For that matter, as the history of the 20th-century scientific revolution reveals, fictionalism can be a defence for obscurantist outlooks *within science.*[8] In brief, fictionalism and open-mindedness do not by any means imply one another necessarily, nor as they causally related to social reality.

A proper test of the often claimed relations, here contested, of fictionalist philosophy of science and "experimental (i.e., antidogmatic) orientations" in the religious and cultural spheres would evidently require a series of case studies "in depth" of the lives (civic, professional, and other) of the major scientists and philosophers of science of the modern era. Surely at the very least one would wish to have studies of the careers of men of the stamp of Mach, Poincaré, Duhem, Planck, Einstein, Bohr, and others. Clearly we must forego an effort to meet the requirement in the present pages.

Here we confine ourselves to developing some evidence on two of the great Pierre Duhem's most celebrated theses, which continue to command extraordinary support from historians and philosophers of science, especially in Great Britain (Crombie 1950, 1953, 1964), and the United States (Clagett 1959a, 1959b). In the first statement, Duhem proclaims the decisive importance of the year 1277 for the birth of modern science. He writes:

> If we had to assign a date to the birth of modern science, we should undoubtedly choose the year 1277, when the Bishop of Paris solemnly proclaimed that there could exist many worlds and that the ensemble of

celestial spheres could, without contradiction, be moved in a straight line. (1909, II, p. 412; cf. Weisheipl 1959, p. 63)

In the second, he offers us an unambiguous assertion of the superiority of Osiander and Bellarmine over Kepler and Galileo in respect to matters of logic and experimental method. Duhem writes:

Logic was on the side of Osiander and Bellarmine [who] grasped the exact significance of the experimental method; . . . despite Kepler and Galileo, we belive today with Osiander and Bellarmine, that the hypotheses of physics are only mathematical devices designed to explain the phenomena. (1908, pp. 587–588, 592 [1969, pp. 113, 117])[9]

We turn now to the year 1277.

III. THE YEAR 1277: BIRTHDAY OF MODERN SCIENCE?

If present trends in the history and philosophy of science continue, we soon will need a series of exhaustive monographs on each of the reputedly lucky outcomes of the extraordinary condemnations which occurred in the year 1277. In that year, as some may remember, Bishop Etienne Tempier of Paris denounced 219 propositions as "obvious and loathsome errors" (see Lerner and Mahdi 1965, for the English translation, and the Latin in Mandonnet 1911, vol. 2, pp. 175–191). Many of the most notable of these propositions were of Averroist origin and it has long been known that a certain number of the Averroist propositions are actually to be found in the writings of St. Thomas Aquinas (Van Steenberghen 1955). The principal Averroist under attack is said to have been Signor de Brabant, who is accorded a place of honor in Dante's *Paradiso* (X, pp. 130–135; and see Gilson 1963, p. 257; and Nardi 1958).

One might have supposed that censures of this scope, together with their sequel at Oxford sponsored by Archbishop Robert Kilwardby (Van Steenberghen 1955; Geyer 1928, pp. 494ff), would have come down in history as among the lowest points reached by the repressive wing of the medieval church in dealing with great university scholars who were exercising their right of philosophizing freely on questions not yet authoritatively determined by faith (Sutton 1953).

Not so. Paradox never fails to claim its quota of followers amongst historians. As bad as 1277 may have been for philosophy, these spokesmen have contended, so good was it for science. Duhem, pious son of the Church as well as physicist, historian, and philosopher of science, was the principal architect of this view.

Duhem celebrates the triumphs which resulted from the condemnations of 1277 again and again in the course of his writings. They were, he asserts, the starting-points of modern science. The reason given for this conclusion are many and varied, but the principal factor for Duhem appears to have been the implicit support given to the fictionalist view of theory through the rejection of a realist natural theology (Duhem 1909, vol. 2, especially at pp. 75-76).[10]

Two younger American writers, Professors Edward Grant and Francis Oakley, have recently put forward exciting versions of Duhem's view with distinctive stresses of their own. The actions of 1277, Grant has remarked, spurred philosophers and scientists to develop modes of probable inference and argument. This probabilism, he has contended, underlies the advent of the modern scientific movement which culminated in Copernicus and Galileo (Grant 1962a).

Professor Oakley finds yet another virtue in the condemnations of 1277. These censures, he declares, provided the basis for the successful assertion of the Occamist view of the absolute sovereignty of God's will. It was this notion, he contends, which is the ultimate source of the concept of a law of nature upon which the whole of modern science rests (Oakley 1961, pp. 433–457).

I have elsewhere indicated (Nelson 1962a) my reservations concerning a number of the main constructions placed upon the censures of 1277. Even though the outcomes of 1277 have yet to be summed up by responsible scholars, I am convinced that the claims of Duhem, Grant, and Oakley go far beyond the available evidence. It is not the case, for example, that probabilism waited for the censures of 1277 to come into being. Probabilistic modes of argument have their origin in Aristotle and Stoics and were undergoing elaborate development in the 13th century by the time of the *Summulae logicales* of Petrus Hispani (Pope John XXI, d. 1271). In any case, not one of the pioneers of early modern science and philosophy was a probabilist, though for evident reasons, Galileo tried very hard in his *Dialogue of the Two Chief World Systems* to sound like one.[11]

As for Professor Oakley's view, we know that Occam regarded the condemnations of 1277 with horror (see Occam in Geyer 1928, p. 582).[12] To him they represented a perverse invasion of the right of philosophers to speculate freely about open physical questions.[13] Nor, for that matter, do I see decisive evidence that it was the Occamist variant of the idea of God's sovereignty which lies at the base of modern science.[14]

Are any clues to be found in an overview of later medieval probabilism and Occamism?

IV. FICTIONALISM, PROBABILISM, AND THE THEOLOGICAL ESTABLISHMENT

We are only in the first stages of our documentary explorations of the sources of medieval and early modern probabilism.[15] Even now, however, I would propose an initial caution suggested by this research. Assume what one will, probabilist argument has not always had the same sociological meanings or correlates throughout its history. I should, in fact, venture the provisional impression that the medieval origins of probabilism in speculative and moral matters are associated with a new approach adopted by those inclined to greater liberalism of thought and expression. The earliest probabilists of the Middle Ages were those who wished to dispel the confusion of revelation with natural theology and philosophy. Another of their main aims was to devise a dialectic of moral decision offering expression to complex and apparently discordant points of view (see Kirk 1927).

Thirteenth-century probabilists and their allies among the logicians continue in the main to have this liberal point of view. *Eventually*, however, this very probabilism came to be the defensive underpinning of the theological status quo. Already in Nicholas of Autrecourt (fl. 1335; Weinberg 1948), however, we detect a sympathy for probabilist arguments serving to promote theological fideism (O'Donnell 1931; de Gandillac 1933). Nicholas, whom Hastings Rashdall called the "medieval Hume" (1907), takes a maximalist view of certitude and natural knowledge which allows him to expand the claims of revelation and moral law. Many probabilists are found to reveal this particular pattern in the 14th and 15th centuries. One of the exceedingly few careful studies of the use of probable arguments in the late Middle Ages shows how nicely probabilism and fideism support one another in the work of cardinal Pierre d'Ailly (de Gandillac 1933).

Here in outline we can already perceive the constituent elements of the intellectual policy of all who were dedicated to the defense of hierarchical institutions and principles. Fictionalism, probabilism, and fideism in varied blends were the bases of the programs of Bellarmine and other high church officials in the 17th century. Even if, in short, the aforementioned pioneers of early modern science and philosophy had wished to adopt an instrumentalist or fallibilist theory of scientific law, the choice was not really open to them. The fictionalist and probabilist positions had been preempted by the rulers of the establishment.

The situation was not very different among Lutherans. The Lutheran authorities met the challenge of Kepler's *Mysterium Cosmog-*

raphicum (1557) by appealing to "fictionalism" in Osiander's mode, thus clinging to their own variety of fideism.[16] It is interesting to note, however, that the authoritative Lutheran theologians were not attracted to the new probabilism *(minus probabilismus)* which gripped the Catholic world in the 16th and 17th centuries (issues which Mosse [1957] has confused, in contrast to Wood [1952]).

What if the pioneers of modern science and philosophy were not probabilists or Occamists? *Were they not, then, skeptics?* The latest literature on this issue tells an interesting story.

V. THE ROLE OF SKEPTICISM IN THE EARLY MODERN REVOLUTION

Is it the case, as many since the mid-18th century have supposed, that early modern science and philosophy owe their origin to a great forward thrust of skepticism? Luckily, critical research on this issue (especially Popkin 1960) is moving forward today along a number of fronts, but scholars are still far, however, from matching the analytical finesse illustrated by the lamented A. O. Lovejoy in discriminating the varieties of romanticisms, pragmatisms, primitivisms (Lovejoy 1955, 1963; Lovejoy and Boas 1936), or the uses of the word "nature" (especially Lovejoy 1936). We cannot yet tell what versions of skepticism and what schools of skeptics exerted what particular influence, if any, on the modern movement in science and philosophy.

A number of points may, however, be offered with assurance even at this time. It is apparently not the case that those who were Averroist philosophers whether they were connected with France or Italy (see Van Steenberghen 1955; MacClintock1956; Nardi 1958; Kristeller 1956), were consistently devoted to the program of scientific experiment or mathematical formalization (Koyré 1957b). The evidence on the relation between Galileo and Paduan Avveroists does not clearly support John Herman Randall's efforts to trace out a consistent anticlerical Aristotelian and Averroist tradition. The positive contributions of the Averroists to the leap forward of 17th century science and philosophy have yet to be clearly demonstrated (now see Gilbert 1963).

The renewed vitality of classical pyrrhonism and other variants of ancient skepticism made little or no impression on scientists intent on uncovering scientific knowledge. Copernicus, Galileo, Descartes and Pascal were not especially involved in the advancement of the skeptical tradition, and were not notably influenced thereby.

Richard H. Popkin (1960), who has recently gone to great pains to review this evidence of the Renaissance resurgence of classical skepticism, has had to acknowledge that neo-skepticism appeared in two chief guises: The first variant, which Popkin identifies with Agrippa of Nettesheim, Montaigne and Montaigne's disciples, is admitted to have been destructive of all possible claims to knowledge; all science and scientists, contemporary as well as ancient, came equally under its suspicion.

It was only the *constructive* skeptics—as Popkin describes them (see Popkin's Preface to Van Leeuwen 1963, p. ix)—Francisco Sanchez, Mersenne, Gassendi and their followers—whose thought allowed a slight opening for the prosecution of scientific work. And so Popkin comes in his own way to a conclusion—agreeing in result but not in spirit with our own—that not one of the men universally acclaimed as a pioneer in the emergence of the early modern revolution in science and philosophy—not Copernicus, not Galileo, not Descartes, not Francis Bacon—was a skeptic or friend of the skeptics.

The two decisive turns in Popkin's arguments flow from and assume this argument. Thus he writes:

> The constructive scepticism of Sanchez, Mersenne and Gassendi led to a type of theoretical empiricism and positivism that was probably too complex for the state of science of their day . . . it led to little of importance in experimental or practical results, and it was overshadowed by the dogmatic metaphysical theory of their great contemporary, René Descartes. Their constructive scepticism was brushed aside, to remain alive mainly in the writings of the anti-Cartesians, especially among the Jesuits. (p. ix in Van Leeuwen 1963)

But, continues Popkin, somewhat unexpectedly:

> . . . *A new version of constructive scepticism was developing in England among the theologians and scientists of the Royal Society, and with the scientific success of the Society, its theoretical outlook also triumphed.* (Emphasis added)

This last thesis has the look of a *deus ex machina.* Fully to grasp this thesis, we need to turn to an interesting, if inconclusive, study by Henry G. Van Leeuwen, *The Problem of Certainty in English Thought: 1650–1690* (1963). Van Leeuwen concurs with the view of Popkin that, far from being a major spokesman for modern scientific outlooks, Francis Bacon was a dogmatist, who needs in the end to be grouped with Aristotle. The true initiators of the scientific point of view, Van Leeuwen discovers, prove to have been two Anglican controversialists of essentially traditional outlooks, William Chillingworth and John Tillotson, who were dedicated to the exploration of the meanings and limits of certainty in the several spheres of knowledge, opinion, faith. In Van Leeuwen's pages, Boyle and Newton are the legatees of these religious traditions in the secularized form which had been transmitted to them by John Wilkins and Joseph Glanville.

In short, to quote Van Leeuwen:

The solution to the problem of certainty arose from a controversy between Catholics and Protestants concerning what beliefs are necessary for salvation. Each side claimed its doctrines to be the necessary ones, and the question then arose by what criterion or rule one could decide with certainty which doctrines were necessary . . . (1963, p. xiii; and compare p. 81).
. . . It is shown, as a subsidiary thesis, that the view that Francis Bacon is the spiritual father of experimental science with respect to aims and method, though initially plausible, must be rejected as false. Though the early members of the Royal Society emulate his view that knowledge is for the improvement of human life and speak highly of his passion for observation of fact, they reject the view that scientific investigation yields absolutely certainty about the real structure of nature. Bacon's views concerning certainty of scientific knowledge are thus shown to be at odds with those of the more influential of the early members of the Royal Society. (p. xiv)

If we would believe Van Leeuwen and Popkin,[17] it was not Galileo, Descartes, or even Bacon who initiated the philosophy or methods of modern science, but Chillingworth and Tillotson.

This is not the place in which to initiate a full-scale analysis of the evidence and claims of Popkin and Van Leeuwen. I must be content to say here that I am unable to accept the theses, and this is especially so in the case of Van Leeuwen. I am, however, pleased to find the view I had independently reached confirmed by their *ex professo* studies of the birth of skepticism.

VI. THE QUEST FOR CERTITUDE: A RETROSPECTIVE VIEW

The founders of modern science and philosophy were anything but skeptics. They were, instead, committed spokesmen of the new truths clearly proclaimed by the Book of Nature which, they supposed, revealed secrets to all who earnestly applied themselves in good faith and deciphered the signs so lavishly made available by the Author of Nature. Nature's Book, in their view, was written in numbers and never lied, whereas the Testaments were written in words which were both easy and tempting to misconstrue. Men like Galileo and Descartes were vastly more certain about the truth *revealed* to them by number than they were by the interpretations placed upon Scriptures in the commentaries of theologians (cf. Koyré 1957b; Crombie 1959c; [and chapter 9]).

Established stereotypes notwithstanding, the source of the main opposition to the innovators was not the superstitious country folk or a narrow-minded minority among the extremists in the priesthood. Our image must not be that of disbelieving Everyman being upset by the fallibilist experimental temper of scientists. The enemy was not the simple man. The enemy was the suave and flexible elite of the Ecclesiastical Establishment who followed the works of scientists and philosophers with considerable interest and did not raise objections so long as the innovators made no inappropriate claims to truth or certitude which openly challenged received doctrine. Andreas Osiander, a Lutheran official and theologian, the author of the false preface to the posthumously published work of Copernicus, did not need to reject Copernicus's treatise so long as he could pretend that Copernicus made no claim as to the truth of his hypothesis (see his Preface to Copernicus [1543] in Rosen 1959, pp. 24–25). The Roman Church would have been content to let astronomy make progress within its own area so long as neither physicists nor philosophers asserted that they had demonstrative *knowledge* of anything in the physical or moral realms (revealed, for example by Bellarmine's letter to Foscarini; see Blake in Madden 1960, p. 44).

The restraints which were eventually imposed on the Copernican view and the *Dialogue* of Galileo stemmed from the fact that the new and urgent claims of truth, which challenged the dominant fictionalism, evidently raised questions about the more or less accredited interpretations of scriptural passages by the theologians. There might never have been a great crisis if the scientists and philosophers had been content to desist from advancing truth claims for their propositions.[18] The fundamental issue at stake in the struggle over the Copernican hypothesis was not whether the particular theory had or had not been established, but whether in the last analysis the decision regarding truth or certitude could be claimed by anyone who was not an offically authorized interpreter of revelation.

The 16th and 17th centuries saw an unrelenting contest over the issues in the logic of decision in all regards and spheres, speculative, moral, physical, natural, mathematical, etc. The pioneers to whom we ascribed the breakthrough were not those who took the fallibilist view but those who opposed fictionalism and probabilism, claimed certitude, and insisted that the one and only proper interest of man and science was truth. This theme appears in a somewhat confusing way in Galileo's *Dialogue Concerning the Two Chief World Systems,* where he tried as hard as any man could to play the Church's game, indeed, to beat the Church at its own game,[19] but it is unmistakably clear in the writings of Descartes, Pascal, Newton, and others (as noted in chapter 7; and see Strong 1959). Modern popularizers notwithstanding, the pioneers of modern science and philosophy were not fictionalists, not conventionalists, not probabilists, not plausibilists. They felt certain that they had established a truth. And they were reluctant to hide it.

VII. FURTHER ISSUES AND PROSPECTS

Here several final questions arise: Have not a number of the major scientific developments of our own century, especially those associated with the names of Einstein, Heisenberg, and Bohr, served to give renewed importance to Duhem's theories of science?[20] Perhaps, for all one knows, most active physicists are currently prepared to say with Duhem: "Despite Kepler and Galileo, we believe today with Osiander and Bellarmine, that the hypotheses of physics are only mathematical devices designed to explain the phenomena." There are even those who would go further with Duhem and say "Logic was on the side of Osiander and Bellarmine."

Granting the fact that contemporary physicists and philosophers of science find many compelling reasons for preferring fictionalist and probabilist interpretations of hypothesis and law, must we not ask whether they or their historian allies are entitled to read their views back into history? How far are they ready to go in support of Duhem? Are they ready to follow him to the point of saying that Osiander and Bellarmine "grasped the *exact significance* of the *experimental method?*" Must we agree that Bishop Etienne Tempier and Bellarmine set men on the path of modern science and philosophy?

Whatever one may think of the ultimate outcome of the views of the pathfinders of modern science and philosophy, it is surely not accurate to charge them with having had a naive relation to the scientific and philosophical situations *of their time.* We seriously misjudge the realities with regard to the *intention* and *effect* of plausibilist theories of *science*—we do not write "religion" or "theology"—of Osiander and Bellarmine, if we construe them from our latter-day perspective, that is, if we suppose Osiander and Bellarmine to have held views that were the prototypes or the counterparts of the views of Einstein, Heisenberg, and Bohr.[21] Unhelpful ideologies and analogies are no base on which to rear an accurate historical account of the consensual validation of scientific theories and systems. One does not need to exhaust the archives in order to know that the central aim of Osiander and Bellarmine was to preserve the theological and ecclesiastical *status quo.* Astronomy and physics counted little in either their private or their official lives. Surely these were not the prototypes of Einstein, Bohr, and Heisenberg.

In fact, we need to remember as stressed earlier, that we should never have arrived at contemporary science and philosophy had the pathfinders from Copernicus to Newton not pressed forward vigorously against the shaky foundations of the ruling versions of plausibilist conjecturalism. This is not to deny that a good many constructions placed upon their teachings by those who carried these methods into areas imposing notable obstacles to experimental tests and mathematical procedures and resisting easy or full logical formalization such as history, psychology, and the social sciences generally leave a great deal to be desired. These and innumerable other questions bound up with the history of the logic of decision in scientific and moral affairs comprise an area sadly in need of intensified study by disinterested historians of ideas.

We do not yet know, indeed, when it will become possible to list, to say nothing of measure, the multiple factors involved in the transition from the late medieval to the distinctively modern scientific and philosophic systems. If we are to identify the central sources, prototypes, and circumstances associated with the 17th century revolution, we need to adopt many fresh perspectives and explore strangely neglected contexts. Thus, we must expand our knowledge of the careers of given concepts by engaging in well-designed studies in the direction of a polyphonic depth history and sociology of cultural expressions and experience. Otherwise, we readily fall into the trap of overstating the importance of one or another idiosyncratic link in the careers of particular notions in the complicated histories of ideas across centuries and civilizations. Thus, I would hope that in any discussion such as the present one, we would not be oblivious to the itineraries of the concepts of hypothesis, knowledge, opinion, certitude, ignorance, etc., within the medieval Islamic and Hebraic traditions, as well as the history of Western thought. Of particular importance in the present discussion is an understanding of the ways in which medieval "logical empiricism" (called nominalism in Western Europe), "skepticism," and "fideism" supported and reinforced one another in *kalam* (Wolfson 1964), and in the so-called nominalist philosophies (known to contemporaries as the *via moderna* [Michalski 1922, 1924; Geyer 1928, pp. 591–597]) of the 14th and 15th centuries. In this spirit, I must confess that I have long felt that one of the major desiderata of research in this field would be a close *morphological* as well as *historical* examination of the systems of al-Ghazali (cf. Van de Bergh in Averroes 1954; Munk 1927; Geyer 1928, pp. 59–94; Jabre 1958), Nicholas of Autrecourt (Weinberg 1948), David Hume (Rashdall 1907), and the Viennese positivists of our own era (Kraft 1957; Passmore 1957, pp. 369–393; Ayer 1959).

The fact that the scientific movement and outlook eventually gained vastly greater ascendancy in Protestant rather than in Catholic lands should not blind us to the very great part played by Catholic "culture areas" in initiating the scientific and philosophic revolution of the 16th and 17th centuries. It was not Francis Bacon or the members of the Royal Society, nor even Newton who set the wheels in motion; it was Copernicus, Galileo, Descartes, Pascal, and many others who had been bred in Catholic schools and had to struggle to win their way to a conviction that they had discerned new *truths* about the Book of Nature's Revolution (cf. Koyré 1957a; and Galileo's "Letter to the Grand Duchess Christina" in Drake 1957).

As the instances of Galileo and Pascal were to show, fictionalism, probabilism, and fideism were to triumph wherever the Catholic establishment was well entrenched. The implications of this Establishment victory for the future development of science and cultural development in Catholic lands have yet to be studied. We are better informed about the march of science, engineering, and the various versions of a scientific outlook in Protestant culture areas. That a conventionalist theory of science principally promoted by a devout Catholic physicist, philosopher, and historian of science was to be taken up and given world currency by the proudly anti-theological and anti-metaphysical Vienna positivists is one of the many ironies of this development.

A final thought arises as one contemplates the ground thus far covered. Thomas Kuhn (1962) and others (e.g., Merton 1970; Bochner 1966) have recently been seeking to chart the paths of scientific development, especially the onset and meaning of a "scientific revolution." In Kuhn's view, the central fact is the emergence of a "paradigm" which is felt by many scientists to give a coherent explanation to the masses of data clamoring for explanation. The "paradigm" also has the character of a "compelling metaphor," which stamps itself on the minds of a great number of thinkers working at any given time.

The evidence here adduced implies that one element needs to be added to Kuhn's discussion. A scientific paradigm is one and only one element in a set of cultural perspectives which need to undergo change if the mental contents of a given period are to receive compelling re-orchestration. New paradigms can only succeed if the cultural climate favors. The present paper has suggested that the scientific innovations of the 16th and 17th centuries could not attain their necessary momentum and consequence until the philosophical foundations of an earlier era had been broken through. "Scientific theories" having the power of paradigms can rarely be treated apart from "philosophic" foundations (Lovejoy 1936; Pepper 1957).

There is much ground here waiting to be explored by resolute scholars.

APPENDIX A: KOESTLER, CROMBIE AND DUHEM ON GALILEO

The Sleepwalkers (1959) is the acknowledged summing-up of Koestler's lifelong interest in science, history, and artistic creation. Here the author of *Darkness at Noon* allows himself to take a long look at the author of the *Dialogue Concerning the Two Chief World Systems.*

The results of the inquiry can be put in a single—surprising—sentence. Arthur Koestler has written the most bitter attack on Galileo, and offered the most generous praise of Galileo's ecclesiastical opponents, to come from the pen of any noted writer or scholar in recent decades. Koestler renders such hard judgment as to be called to account by Father James Brodrick, the distinquished Jesuit historian and biographer of Bellarmine, well known for his lively and skilled defenses of his order and Church (see Appendix B for Father Brodrick's criticisms of Koestler). We would need more space than we are allowed here to explore and explain the asperity of Koestler's attack adequately.

In any case, first things must come first.

Whatever else *The Sleepwalkers* may be, it is a work of serious—and at times of penetrating—scholarship[22] and needs to be treated with scholarly respect. Our main task here, therefore, must be to review the philosophical and historical arguments which Koestler put forward to sustain his portrait of Galileo's personal trials and his trial by the Roman Inquisition (cf. de Santillana 1955; Abbé Léon Garzend 1912; and the original documents in Berti 1876, esp. pp. 111–125; Favaro 1907).

Koestler's argument will be recorded here against the background of the works of Crombie and Duhem since they offer the main support for his case against Galileo.

Galileo's views are under implicit discussion in all of Crombie's writings. The fullest statement from the point of view of this Appendix will be found in Crombie (1950). Oddly, Crombie's work and name are not mentioned in Koestler's pages although there is evidence that he did make use of Crombie's writings. Reference is made, however, to Butterfield, who also introduced Koestler's book.

A neglected but exceedingly interesting summing-up of Duhem's views will be found in his preface to the first edition (1911) of A. Maire's annotated bibliography of Pascal (especially, pp. v-vx in Maire, I, 1925).

Even though all three writers [i.e., Koestler, Crombie, Duhem—Ed.] go to great pains to place Galileo in the proper contemporary contexts, all three, in my view, miss several important features of the intellectual and spiritual situations.

Galileo, explain his critics, was as mistaken as he was irrational in looking for *truth* where *certainty* was not then—is not now, and, indeed, *is not ever*—to be had (Crombie 1950, pp. 135–138; cf. Weisheipl 1955). It was Galileo and not the spokesmen of the church, Crombie and Koestler observe, who violated the most advanced teachings of that day (Duhem 1969, pp. 113–117; Blake in Madden 1960, chapter 2, esp. pp. 47–49)—and indeed of our time (Koestler 1959, p. 533; Feyerabend 1962, esp. pp. 260–262)—with regard to mathematical and physical inquiry. Far from being required to adopt a restricted parochial position, Galileo was treated with attentive respect by both Cardinal Bellarmine and Pope Urban VIII (Koestler 1959, pp. 432, 442, 445, 460). Indeed, these two princes of the Church were truly conversant with and sympathetic to the advanced philosophical positions of their day.

It was Galileo and not Bellarmine or Pope Urban, Crombie and Koestler add, who destroyed the hopes for a peaceful development of science and its productive relations with philosophy and theology. Koestler writes:

> After Galileo, fragmentation of experience sets in again, science is divorced from religion, religion from art, substance from form, matter from mind. (1959, pp. 389 and 537–542 for stronger statements)

Galileo had taken the offensive, and the spokesmen for the Church had to respond to his challenges (Crombie 1950, pp. 114–115, 132, 134; Koestler 1959, esp. pp. 476–479). In defending its traditions, the Church was representing not only its own interests as an institution, but, as it chanced, the central interests of undogmatic philosophy and science. [Hence] Koestler's ringing conclusion:

> His bones, unlike Kepler's, were not scattered into the wind; they rest in the Pantheon of the Florentines, the Church of Santa Croce, next to the remains of Michelango and Machiavelli. His epitaph was written for him by posterity: *eppur si muove*—the famous words which he never uttered at his trial. When his friends wanted to erect a monument over his grave, Urban told the Tuscan Ambassador that this would be a bad example for the world, since the dead man "had altogether given rise to the greatest scandal throughout Christendom." That was the end of the "perilous adulation," and the end of one of the most disastrous episodes in the history of ideas; for it was Galileo's ill-conceived crusade which had discredited the heliocentric system and precipitated the divorce of science from faith. (1959, p. 495; cf. pp. 537–542; but see Einstein's [1953] remarks, pp. x–xvi)

Bellarmine sought to recall Galileo to the established notions of "hypothesis, saving the phenomena," which had the support of leading thinkers, whether they were Paduan professors (Duhem 1969, pp. 106ff) or Jesuit theologians (Blake 1960, pp. 32–35, 38; Crombie 1950, p. 133). All of the responsible scholars to whose person and work Bellarmine had ready access, the argument continues, adopted one or another version of the dominant positions[23] in the logic and philosophy of science. Father Clavius and Agostino Nifo are especially critical witnesses here (Crombie 1959a, vol. 2, pp. 26–27, 208–209; Blake 1960, pp. 32–35, 8). As for the Pope himself, Koestler and the others add, Urban VIII's seemingly theological suggestions were meant to bring home to Galileo a version of fallibilism indispensible to the advancement of empirical science (Santillana 1955, pp. 165–166; cf. Koestler 1959, p. 533 for a strong endorsement of Urban VIII's argument in the light of current physical theory).

To these points and to others in this vein, the following replies may be made:

(1) After 1616, by the canons of strict theology, Galileo did not have a single congenial option unambiguously open to him. Once the Holy Office had spoken and Galileo had been admonished by Bellarmine and the Commissaries of the Inquisition,[24] once the Congregation of the Index had issued its decree against the work of Copernicus (Favaro 1907, pp. 62–63; Crombie errs in tracing the 1633 proceedings to this decree alone), Galileo ran the risk of being denounced as a heretic for ascribing *any* measure of probability whatever to the Copernican view.[25] None of Galileo's critics stresses this fact sufficiently. As I shall seek to show in detail elsewhere, Galileo *did* hope to avail himself of an ambiguity in the contemporary teachings about probability, *but* the feigned compromise was flatly rejected by the Qualifiers and the Signatories of the Sentence of 1633. There *were* ambiguities in Galileo's situation when viewed from the standpoint of the highest level of moral theology. It is hard to account for Galileo's acts after 1616, except on the hypothesis that Galileo and his learned advisors in the clergy thought that there were theological loopholes available to Galileo so long as he adopted the *probabilist* stance. (Cf. the passage cited from Cardinal Caramuel's *Theologia Moralis Fundamentalis* [in chapter 7, p. 117 above]; and the *Anti-Aristarchus* of Galileo's contemporary, Fromond of Louvain [Ward 1865, p. 40].) In short, there was no clear-cut doctrine as to whether a view denounced by a Congregation of the Index as opposed to Holy Scripture did or did not preserve any measure of probability of one or another sort (see Descartes's notable . . . remarks on the issue in his letter to Mersenne of April 1, 1634, in Crombie 1959, 2, pp. 218–219). The Qualifiers closed off Galileo's routes in two ways: (1) They insisted

that he had discussed matters *assertively* rather than *hypothetically;* (2) they charged that it was heresy to impute probability to a proposition which had been declared to be contrary to Holy Scriptures (see Favaro 1907, pp. 87–96). Koestler misconstrues the reference to probability in the sentence of the Inquisitors of 1633. (W. G. Ward's Victorian discussion [1865] of the issue of probability is much better informed; especially pp. 412–414 where he discriminates between the meanings of probability in strict theology and in ordinary usage. He also offers interesting sidelights on theological notions of "hypothesis" and "scientifically grounded." I am not, however, able to accept the details of Ward's reconstruction, pp. 414–415, of the meaning and intended effect of the 1633 Decree.)

As the 1633 Sentence clearly reveals, the act of putting forward the Copernican hypotheses *as a probable opinion* was itself deemed of heretical nature (translated in Ward 1865 and Santillana 1955, pp. 306–314). Copernicus's teachings, the Qualifiers held, being judged erroneous, absurd, and even heretical, were therefore devoid of probability. In their view, Galileo was no more entitled to allege probability for the Copernican theory than he could claim certitude.[26] (3) If after the 1616 injunctions, any permission to speak on the subject was still open to Galileo by virtue of the Pope's personal favor, it was the permission to present his views in the guise of a *suppositio* (a fact which clearly emerges in a study of the documents of 1633) or a fictitious,[27] not—as Koestler says—a "working hypothesis,"[28] one already denounced by the Church as false and absurd. This Galileo would not do.

Why? Our answers to this question will help us understand why Galileo could not, so readily as Crombie and Koestler would have wished, accept the "fictionalist" compromises offered by Bellarmine and Pope Urban. In our view, the answers are as follows—in order of ascending importance:

(1) The adoption of a purely hypothetical mode of reasoning was not natural to Galileo, who was a physical astronomer and a physicist in the new style, rather than a speculative astronomer in the older manner (Crombie 1950, 1959, vol. 2).

In any case, the older views about "hypothesis saving the appearances" had become sadly compromised by Kepler when he exposed Osiander's authorship of the anonymous preface to Copernicus and mocked Osiander's statement of the issues (Rosen 1959a, p. 24 n68). Osiander had opened the gates to doubt and confusion himself. It was one thing to say that a mathematical formalism could prove immensely useful even if its "physical truth" could not be established; it was another to assert that a *false* hypothesis could be as useful and effective as true ones in explaining the phenomena.

(2) Pope Urban was in truth proposing a theological position to Galileo which the latter must have found demeaning to his convictions as a scientist. The Pope urged him to acknowledge the possibility, that at any time it suited the Lord's convenience, He could rearrange Nature in any way He pleased (cf. Duhem 1969, pp. 110–111; Crombie 1959, 2, pp. 214–216; Koestler 1959, p. 473 for modern interpretations of this passage). This must have seemed as implausible to Galileo as in earlier days the views of Peter Damiani (see his *De divina omnipotenia,* discussed in Geyer 1928, pp. 187–188) had seemed to the logicians of the Little Bridge in Paris (Geyer 1928, pp. 186–190, esp. 187; Gilson 1930, pp. 37–38; and cf. the discussion of Walter of Saint Victor [d. 1180] and his treatise, *Against the Four Labyrinths of France,* in Gilson, p. 79). Interestingly, the compromise proposed to Galileo by Urban VIII was a commonplace of skeptical thought in the 17th century (Van Leeuwen 1963; Popkin 1960).

(3) In the last analysis, both Cardinal Bellarmine and Pope Urban VIII—both Professor Crombie and Mr. Koestler—seem to have misjudged Galileo. The Florentine innovator was not content to adopt a purely professional stance inside mathematics or physics. He struggled to know and proclaim the truth. The compromise offer of fictionalism proved unacceptable in the end because his conscience hungered for both certainty and certitude and *surely his faith must have been one source of inspiration which taught him to act in the light of his committed conscience.* [See chapter 7 above for discussions of these issues].

To grasp the force of motives and belief in Galileo, we need to work our way back to the issues vital to the men of Galileo's age. Galileo's contemporaries were under continuing pressures to relate to issues of conscience with exceptional seriousness. It was, indeed, the great contemporary struggles over conscience which pointed up the claim that certitude—and certainty—were necessary for true faith, true science and true virtue. Galileo was in this regard a member of the family of Descartes and Pascal. Our point here is not rendered weaker by referring, as Duhem and his followers do, to Pascal's queries on Copernicus. I hope one day to study the ideological cross-currents in the polemics against Galileo and Descartes by the French "Pascalians" among the historians and philosophers of science (see Duhem 1911; and Picard 1922). The present paper is the discussion of the opposition of these thinkers by German and other scholars sharing a partiality for Kepler *(The Sleepwalkers)* against Galileo in this regard.

Lest an impression be left that Galileo was, *after all,* a *probabilist,* it remains to be added that Galileo—like Descartes, Pascal, and Newton—*had to strike out against the vast storehouse of probable arguments received from the past* (see *Dialogue,* Day 2, as cited in Crombie 1950, p. 117).[29]

APPENDIX B: FATHER BRODRICK ON GALILEO AND BELLARMINE, DUHEM AND KOESTLER

A striking contrast to Koestler's discussion of Galileo is provided by Father James Brodrick, who has devoted his life to learned studies of the history and the notable figures of his order.

Father Brodrick's detached asides about Koestler's recent book in his own popular biography of Galileo (1964) need to be read in the light of two of his important discussions of Galileo, which have been unduly neglected:

(1) His heavily documented two-volume biography of Cardinal Bellarmine first published in 1928 and then reprinted in 1950;

(2) A greatly re-worked and updated biography of Bellarmine in one volume, published in 1961.

It is in the latter work that we will find Father Brodrick's professional assessment of Koestler. It is here, too, that we can read Father Brodrick's candid and wittily self-critical account of the development of his own views on the extraordinary issues involved in the Galileo matter during the years which elapsed between the original two-volume biography and the 1961 study. The spirited tone of Father Brodrick's self-criticisms is struck at the very outset:

> The most disconcerting chapter in the old volumes re-read thirty-three years later was that entitled "The first Troubles of Galileo." Myself when young spent many weary months in the Reading Room of the British Museum poring daily over the twenty volumes of Favoro's *Opere di Galileo Galilei,* but the labour did not cure my romantic determination to justify Roberto even in the esoteric realms of physical science where I see now rather ruefully that he possessed no competence whatever. (Brodrick 1961, p. ix)

The more, Father Brodrick explains, he pondered the issues, the more he regretted his initial endorsement of Duhem, in respect to both philosophy and history (pp. 365–366). Freely admitting the limits of his earlier statement, Father Brodrick explains:

> Alas, thirty-two years ago the writer of these present lines was misled by Duhem's eminence as a practical scientist in thermodynamics into embracing his views as a historian of science. Thirty-two years is a long time, and one almost of necessity learns a little wisdom with greying hairs. (p. 365n 3)

In the end, as Father Brodrick tells us, he had come to realize that Duhem had overstated his case. Anachronism and paradox had won the day when he claimed that Osiander, Bellarmine, and Pope Urban, rather than Galileo, had grasped the "exact significance of the experimental method." Hoping to set straight the historical situation, Father Brodrick writes:

> . . . To attribute to Bellarmine and Urban VIII a grasp of scientific principles in the modern sense, while denying them to Kepler and Galileo, was surely an appalling piece of anachronism. Robert Bellarmine was a lovable saint and a phenomenally learned theologian, but he knew next to nothing of science in Kepler's and Galileo's understanding of the word, and perhaps cared just as little about it. He was aware, because the experts of the Roman College had so assured him, that Aristotle's unchanging heavens had been laid in ruins by the observations of Tycho Brahe, Kepler and Galileo, and that Galileo had also irreparably damaged much of Aristotle's physics as well as his astronomy. But at seventy-four and with his cast of mind, it was psychologically impossible for him to accept the revolutionary ideas, which Copernicus, Kepler, and Galileo had intuitively grasped, that the solid earth under his feet was a planet in rapid motion around the sun. (pp. 365–366)

And once again with evident respect for Galileo and science, he continues:

> Galileo was perfectly justified in rejecting Cardinal Bellarmine's naive invitation to regard the Copernican theory as a mere hypothesis to account for the face of the heavens. Besides being a great scientist, he was the first great theoretician of science, and knew in his bones that its business was to attain to the structure of reality, that hypotheses are formulated, not to save appearances, but to be verified or rejected by experiment, and also to try and bring a great unifying principle into the diversity of phenomena. By his own telescopic discoveries he had already made an enormous contribution to the unifying principle by proving that the material of the heavens, of the sun and of the moon, was no privileged fifth incorruptible essence, but the same changeable elements as those that composed the earth. (p. 367)

Though not relaxing in his regard for Bellarmine, who had, indeed, been canonized in the interval between the two biographies, Father Brodrick admits that Bellarmine's ways of relating Scripture and natural science left a great deal to be desired. It was Galileo's *Letter to the Grand Duchess Christina,* he explains, and not Bellarmine's exegetical method which is confirmed by *Providentisimus Deus* of Leo XIII (pp. 364–365).

Then, turning to a deeper analysis of Bellarmine's, Father Brodrick writes:

> Bellarmine's rather "fundamentalist" views were not special to him. They were widespread at the time and, in a sense inevitable, owing to the cautionary and defensive attitude with regard to the Scriptures, forced on the Church by the Protestant revolution. (p. 365)

Against this background, it is not hard to understand that Father Brodrick is no more able to accept Koestler's account of the developments than he can approve of every detail of Koestler's scholarly practice:

> In his exhilarating and most informative book, *The Sleepwalkers* . . . Arthur Koestler, who admits to a bias against Galileo because of his scurvy treatment of Kepler, pretty well revives Duhem's idea of Bellarmine's rightness and Galileo's wrongness. In the process, he makes Bellarmine General of the Jesuits and head of the Roman College, both erroneous statements. All his information about St. Robert is borrowed confessedly from Professor Giorgio de Santillana, who unconfessedly borrowed it all, including its mistakes, from Brodrick's Life of Cardinal Bellarmine in two volumes, London, 1928. (pp. 365–366 n3)

Scholars who may wish to dig further into the development of Bellarmine's own views on issues related to his part in the procedures against Galileo are advised also to look into Father Brodrick's just-mentioned two-volume biography of Bellarmine. Here he quotes extensively from Bellarmine's writings, including the *De ascentione mentis in Deum* which appeared in Douai in 1616, the very year in which Bellarmine was supervising the restraints on Galileo. The truly extraordinary relevance of the following autobiographical passage to the present theme will escape no one:

> Afterward, the same Prophet (the Psalmist) doth extol the course of the Sun, which also is very admirable. *He hath rejoiced* (saith he) as a giant *to run the way.* A giant truly if he extend his steps according to the greatness of his body and even as fast as his strength will afford, will in a short time pass a long way. And indeed the Prophet having compared the Sun unto a bridegroom thereby to declare the beauty thereof, also compareth it to a giant that by that resemblance he might in some sort shew the most speedy course thereof.
>
> But albeit he had not compared it to a giant but to the flight of birds and arrows, or to the winds and lightning; yet should it have been far from the thing indeed. For if that be true which with our eyes we see, to wit that the Sun in foure and twenty houres passeth about the whole compass of his orb; and if the compass of the Sun's orb exceedeth almost without comparison the compass of the earth; and if the compass of the earth containeth 20,000 miles; all which is most true: it must then needs follow that the Sun every houre runneth many thousand miles.
>
> But why say I every houre? Nay, every quarter of an houre; yea, almost every minute. For whosoever shall observe the rising or setting of the Sun in an open horizon as at sea, or in a plain field, shall perceive the whole body of the Sun to ascend above the horizon in less space than the eighth part of an houre. And yet the diameter of the Sun's body is much greater than the diameter of the earth, which notwithstanding containeth 7,000 miles.

I myself being once desirous to know in what space of time the Sun set at sea, at the beginning thereof I began to recite the Psalm *Miserere*, and scarce had read it twice over before the Sun was wholly set. It must needs be, therefore, that the Sun in that short time in which the Psalm *Miserere* was read twice over, did run much more than the space of 7,000 miles. Who would believe this unless certain reason did demonstrate it? (Bellarmine in Brodrick 1928, 2, p. 335)

It only remains to be added that there is a great deal more to the encounter of Father Brodrick and Arthur Koestler and the worlds they have depicted than we can include in these pages.[30]

NOTES*

1. For general guidance on the research and issues, see Burtt (1925); Butterfield (1951); Crombie (1952-53); Crombie and others (1963, 1964); Dijksterhuis (1961); Dugas (1958); A. R. Hall (1954, 1963a); Marie Hall (1962); Hanson (1958); Randall (1957a, 1957b); Rosen (1959; 1964, 1965); Taton (1964). Also see the instructive papers in two newly issued books on Galileo, especially Hanson in Kaplon (1965) and Moody and Spini in Golino (1966).

   The present study deals mainly with the interplay of central institutional and intellectual patterns in the transitions from medieval to modern science, philosophy, and religious life. For this reason, particular prominence is given to the work of Pierre Duhem (1906-13, 1908, 1913-59). With rare exceptions, the current British and American generation of historians of science have tended to take Duhem's philosophical, as well as his historical views, as points of departure for their own research. See, e.g., A. C. Crombie (1950-1952); M. Clagett (1959a); E. Grant (1962a, 1962b).

   This trend has been vigorously opposed by Edward Rosen. Though his work is mainly historical in emphasis, Rosen is evidently also unsympathetic to Duhem's philosophical views (see Rosen 1959 and 1964). The late Alexandre Koyré was another outstanding critic of Duhem. See, for example: Koyré (1939, 1957a, 1964). Limitations of Duhem have also been remarked by A. Maier (1949), Father Joseph Clark (1959).

   Again with only rare exceptions, leading philosophers of science have in the main, taken Duhem's *historical* analysis for granted. Here the outstanding dissenter has been Karl Popper. Sir Karl carries his learning in this area so lightly that it is easy not to notice that his essays are as original in their scholarship as they are acute in their logical analysis. No

*In preparing this paper for the present volume, many of the footnotes which contained material integral to the argument of the text were inserted in the text. This seemed to enhance the reading of the paper and to strengthen the argument, especially in the Appendices.—Ed.

fewer than four of his essays, now happily gathered together in his *Conjectures and Refutations* (1962, 1965), could not profitably be made required reading for students interested in the *History* of warring philosophies of science from the 16th century to our own day. Luckily the special relevance of these essays for my own research became apparent to me before it was too late—while the last drafts of this essay were in progress. The four essays of critical relevance to the present paper include. "On the Sources of Knowledge and of Ignorance" (pp. 3–30); "Three Views Concerning Human Knowledge" (pp. 97–119); "A Note on Berkeley as Precursor of Mach and Einstein" (pp. 166–174); and "On the Status of Science and of Metaphysics" (pp. 184–200). I call attention in the present context to Popper's remarks on pp. 97–100 on "instrumentalist" views of physical theory. Paul Feyerabend (1962, especially pp. 260–261) takes exception to certain of Popper's views and offers an interesting comparison of the current state of microphysics with the situation which prevailed in Galileo's days. I shall not be able in these pages to do justice to all the detail of Feyerabend's statements. I would, however, call attention to Bellarmine's views and reasonings in cosmology and the characterization of his methods by his learned fellow-Jesuit and biographer. See Appendix B below, "Father Brodrick on Galileo and Bellarmine, Duhem and Koestler," which includes a fascinating excerpt from the Cardinal's *De ascentione mentis in Deum*, a work issued at Douai in 1616, the year in which Bellarmine first officially imposed restraints on Galileo. These newer writings of Popper, Hanson, Feyerabend, and others, extend our knowledge beyond Duhem in respect to conflicting logics of scientific inquiry, including rationales of theory construction and designs of experimental research. In this connection, the learned chapters of the late professor R. M. Blake—see Madden and others (1960)—deserve more attention from historians of science than they have so far received.

2. The phrases above from Rostow appear in my first paragraph because I wish to lose no time in suggesting the two sides of my outlook: (a) medieval innovations did play a part (not so great as Duhem and some of his followers have at times contended, but an important part nonetheless) in the revolution of scientific interests and outlooks; (b) the discoveries of the 16th and 17th centuries do constitute a "breakthrough" and "take-off," one which continues at a vastly accelerated rate in our own century. For a related view, see now Moody (1966). Also cf. Price (1962).
3. This distinction hinted at between the emergence of modern science and its *subsequent institutionalization* is assumed in works of informed theorists like Merton and Parsons, but their almost exclusive interest in Protestant developments has tempted many less learned colleagues and students into faulty views on the Catholic developments. See, e.g., Merton (1970); Bellah (1964); Eisenstadt (1965); Ben-David (1960). Although all the necessary qualifications and cautions appear to be made by Merton, it remains true that his study makes no reference to the names of Duhem, Gilson, and many others who have done intensive research in intellectual

life of Catholic culture areas from the 12th to the 17th centuries. Almost nothing is said of medieval technological development, which has now been presented in a new light in extensive research summarized by Lynn White (1964). The only references actually made to medieval situations—those to St. Peter Damiani and the prohibitions of the Council of Tours (1163)—give a quite negative view of science of the Middle Ages (Merton 1970, p. 77 and in notes). One wonders whether Merton's position may be correlated with his high valuation of Francis Bacon's view of science. On this point and on Merton's historical views, see now A. R. Hall (1963b, esp. pp. 12–13). For another unenthusiastic view of Bacon's role in the history and philosophy of science, by an earlier writer, see M. R. Cohen (1926). A superb depiction of the role of Baconian ideology in the promotion of the scientific movement will be found in Jones (1961).

4. Feuer writes:

   In this study, I shall try to show that the scientific intellectual was born from the hedonist-libertarian spirit which, spreading through Europe in the sixteenth and seventeenth centuries, directly nurtured the liberation of human curiosity. Not asceticism, but satisfaction; not guilt, but joy in the human status; not self-abnegation, but self-affirmation; not original sin, but original merit and worth; not gloom, but merriment; not contempt for one's body and one's senses, but delight in one's physical being; not the exaltation of pain, but the hymn to pleasure—this was the emotional basis of the scientific movement of the seventeenth century. Herbert Butterfield has spoken of "a certain dynamic quality" which entered into Europe's "secularization of thought" in the seventeenth century. What I shall try to show is how the hedonist-libertarian ethic provided the momentum for the scientific revolution, and was in fact the creed of the emerging movements of scientific intellectuals everywhere. (1963, p. 7)

5. An exceptionally revealing account of the decisive importance of the notion of certitude in the debates over probabilism is given in Fagnani (1661, written in 1656-7). The probabilist background of the controversies connected with the names of Galileo, Descartes, and Pascal are overlooked by Popkin (1960), who allows himself only one sentence for the analysis of (subjective) *certitude* and (objective) *certainty*. Fagnani and other probabilist also go unmentioned in the pages of Van Leeuwen (1963). For preliminary insights into the outlooks of the intellectual innovators of the 16th and 17th centuries, see Febvre (1947), Blake (1960).
6. The discussion above in chapter 3 will help to explain why I compare men discussed here to "prophets" rather than to "sages," and why I speak of Catholic prophetism. As I construe their sense of mission, Galileo, Descartes, Pascal and Newton saw themselves as having a *prophetic* call. In the illuminating phrase of George Santayana, the ethic of the sage is *post*-rational, as a study of Western literature and religion confirms. See Santayana (1905, chapter 10, pp. 262–300) and Pfeiffer (1941).
7. For a critical discussion of the contrary view recently stated with great force by Koestler (1959) on the basis, largely, of the materials developed by Duhem and Crombie, see Appendix A below.
8. I am happy to find confirmation for this observation in Popper (1965, p. 100) who instances Mach's views on atomism. Also see Mach's remarks on relativity, in Mach (1960).
9. The oft-quoted lines given above condense two longer passages, one

from the beginning and the other from the end of the striking conclusion of Duhem's remarkable study. A careful reading of the full text of the passages given below will reveal that the phrases usually omitted, especially those appearing as the last in Duhem's essay and here, add a dimension to the argument:

> Many philosophers since Giordano Bruno have taken Osiander harshly to task for the preface he placed at the head of Copernicus' book. And Cardinal Bellarmine's and Pope Urban the Eighth's counsels to Galileo have been treated with hardly less severity since the day they were first published.
>
> The physicists of our day, having gauged the worth of the hypotheses employed in astronomy and physics more minutely than did their predecessors, having seen so many illusions dissipated that previously passed for certainties, have been compelled to acknowledge and proclaim that logic sides with Osiander, Bellarmine, and Urban VIII, not with Kepler and Galileo—that the former had understood the exact scope of the experimental method and that, in this respect, Kepler and Galileo were mistaken.
>
> Yet in the history of the sciences Kepler and Galileo are ranked among the great reformers of the experimental method whereas Osiander, Bellarmine, and Urban VIII are passed over in silence. Is this history's supreme injustice? Could it not be the case that those who ascribed a false scope to experimental method and who exaggerated its worth worked harder and better at perfecting it than did they whose estimate was from the start more measured and exact? (1969, p. 113)
>
> Despite Kepler and Galileo, we believe today, with Osiander and Bellarmine, that the hypotheses of physics are mere mathematical contrivances devised for the purpose of saving the phenomena. But thanks to Kepler and Galileo, we now require that they save *all the phenomena* of the inanimate universe *together.* (p. 117)

10. The neglected essay of Maire (1911, vol. 1, pp. X–XII) is also of interest in this connection. One wonders why neither Duhem nor his followers troubled to discuss this document and the 219 errors in a systematic way. Had this been done, there would have been a more reliable basis for proving or disproving the claim that 1277 was the starting-point of modern science. I call attention especially to the propositions numbered in the translation appearing in Lerner and Mahdi (1965, p. 335ff.).
11. Another view of the contexts governing Galileo's trial and sentence are sketched above in chapter 7, and which includes a summary of Cardinal Caramuel's review of the trial. Cf. Einstein's remarks (1953).
12. See Occam's *Dialogue,* Pars. 1, lib. 2, c. 22: "Especially physical assertions, which do not pertain to theology, are not to be solemnly condemned or interdicted by anyone; because in such assertions everyone ought to be free, and say freely what he pleases. And clearly, because the said archbishop damned and interdicted grammatical and logical as well as physical opinions, his judgment was considered rash. He condemned the opinion of Thomas concerning the unity of form in man, among others, which nonetheless you know many in Paris publicly hold and defend and teach, and so of many others." Cited in Geyer (1928, p. 582).
13. A new longitudinal and depth analysis of the notion of *libertas philosophandi* is needed. Meanwhile, see Sutton (1953), McLaughlin (1964), and Hofstadter and Metzger (1955).
14. For evidence that the absolutist theory of God's sovereignty may have contributed to the development of skepticism, see Popkin (1960).
15. The reader can now see the texts and discussions in De Rijk (1962), Petrus Hispani (Bochénski 1947), and Deman (1936).

16. In a letter sent to Kepler from Tübingen on April 12, 1598 (old Style), Matthias Hafenreffer wrote: "I firmly hope that there is room for my brotherly counsel; for in demonstrating hypotheses in this way you proceed as the pure mathematician *(nudum Mathematicumages)*, caring nothing whether or not they respond thus to created things. For the mathematician, if he offer such hypotheses, to which the phenomena most exactly correspond, I judge him then to follow his true purpose: and I think you yourself would yield to him who would be able to put forward better [hypotheses]. Nor, however, does it follow that the truth of things at once conforms to the hypotheses fashioned by art and thought (Nec tamen consequitur, uniuscuiusque Artificis Meditatis hypothesibus, rerum veritatem confestim conformari). I do not wish to mention those unconquerable things I could utter from the sacred. I believe there is need for brotherly advice, not disputation. But if you will acquire these things (and I am sure you will) and do abstract mathematics, I have no doubt that your thoughts will be agreeable to many (as certainly they are also to me). But if (which great and good God avert) you should wish publicly to reconcile these hypotheses with holy scriptures, and fight on their behalf, I fear that the matter will certainly erupt in dissension and conflict: in which case I would wish myself never to have seen these your thoughts, excellent and noble [though they be], considered in themselves and mathematically: for a long time now in the Church of the Lord, there is more of contention than aid to the weak." See Kepler (1938, vol. 13, pp. 202-204, esp. 203). Professor Edward Rosen graciously placed a copy of this passage at my disposal in response to my request for texts bearing on the fictional view of hypothesis.
17. Popkin's and Van Leeuwen's lack of reference to the long medieval record of development of these ideas can hardly be explained. See the hints in Kirk (1927) and Wood (1952).
18. The elements of truth in this view are expanded to the utmost limit by Koestler (1959) when he concludes that Galileo's aggressive personality is the sole explanation of the clash. See Appendix A.
19. Einstein's remarks (1953) on this point are well taken if not literal or precise.
20. For the affinities of relativity theory and Duhem's views, see Cohen (1949, p. 229). See also the discussion of Einstein and Bohr in Schilpp, ed. (1946), reprinted in Bohr (1958, pp. 32–66); Heisenberg (1958). Especially important discussions on the philosophical and historical contexts of the Galileo trial will be found in Popper (1965) and Feyerabend (1962). The latter's striking statements about the modernity of the church's outlook on the scientific issues at stake with the trial are reserved for more detailed discussion in Appendices A and B below.
21. Feuer (1963, pp. 158–159) cites critically a telling passage from Cohen's *Faith of a Liberal* (1947) and a number of other authors generally assumed to be partisans of intellectual freedom. The most striking recent assessment of Galileo is that offered by Koestler in *The Sleepwalkers* (1959).

    For Bellarmine's views, see Broderick (1961), and for contemporaries, Bohr (1958), and Heisenberg (1959); and Appendix B below.
22. A number of writers, including Popper, have praised Koestler for his

pages on Kepler. Popper adds, however, that Koestler fails to treat Galileo with the same understanding and sympathy as he does his German "Sleepwalker" hero. (Popper 1965, p. 188n).

23. As we note elsewhere (see below, note 25) Koestler and other writers do not discriminate the shades of the arguments directed against Galileo. The issues are of the greatest moment from the points of view of both theology and philosophy. Fictionalism and logical probabilism had different sources, foundations and contexts. Bellarmine came down consistently on the fictionalist side, but probabilist motifs are given greater prominence by some other writers. At no point, however, did Bellarmine ascribe *probability* to the Copernican hypothesis. If he had done that, there could have been little defense later against Galileo's effort to claim probability for his views. Bellarmine only refers to *suppositio* in the sense of a "counterfactual" *hypothesis,* thus remaining in the orbit of mathematical fictionalism. Osiander's Preface also implies a sharp distinction between mathematical fictionalism and probable opinions or true knowledge.

24. For the texts, see Favaro (1907, pp. 61–62). A vast literature has accumulated on the discrepancies between several 1616 actions—notably those of February 25 and February 26—especially as these arise in and qualify our understanding of the 1633 Sentence. The February 26, 1616 protocol has, indeed, been declared a forgery by many scholars because of its odd twists and turns. Koestler's textual discussion of the juridical and theological issues reveal lapses of command and lacks of precison. See the criticisms by Father Brodrick (1964, pp. 366, 376–377 in notes). I hope to discuss the relevant procedural and moral theological questions from fresh points of view in later essays.

25. Koestler's references to "probability" in the 1633 Sentence are in manifest contradiction with the argument of Father Inchofer's official protocol as Qualifier to the Inquisition (Favaro 1907, p. 89, p. 95). Yet Koestler's errors are easier to overlook than it is to understand Crombie's highly personal reconstruction of Bellarmine's life and mind:

> A student of astronomy in his youth, it had been Bellarmine's unhappy task to frame the decision that led Giordano Bruno to his death at the stake in 1600. Undoubtedly his policy over Galileo was based on a determination never to let that episode be repeated. Over seventy years old, he aimed at administrative peace, and his method of achieving it was to take the alternative way to Galileo's in order to escape the conflict between astronomy and Scripture. His policy was to weaken the conclusions of natural science and to accept the new astronomy as in no sense established with "indubitable certainty" but ony as "probable opinion and plausible conjecture," to accept it only in a form that would leave undisturbed the literal interpretation of Scripture and the Aristotelian cosmology which historical accident had married with it. He shut his eyes to the respects in which the union was less like marriage than living in sin. Yet although primarily administrative in their aim and limited in their application, it cannot be denied that Bellarmine's arguments succeeded in making a philosophical point against Galileo. Their two philosophies represent a classical polarisation of opposities, an antithesis in the conception of the discoveries and inventions of theoretical science that is at once ancient, persistent, and easily misunderstood. (Crombie 1959, 2, pp. 207–208)

The moral of all this is clear: Scientists and scholars now working in the history of science need to become more careful in their rendering of certain important expressions. If strict fidelity to logic and history is sought, *probabilis, probabilitas, probabilismus*—and only these words—should be rendered as "probable," "probability," or "probabilism." In point of tradition, terms such as *verisimilis* and notions like "hypothesis saving the phenomena," or, "statistical frequency," are at a great remove from *speculative* and *moral probability,* which relate to the technically elaborated dialectics of different shades of probable *opinion*. (Professor Popper is one of the very few writers on the subject who have recognized the pitfalls of the word *verisimilis.*) Even exceptionally reliable writers as Rosen and Blake waver now and then on this point (See, e.g., Rosen 1959, p. 25; Blake 1960, p. 32).

26. Despite the studies of Van Leeuwen (1963) and Popkin (1960), we are in great need of a critical history of the linked ideas of certitude and certainty. No such study can be done without an exhaustive study of the treaties and polemics on conscience and probability, where the matters of certitude receive their most elaborate analysis. It is not surprising, therefore, to note that Koestler and Crombie almost entirely do little by way of exploring these issues.
27. Cf. the careful statements of the Qualifiers, esp. those of the Jesuit Melchoir Inchofer (Favaro 1907, pp. 89–95). Inchofer explains: "Sed non est hypothesis mathematica quae conclusionibus physicis et necessariis stabilitur" (p. 92).
28. Koestler writes: "This legend and hindsight combined to distort the picture, and gave rise to the erroneous belief that to defend the Copernican hypothesis entailed the risk of ecclesiastical disfavour or persecution. During the first fifty years of Galileo's lifetime, no such risk existed and the thought did not even occur to Galileo" (1959, p. 358; cf. p. 449).
29. In the final note of this Appendix it is a pleasure to be able to add that in the 1959 revised edition of his valuable book, Crombie offers a fresh review of the influence of the struggle over hypothesis in the life and work of Galileo, Newton, and in the 17th century science generally (Crombie 1959a, 2, pp. 215–290, 315–333). This discussion carries Crombie well beyond his own 1950 essay, beyond Duhem's writings of 1905–1914, and for that matter, beyond Koestler, who, as pointed out above, strangely never mentions Crombie.
30. *Addendum* (10 July 1967): While this essay was going through the press, I have studied the following relevant writings with profit: the two chapters on Scheiner and Maffeo Barberini in A. Favaro's series on the "Oppositori di Galileo" published separately at Venice in 1919 and 1921 (cf. *Atti del R. Ist. Veneto di sc., lett. ed arti* 78, pt. 2a and 80, pt. 2a); W. Risse, *Logik der Neuzeit,* I (Stuttgart-Bad Cannstatt, 1964); H. Blumenberg's moving introduction to the German transl. of the *Siderius nuncius* and other selections from Galileo (Frankfurt: Sammlung Insel, 1965); and Mach's unnoticed Foreword to F. Adler's German translation (1912) of Duhem's (1905).

NINE

Certitude and the Books of Scripture, Nature, and Conscience

The presence of Edward Rosen in our midst today encourages me to begin by explaining that some pages in Rosen's lively and learned book on *Three Copernican Treatises* served as one of the main spurs that encouraged me to quicken my pace in a research program that has now carried me to new borders of the comparative historical *differential* sociology of culture and science—*in civilizational perspective.*

Having had a rather extensive involvement in the study of the jurisprudence and moral theology of the 16th and 17th centuries, and having worked rather intensively on crises over disputed logics of variant casuistries of conscience and probabilities of opinion, I found myself stirred to curiosity over theological, legal, and forensic philosophical backgrounds of critical documents and episodes in the history of science. I focused especially on controverted texts central to issues at stake in the struggles associated with the teachings of Copernicus and Galileo:

1. The actual and the claimed intention of Copernicus's masterpiece.
2. The identification of Osiander as the author of the anonymous preface.
3. The interpretation of the meaning of Osiander's preface.
4. Bellarmine's proposal to Foscarini and Galileo.
5. The 1616 decree against Copernicanism by the Congregation of the Index.
6. The cultural contexts and outcomes of the 1633 Sentence imposed on Galileo.

Reprinted from *The Nature of Scientific Discovery: A Symposium Commemorating the 500th Anniversary of the Birth of Nicolaus Copernicus,* edited by Owen Gingerich (Washington, D. C.: Smithsonian Institution Press, 1975), pp. 355–372. The original paper was dedicated to Professor Paul Oskar Kristeller of Columbia University. Reprinted by permission of the Smithsonian Institution Press. The illustrations on pp. 159 and 161 are reproduced by permission of the Houghton Library, Harvard University.

Early in my investigations I found myself noting that many available accounts of the prohibitions and the injunctions and the trial did not seem to me to explain what was happening. I could not grasp, for example, how it was possible for Galileo to think of publishing his *Dialogue Concerning the Two Chief World Systems* after such injunctions as those of 1616, nor could I understand how less than a quarter century after the Sentence against Galileo it would be possible for a theologian-scientist of such eminence as Cardinal Caramuel y Lobkowitz to say that the sentence against Galileo really was not in any sense a condemnation on the grounds of heresy, but that it was basically a sort of practical interdiction. Least of all could I grasp why so few present-day scholars had detected and clearly set forth the powerful workings in these developments of such influential ideas as probabilities of varied strengths, certainty, and certitude of conscience.

These developments brought into sharp focus some of the different cultural, theological, historical and sociological backgrounds for the development of science, and, as it were, the histories of how people actually experienced and ordered their worlds as they went about "sciencing."

Let me turn now to the main issue of the present discussion.

Edward Grant has called our attention to *"the great quest for reality initiated by Copernicus."* He explains:

> Modern science has shown a greater affinity with the XIVth century than with the century of Galileo and Newton. In the judgment of Pierre Duhem medieval scholastics had a truer conception of science than did most of the great scientists of the Scientific Revolution. Duhem even saw the Parisian nominalists as Christian positivists—forerunners of the positivist movement of his own day. He could not hide his scorn for the naïveté of some of the greatest figures of XVIIth century science who confidently believed they could—and should—grasp and lay bare reality itself. Most of their basic errors, Duhem insisted, derived from their delusive search for reality which served only to corrupt the theoretical structure of science.
>
> Duhem is, in general, quite right. Scholastics were most sophisticated and mature in their understanding of the role which an hypothesis must play in the fabric of science. They were not, as we have seen, deluded into believing that they could acquire indubitable truths about physical reality. But it is an historical fact that the Scientific Revolution occurred in the XVIIth century—not in the Middle Ages under nominalist auspices. Despite the significant achievements of medieval science—which Duhem himself did so much to reveal—it is doubtful that a scientific revolution could have occurred within a tradition which came to emphasize uncertainty, probability, and possibility, rather than certainty, exactness and faith that fundamental physical truths—which could not be otherwise—were attainable. *It was Copernicus*

who, by an illogical move, first mapped the new path and inspired the Scientific Revolution by bequeathing to it his own ardent desire for knowledge of physical realities. (1962b, pp. 219–220; emphasis added)

As I see it, Copernicus was not the first to get men started on the *quest for Reality.* In my view, the central issues of the quest in which the pioneers of early modern scientific revolution were engaged had another name in the days of Copernicus and Galileo. It was the quest for simultaneous (objective) certainty and (subjective) certitude of conscience, and this double-edged quest for truth had been insistent through the medieval era and, indeed, recurrently throughout the history of Western thought (see Schmitt [1972, e.g. p. 260 n92] on the discussion of *certitudo mathematicanum*). Once this has been admitted, two linked questions instantly come to the fore: Why did the quests for certitude and certainty have the great thrust and significance they reveal in the 16th and 17th centuries? And why did the thrusts in this direction prove to have so much more significance in the West than they proved in China? In asking the last question I give expression to my conviction that few—if any—developments in the way of sciencing and science in the world are "natural" (in the sense of invariant). Rather, these shifts in orientation occur within civilizational contexts that have their own particular complexities.

References to the two kinds of certainties, (objective) *certainty* of proof and the (subjective) *certitude* of conscience, regularly recur in discussions among Catholic thinkers on the distinctive features of the works of the speculative and practical intellect and the mathematical and physical demonstration. Finding themselves unable to acquiesce in counsels of prudence and accommodation, the pioneers aspired to a knowledge of truth that was secure and that guaranteed access to reality. They wished to know the order of the world. They wished to uncover the *machina mundi.* They wished to be able to see the design and laws of the physical universe. . . .

Another issue is very familiar to those who have read the writings of Joseph Needham. He argues that Chinese science and technology—and there I think he is mistaken in too often treating the two as though they were the same—were far ahead of the West until the era from Leonardo to Galileo, when the great quantum jump to Galileo's mathematical physics occurred. The turn came in the West, Needham insists, because of the breakthroughs of mercantile capitalism in the 16th century (Needham 1969a, chapter 6 [and see chapter 10 below]). . . .

Many were the differences of orientation, outlook, and organization at all levels, which made it unlikely that a science of a universalistic character could openly establish itself in China. You will recall that Matteo Ricci was the principal figure in one of the greatest missions in the history of Christendom, that which brought a number of very distinguished Jesuit scientists and engineers to China (see Sivin 1973). In his journal, Ricci gives a critical clue about some of these questions:

> It was during this time, when they had settled down in their new residence, that the Fathers undertook a work which at first sight might not seem to be wholly in keeping with the purpose of their mission but, once put into practice, proved to be quite beneficial.
>
> Dr. Hsu Paul had this one idea in mind. Since volumes on faith and morals had already been printed, they should now print something on European sciences as an introduction to further study in which novelty should vie with proof. And so this was done.
>
> But nothing pleased the Chinese as much as the volume on the Elements of Euclid. This perhaps was due to the fact that no people esteem mathematics as highly as the Chinese, despite their method of teaching in which they propose all kinds of propositions but without demonstrations. The result of such a system is that anyone is free to exercise his wildest imagination relative to mathematics without offering a definitive proof of anything. (Ricci 1953, p. 476)

I consider this passage an exceptionally revealing illustration of the wider horizons of what I have insisted on as the "quest for certitude."

I turn to another distinction that I have already sought to clarify elsewhere. It names the two key—but different—targets of the innovators: *the fictionalist theory of hypothesis or "saving the phenomena" and the probabilist casuistry of conscience and opinion.*

"Hypothesis" and "saving the phenomena" [as noted earlier, chapter 8, p. 124] marked the approach that Osiander sought to impute to Copernicus, and Cardinal Bellarmine later urged this approach upon Foscarini and Galileo. The resistance to this fictionalist concept of hypothesis by Copernicus, Galileo, Kepler, and Newton is now well known.

The second tradition is no less important in understanding the details of the various actions centering around the condemnations of Copernicanism and, of course, Galileo. It originates and develops independently of the status of mathematical schemas in astronomy. I refer now to the neglected tradition of the probabilist logic and epistemology loosely called probabilism. . . . Let me simply remark that in the 16th century, thanks to a very critical work by the Dominican, Bartholomeus de Medina, there was formulated a special kind of probability called *minus probabilismus* that altered the cast of all issues embraced under the rubrics, conscience, opinions, demonstration, proofs. It constituted one of the most critical contexts of many of the developments of the 16th and 17th centuries. . . . As I elsewhere argued:

Whatever their hopes—whether they wished to advance pure science, logic, secular progress, true or "natural" religion—the pioneers of early modern science and philosophy sought to establish new foundations for exact and compelling knowledge and belief. The prevailing probabilist dialectic that inhibited the quest for new data, the recourse to well-designed experimental research, strict logical proof, and mathematical formalization inevitably became a major target of the 17th century writers. (1962a, p. 615)[2]

. . . The excitement that was generated by the works of innovating scientists and philosophers was great, but the battle was not felt to have been drawn until forthright claims to truth or certitude occurred, claims that openly challenged entrenched (received) doctrine threatening the very foundation upon which all vested authority rested: The claim that there could be some set of truly trustworthy assurances based upon the evidence of the senses concerning the plan and pattern of the "Book of Nature". . . .

The time has now come to admit that I was myself startled one day to realize that I did not truly know the full contexts of two of Galileo's notable passages, which I had incorporated into the following statement [above, p. 132]:

The founders of modern science and philosophy were anything but skeptics. They were, instead committed spokesmen of the new truths clearly proclaimed by the *Book of Nature,* which they supposed revealed secrets to all who earnestly applied themselves in good faith and deciphered the signs so lavishly made available by the Author of Nature. Nature's Book, in their view, was written in numbers and never lied, whereas the Testaments were written in words which were easy and tempting to misconstrue. Men like Galileo and Descartes were vastly more certain about the truth revealed to them by number than they were by the interpretations placed upon Scriptures in the commentaries of theologians.

The two passages which now intrigued me were statements offered by Galileo at critical moments in the controversies with his opponents, especially the theologians. The first statement had to do with what was usually called the "Book of Nature"; the second related to what was usually called the "Book of Revelation."

1. The Universe, that is, the Book of Nature, Galileo explained (in *Il Saggiatore*), was written in numbers and as such was a sort of "universal manuscript" available to those who knew numbers, a book not readily falsified by any for their own interests. Galileo writes:

Philosophy is written in this grand book, the universe, which stands continually open to our gaze. But the book cannot be understood unless one first learns to comprehend the language and read the letters in which it is composed. It is written in the language of mathematics, and its characters are triangles, circles, and other geometric figures without which it is humanly impossible to understand a single word of it; without these, one wanders about in a dark labyrinth. (In Drake 1957, pp. 237–238)

Numbers, as we would say today, were the medium and numbers were the message—a medium and message so constituted deserved more credence than the conjectural and often contradictory interpretations of Scriptural texts, the texts of the so-called Book of Revelation on the part of rival theologians. All who knew numbers were free to read the Book of Nature; they required no special theological credentials and could test each other's readings. This was not the case with the Book of Revelation.

As everyone knew, any text was open to multiple constructions and yielded up its meanings only when these multiple constructions were put into play. Any phrase in Holy Scriptures or the Book of Revelation admitted of meanings at every level of interpretation, e.g., the literal, historical, analogical, anagogical, tropological, etc. All agreed that mistakes were bound to occur whenever an inappropriate meaning was imposed upon a text by a theologian insufficiently versed in the modes of interpretations.

2. The second passage in Galileo that now drew me forward into more intensive research was ascribed to Cardinal Baronius in his *Letter to the Grand Duchess Christina.*

Cardinal Baronius was correct, said Galileo, when he declared that the Holy Spirit had given us Scriptures—that is, the Book of Revelation—not to tell us how heaven goes, but how to go to Heaven. Galileo's words on this score are memorable:

From these things it follows as a necessary consequence that since the Holy Spirit did not intend to teach us whether heaven moves or stands still, whether its shape is spherical or like a discus or extended in a plane, nor whether the earth is located at its center or off to one side, then so much the less was it intended to settle for us any other conclusion of the same kind. And the motion or rest of the earth and the sun is so closely linked with the things just named, that without a determination of the one, neither side can be taken in the other matters.

Now if the Holy Spirit has purposely neglected to teach us propositions of this sort as irrelevant to the highest goal (that is, to our salvation), how can anyone affirm that it is obligatory to take sides on them, and that one belief is required by faith, while the other side is erroneous? Can an opinion be heretical and yet have no concern with the salvation of souls? Can the Holy Spirit be asserted not to have intended teaching us something that does concern our salvation? I would say here something that was heard from an ecclesiastic of the most eminent degree: "That the intention of the Holy Spirit is to teach us how one goes to Heaven, not how heaven goes." (Galileo [1615] in Drake 1957, pp. 185–186 with slight modification from the translation)

On first seeing this passage, I became convinced that it represented the sophisticated point of view of a theologian who was already possessed of a rationalized[3] structure of consciousness. Being so persuaded, I decided to look into the earlier life of the images to discover how such a distinction had been formed.

With luck, I soon found that an important step on the way to this destination were some passages in the *Natural Theology* or *Book of Creatures* by a Catalan theologian, teaching at Toulouse, generally known as Raymond de Sebonde (Ramon de Sibiude) on whose behalf Montaigne was to write an *Apology* and whose work Montaigne undertook to translate at the behest of his own father (see Montaigne 1948, pp. 318–457). On closely examining the *Natural Theology* by Raymond de Sebonde, I discovered that the main discussion in the body of Raymond's text did not precisely say what Cardinal Baronius had inferred. Indeed, there was a distinction between the constructions we were to derive from two books, the Book of Nature and the Book of Revelation, but Raymond de Sebonde seems to have given the palm in the body of his work to the Book of Revelation, because the Book of Revelation did offer help in getting to Heaven. Raymond considered this knowledge to be of higher form and purpose than the less certain, more ambiguous and equivocal teaching of the Book of Nature. It was hard to say what was the message of that book. Its main message seems to have been to establish the degree of our dependence on God's power and will (see Sabonde 1966 [1436], pars. 2a, tit. ccxii, pp. 312–15).

A quite different sense of the relation between the two books is conveyed in the Prologue and in the first sections of Book One of Raymond's work. In the Prologue Raymond talks about the Book of Nature as absolutely *knowable with complete certitude* and embracing everything that anyone really needs to know for his salvation. What he means here by the Book of Nature is what might be called the combined "Books of Nature and Conscience"—that is to say, the notion of nature and the self (1966, pp. 27, 37).[4]

Raymond speaks about the Book of Nature in a manner that might easily have served as a critical context for those disposed to find revealed truth in the Book of Nature. It is not claimed here that Raymond's discussion is the source of Galileo's remark.[5]

Being convinced that a "provincial" theologian of the 15th century was hardly likely to have initiated this cluster of notions, I proceeded to look into what must have been his ground sources, the authors of the High Middle Ages, when the nuclear metaphors of pre-Reformation Christian thought achieved their classical form. I turned to Grosseteste, Roger Bacon, Bonaventura, and other writers of the 12th and 13th centuries. I was not surprised when I found that these images were simply everywhere, especially among the Franciscan writers.

Everything in the world was seen as being the work of God's hand. Everything in the world was seen as a "book." All actions, images, and artifacts comprised books. As we have noted, there was not only the Book of Nature or Creatures, there were also the Book of Conscience, the Book of Revelation, the Book of Life, and other Books as well (see Tavard 1954, chapters 1-4; and Curtius 1953, for an older survey of the range of references to the "Books"). Everything in the world (not evidently the work of a human hand) was somehow aided by the creative spirit. As the work of God's hand, it was directly revealed as incarnate Nature. It conveyed its own image directly.

Therefore, the notion that the Book of Nature was somehow more ultimate and available than the Book of Revelation was not the expression of a wholly new idea in Galileo's day but a new accent placed upon an old idea. There was a shift, so to say, in some of the theological and philosophical notions.

Let me mention briefly a noteworthy related case of the linking of theological and philosophical ideas. At the top of the frontispiece of Riccioli's *Almagestum novum* in 1651 we see the outstretched hand of Jehovah. A nearly concealed reference at the ends of Jehovah's fingers to the apocryphal Wisdom of Solomon 11:20 explains that the created world was all made and disposed in number, weight, and measure. The explicit use of Wisdom 11 in artistic representations of the creation intending to stress the ordered creation of the world and

its disposition in number, weight, and measure goes back quite a long way in the Western world, at least as early as the year 1000. A most striking representation of the theme of the Geometer God is found in the Winchester Gospel of about that date in which God is seen with compasses in hand (L. White 1971, p. 189). It is quite remarkable that Riccioli should have thought to place a variant of this symbol as his frontispiece. Was it not a way of responding to the Galilean emphasis on number, weight, and measure? Very interestingly, it is exactly the symbol of Urizen, the Geometer God-Demon of the Compasses which Blake developed in his depiction of "The Ancient of Days," his attack upon mathematics, Newton—and Galileanism (Blunt 1937-38).

In the same breath I allow myself to say that the idea of a "machine of the world" was by no means new in the 17th century. The 17th-century notion of a new *machina mundi* had new elements, new stresses, new force, new impetus, new proofs, but it was not a new idea. The notion of *machina mundi* is classical in origin and it acquires new meanings when there occurs a peculiar junction between the Greek and Christian ways of thinking about the Creation and order of the world (see Koester 1968). More than one medieval philosopher-scientist spoke of unravelling the *machina mundi* and the *machina universitatis* (see Grosseteste 1963; M.-D. Chenu 1968).

I close this story abruptly but allow some final words of clarification and caveat. I shall not be surprised if colleagues and scholars who are expert in the histories of physics and mathematics from the 13th century forward doubt that any such notions could have had any effect in the development of science. Had not the professional theologians and philosophers of the late Middle Ages passed beyond responding in any significant way to symbols and images coming out of these theological and religious traditions?

Even after calm reflection, I discover that I cannot agree with the views of disbelieving friends.[6] To me the evidence seems overwhelming that many of these theological-religious symbols were experienced in different ways, given different kinds of stress and emphasis, but they constitute the backgrounds of even the most highly disciplined theologians and philosophers who in their technical work chose to confine themselves to proofs that were solid in philosophy and in physics; nonetheless, they had these symbols as a ground. A twofold claim is implied here: (1) Modern science has not only metaphysical foundations of Greek origin; it also has *theological and religious*—Hebrew, Christian, and probably Islamic—foundations; (2) a full recovery of the cultural foundations of modern science requires greatly widened access to root images in the substrate of these traditions.

The men of the Middle Ages looked everywhere in the hope of

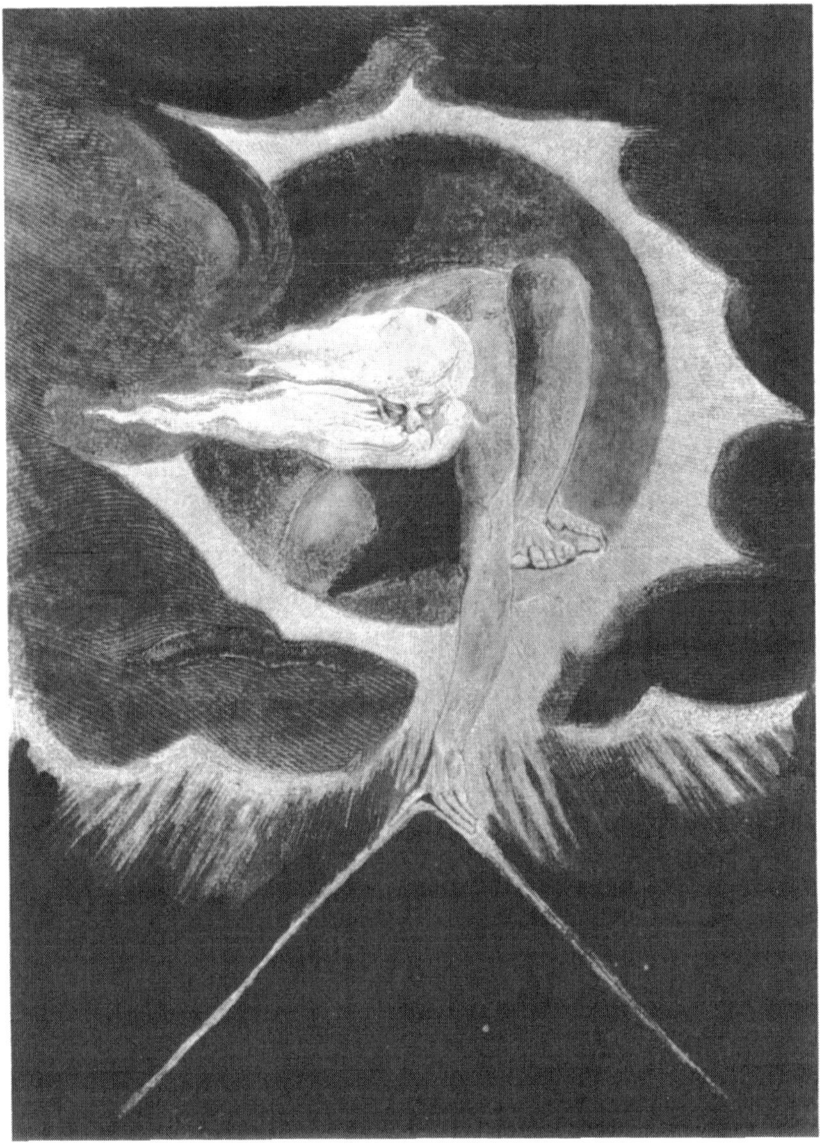

seeing the alphabet, the signature, the characters of the message of the universal manuscript. They were confident that by continuous inspection of Nature they could read the numbers therein. For various reasons and with various outcomes men like Roger Bacon and Raymond Lull looked for numbers everywhere in hopes of achieving a kind of universal language which might, in the view of some, be a vehicle for the conversion of mankind. It was, therefore, I repeat, not altogether a new idea that identified the Book of Nature with numbers.

In sum: The quests for subjective certitude of conscience and objective certainty of proof were powerful forces in the 16th and 17th centuries. The notion that there could have been anything else seems very unlikely to me. My other contention is that the thrust, the intensity of the two, certainty and certitude, had much to do with the fact that the West placed so great a stress upon the notion that the world was the work of God's hand and as such it constituted a Book of Nature, itself a revealed work, revealed to us by God's design. The Book was given to us and it was our task to decipher the pattern and prove the logic and character of the design. My last contention is perhaps a peculiar one whose discussion must be reserved for another day. The world was beyond any doubt a *machina mundi* and it was our responsibility to participate in *co-creation* through the development of mechanisms.

NOTES

1. A posthumously published book by N. R. Hanson, *Constellations and Conjectures* (1973, especially pp. 175–186, 200, 200–235) offers a stimulating alternative way of characterizing Copernicus's central contributions.
2. This citation is altered slightly from the original (1962a) by the author.—Ed.
3. A "rationalized" structure of consciousness in the sense indicated in chapter 5, pp. 95ff.—Ed.
4. When the main text of the *Theologia* was removed from the *Index* in 1564, the Prologue remained. On pp. 40–43 note especially the great stress on certitude.
5. Much research would need to be done before we could establish the actual sources of this saying of Galileo and Cardinal Baronius. A certain similarity in tone and phrasing does seem to exist between the Prologue of the *Theologia naturalis* and Galileo's *Letter to the Grand Duchess Christina*. See, for example, Ramond de Sebonde (1966, p. 36); Galileo, in Stillman Drake (1957, pp. 185–186).

6. This reference to dissent probably refers to the discussion and related exchanged at the Symposium Commemorating the 500th Anniversary of the Birth of Nicolaus Copernicus (Session 11, 24 April 1973 in Washington, D. C.), reported in Gingerich (1975, pp. 372–389).—Ed.

TEN

Sciences and Civilizations, "East" and "West": Joseph Needham and Max Weber

I

For some years now, I have been engaged in a series of "case histories" and "mental experiments" whose main aim has been to strengthen the foundations of an evolving comparative historical *differential* sociology of sociocultural processes and patterns, most notably sciences, in *civilizational perspective*. As I have conceived them, the "case histories" call for a strict study of determinate "civilizational complexes" with a view to analyzing the variations and the shifting balances of their cultural ascendancies resulting from critical contacts. In this same spirit I am continuing to seek to chart the differential impacts and outcomes of specified "intercivilizational encounters."

The title and subtitle of the present paper suggest clearly enough the wider settings to which the present "case" and "experiment" relate.

The working hypothesis on which I shall be proceeding herein is that sciences *as well as other cultural expressions* have taken many different shapes and enjoyed different saliences in the accredited frameworks of orientations and disciplines of different civilizational settings. Most important of all is the fact that the structures of consciousness and conscience associated with the varied sciences of great societies have been characterized by variable mixes of different cultural elements and emphases. This may be seen to best advantage in the study of the changing orientations of societies which have been the vehicles of world religions and the seed-beds of civilizational transformations.

Reprinted from *Boston Studies in the Philosophy of Science*, vol. XI, edited by R. J. Seeger and R. S. Cohen (Dordrecht-Holland: D. Reidel, 1974), pp. 445–493. Reprinted by permission of D. Reidel Publishing Company.

I further proceed on the postulate that the stratigraphy of these shifting ascendancies becomes especially available to view when two complex societies with deep historical imbeddedness come into intense contact with one another in respect to the ultimate structures of their traditions and commitments, above all the rationales legitimating their cultural ontologies and epistemologies. A large part in defining the eventual mixes of cultural elements is played by what I have been calling the "moralities (and logics) of thought" and the "logics (and moralities) of action" enshrined in their symbolic universes by their symbolic technologies (Nelson 1951, and chapters 2 and 3 above).

It is surprising how few have been the historians, sociologists, and philosophers of science who have placed due emphasis on certain basic contexts which must involve noteworthy considerations for all their diverse orientations, however specialized they be. Too few have recognized, for example:

(1) how large was the part played by the varied fortunes of science in shaping the different destinies of East and West;

(2) how large a part sociocultural, including religious, factors have played in the varied fortune of science. (See chapters 7 and 8 above)

II

In the present essay I shall mainly refer to episodes in the East-West relations during two periods of critical intercivilizational encounters: the era of the so-called Scientific Revolution—16th and 17th centuries—and the era of the so-called "Twelfth-Century Renaissance" and the Crusades—the 12th–13th centuries.

I begin with the crucial case of "East"-"West" relations in the era of the Western Scientific Revolution.

On a superficial view, the Scientific Revolution in the West was one continuous breakthrough from Copernicus to Newton. Actually, the story tells differently (Stimson 1917; A. R. Hall 1954; von Gebler 1876). There was no easy road for either astronomy or physics in the 16th and 17th centuries (Rosen 1959, 1965). The *De revolutionibus orbium caelestium* published in 1543 as the great astronomer lay dying at his home in Fromburk (Frauenburg) in Poland, acquired a readership only by being presented to the world in an anonymous introductory Letter to Readers describing the new image of the universe as a contrary-to-fact "hypothesis." It was many years before the author of this fateful Letter was publicly identified as Andreas Osiander, German Protestant Church official and theologian (Rosen 1959).[1] Despite the masking of Copernicus's purposes and claims, his work did not have easy sailing in the first century after its publication either in the Catholic[2] or Protestant worlds (Schofield 1965; Westman 1971; Christianson 1973).

Galileo's early successes did not protect him from later opposition and disfavor. After 1616 he was under admonition to desist from arguing for the truth of the Copernican hypothesis (see chapter 7 above, pp. 139). From this point forward he walked in a labyrinth. Nonetheless, in 1623 he arranged to publish his *Dialogue Concerning the Two Chief World Systems.* Soon he was to undergo trial and humiliation at the hands of the Inquisition.

Despite Galileo's forced self-denunciation, the Scientific Revolution pulsed forward. The new science of Copernicus, Kepler, Galileo, Newton triumphed and made especially rapid headway in the areas marked by a surge toward scientific enlightment. The capstone came in 1687 with the publication of Newton's *Principia.*

What happened in China, Japan, India was another story altogether.

To tell this story we must turn first of all to the renowned mission of Matteo Ricci, S. J. (Ricci 1953; Needham 1954–71, 4, pp. 437–61; Sivin 1973; Hellman 1963). The Jesuits who were associated in the Ricci mission were familiar with the new developments associated with Copernicus. Ricci himself was a student of Christopher Clavius at the Roman College. Many knew Galileo and were themselves Galileo's fellow members at the Academy of the Lynxes. Once, however, the Congregation of the Index had spoken against the Copernican hypothesis, the Jesuits felt bound to obey. They made few references to Copernicus or his work. This situation lasted throughout the crisis associated with Galileo. Publicly the Chinese Jesuits made no reference to the *Dialogue Concerning the Two Chief World Systems* and to the sentence against Galileo.

In short, the Jesuits did not feel under obligation to inform their Chinese hosts of the very stormy developments of Europe. As a result of this silence, the Chinese were kept from learning about the great new breakthroughs in astronomy and physics. Indeed, it was not until the late 18th century that the Chinese and Japanese were appraised of these developments. During most of the early 17th century the astronomy which was taught them and described in books by the Jesuits was that of Tycho Brahe, which represented a third way, the geoheliocentric system which respected the Biblical teaching (Boas and Hall 1959; Hellman 1963, 1970).

Professor Nathan Sivin (1973) has indicated that some Chinese astronomers early saw the limitations of the Tychonic system. The net effect of the Jesuit diffidence, however, was to delay the Chinese reception of Copernicus and Galileo until the late 19th century. The most powerful estimates of the effect of this delay will be found in the following statements by Sivin and Duyvendak:

> The Jesuits were also unable to discuss the wider repercussions of the Scientific Revolution, in particular Galileo's central idea that the only firm basis for knowledge of nature was the work of scientists themselves. The Church's injunction of 1616 against the teaching of heliocentrism was meant to reject this notion. To the very end of the Jesuit scientific effort in China, the rivalry between cosmologies was represented as between one astronomical innovator and another for the most convenient and accurate methods of calculation, rather than between the scholastic philosopher and the mathematical scientist for the most fruitful approach to physical reality. Thus the basic character of developing modern science was concealed from Chinese scientists, who depended on the Jesuit writings. . . . (Sivin 1973, p. 103)
>
> . . . The impossibility for the Jesuits, the mediators of Western science in China, to accept Galileo's heliocentric theory, is a matter of immense cultural significance. . . . It is really with the condemnation of Galileo that the paths of East and West diverge, not to meet again until the nineteenth century . . . . (Duyvendak 1948, p. 328)

III

My direct purpose herein will be to speak to the challenges posed by these developments and to review the historical and sociological theses set forth by Needham in his life-long studies of the dialogues and encounters of East and West. Toward this end I shall be drawing especially on four of Needham's works, his monumental *Science and Civilization in China* (1954–71), his *Grand Titration* (1969a), *Clerks and Craftsmen in China and the West* (1970), and a series of popular lectures which offer many clues to Needham's personal attitudes, *Within the Four Seas* (1969b).

To help illuminate the theoretical frameworks of Needham's hypotheses and challenges I shall mainly draw here upon clues culled from the pages of a small company of titans of sociology: Max Weber principally, but also Sir Henry Maine, Emile Durkheim, Marcel Mauss, and a number of more recent writers. I single out these authors because I am convinced that along with Needham they offer us building blocks on which to rear a comparative depth-historical *differential* analysis of sociocultural processes at the societal, civilizational and intercivilizational levels which we so sadly lack today and will more desperately need tomorrow.

Among other contributions, these authors suggest clues on how to correlate factors involved in the movements toward and away from participations in wider communities of discourage–passages toward and away from wider rationality and fuller freedoms of entry into and exits from frameworks of inquiry; passages toward the elaboration of social and cultural milieux conductive to applications and validations of structures of rationales regarding evidentiary canons and procedures—outcomes expressed in languages marked by logical constancy, abstractions of form yielding universality of opportunity.

IV

I begin with two expressions of Needham's challenges whose current importance for the comparative history and sociology of science and society it is impossible to exaggerate:

Perhaps the most far-reaching issue in Needham's challenge calls out to us from the pages of his monumental *Science and Civilization in China* which sums up the meaning of the famed Jesuit mission in China in the early 17th century. Here Needham asks whether the new astronomy which was in the power of the Jesuits to mediate to the Chinese needs to be thought of as a *peculiarly Western science* as opposed to *Eastern science* or as a *new, universally valid,* world science. Fully cognizant that the entire tenor of his vast undertaking hinges on his answer to this question, Needham replies:

> It is vital today that the world should recognise that 17th century Europe did not give rise to essentially "European" or "Western" science, *but to universally valid world science, that is to say, "modern" science as opposed to the ancient and medieval sciences.* Now these last bore indelibly an ethnic image and superscription. Their theories, more or less primitive in type, were culture-rooted, and could find no common medium of expression. *But when once the basic technique of discovery had itself been discovered, once the full method of scientific investigation of Nature had been understood, the sciences assumed the absolute universality of mathematics, and in their modern form are at home under any meridan, the common light and inheritance of every race and people. . . .*
>
> . . . *And what this language communicates is a body of incontestable scientific truth acceptable to all men everywhere.* Without it plagues are not checked, and aircraft will not fly. The physically unified world of our own time has indeed been brought into being by something that happened historically in Europe, but no man can be restrained from following the path of Galileo and Vesalius, and the period of *political dominance* which modern technology granted to *Europeans is now demonstrably ending.*
>
> In their gentle way, the Jesuits were among the first to exercise this dominance, spiritual though in their case it was meant to be. To seek to accomplish their religious mission by bringing to China the best of Renaissance science was a highly enlightened proceeding, yet this science was for them only a means to an end. Their aim was *naturally to support and commend the "Western"*

> *religion by the prestige of the science from the West which accompanied it. This new science might be true, but for the missionaries what mattered just as much was that it had originated in Christendom.* The implicit logic was that only Christendom could have produced it. Every correct eclipse prediction was thus an indirect demonstration of the truth of Christian theology. The *non sequitur* was that a unique historical circumstance (the rise of modern science in a civilisation with a particular religion) cannot prove a necessary concomitance. Religion was not the only feature in which Europe differed from Asia. But the Chinese were acute enough to see through all this from the very beginning. The Jesuits might insist that Renaissance natural science was primarily "Western" but the Chinese understood clearly that it was primarily "new." (1954–71, 3, pp. 448–449; original emphasis)

Precisely how the Jesuits or the Chinese defined the situation is no easy matter to decide from evidence made available by Needham and others. What matters to us at the moment is how *we* need to assess the elements at issue in this conclusive contact of ultimate rationales.

And now to the second set of Needham's challenges:

> Between the first and fifteenth centuries the Chinese, who experienced no "dark ages," were generally far in advance of Europe, and quite independent of the great ideas and systems of the Greeks. Not until the scientific revolution of the Renaissance did Europe draw rapidly ahead. Throughout those fifteen centuries, and ever since, the West has been profoundly affected not only in its technical processes but in its very social structures by discoveries and inventions emanating from China and East Asia. Not only the three which Lord Bacon listed (printing, gunpowder and the magnetic compass) but a hundred others—mechanical clockwork, iron-casting, stirrups and efficient horse-harness, the Cardan suspension and the Pascal triangle, segmental-arch bridges and pound-locks on canals, the stern-post rudder, fore-and-aft sailing, quantitative cartography—all had their earth-shaking effects on a Europe generally more unstable. Why then did *modern* science, as opposed to ancient and medieval science, develop only in the Western world since the time of Galileo? (Needham 1969a, p. 11)

What shall we say to Needham's challenges and questions? How well does he himself answer them in the light of our present understanding of comparative historical sociology? How secure are the foundations on which Needham's questions rest? Is it truly the case, for example, that:

(a) Between the first and fifteenth centuries the Chinese, who experienced no "dark ages," were generally far in advance of Europe, and quite independent of the great ideas and systems of the Greeks?

(b) Europe did not draw rapidly ahead until the scientific revolution of the Renaissance, more exactly the 17th century?

(c) Throughout the first fifteen centuries of the Christian era and ever since, the West has been profoundly affected not only in its technical processes but in its very social structures by discoveries and inventions emanating from China and East Asia?

I anticipate the discussions to follow by remarking that we will not be able to find a satisfactory answer to the issues Needham poses if we accept without check the statements he has proclaimed facts in the first two paragraphs I have just concluded. The issue of factual accuracy is no small or pedantic matter. The established assurance of an accurate comparative history of science will prove the clue to the analysis we shall describe as comparative historical differential sociology in civilizational perspective.

V

Scholarly courtesy demands that I stand aside at this point and wait upon Needham to develop his own case on the basis of the evidence he knows so well. What does Needham himself see as the explanations of the phenomena and problems under investigation?

First impressions to the contrary apart, it will be noted that Needham is not content to rest his case upon a single or simple formula. Strewn throughout his works are many hypotheses offered with varying stresses scoring the orchestration of elements supposed to constitute the explanations of the phenomena. The main sources which he draws upon for explanatory paradigms are Marx and Engels (Needham 1969a, pp. 123–124, 129, 192–194, 202–303; 1971, 2, p. 291), Karl Wittfogel (1931), Edgar Zilsel (1942a, b), and other less renowned scholars who are believed to have advanced a so-called dialectical materialist, or Marxist interpretation of history. While Needham acknowledges his dependence on Wittfogel's early work, he is very critical of Wittfogel's *Oriental Despotism* (1957; cf. Needham 1969a, pp. 150, 192–194, 203–204). Two scholars particularly praised by Needham are Jean Chesneaux and André Haudricourt (see Needham 1969a, pp. 123–24, 192–94).

It quickly becomes apparent, however, that Needham feels free to range outside the conventional Marxist canon. Needham explicitly rejects fixed stage theories of evolutionary sequence. Some weighty and critical hypotheses suggested to him by the Chinese data are related to other sources. One such critical source is Alfred North Whitehead (see Needham 1969a, pp. 124, 323–24; 1954–71, 2, pp. 201, 291–292, 454). Whitehead is credited by Needham with having developed philosophical views which very closely resemble the central Chinese concepts of process. This leads Needham to ask a very striking question:

The Chinese world-view depended upon a totally different line of thought. The harmonious cooperation of all beings arose, not from the orders of a superior authority external to themselves, but from the fact that they were all parts in a hierarchy of wholes forming a cosmic pattern, and what they obeyed were the internal dictates of their own natures. Modern science and the philosophy of organism, with its integrative levels, have come back to this wisdom, fortified by new understanding of cosmic, biological and social evolution. Yet who shall say that the Newtonian phase was not an essential one? (1954–71, 2, p. 582)

Essentially Needham conceives his main challenge to be double-edged: (1) Chinese science and technology markedly outstripped the West from the second to the sixteenth centuries.

(2) How, in the face of this evidence, explain the fact that the Galilean-Newtonian science did not emerge in China?

Seeking explanations in terms of the sociocultural directions embarked upon by the autochthonous civilizational complexes (a critical element stressed in Needham's comparative historical, differential sociology), he seems, in my reading, to mark out five interrelated systems or complexes in terms of which changes in socio-cultural patterns can be explained:

(1) geographical-environmental settings
(2) socio-economic structures
(3) cultural ontologies and philosophical principles
(4) symbolic (in the narrow sense) technologies, notably linguistic patterns of the written and spoken languages
(5) community and associational patterns and values, including norms of conduct.

Needham himself sums up his findings under four "factors": geographical, hydrological, social and economic (1969a, p. 150).

The key features of these data ask to be arranged under these rubrics as follows:

(1) The *geographical-environmental* conditions of China were marked by an isolation of the Chinese civilization unknown among the Mediterranean or even the Hindu civilization. Furthermore, China was a gigantic land mass connected by rivers, rather than a series of coasts connected by a virtually enclosed sea, and, despite the similarity of the range of climates between China and Europe, only China experienced a monsoon season (Needham 1969a, pp. 190–191).

(2) In its *socio-economic structures,* medieval China was a non-hereditary "feudal bureaucracy" founded on an agrarian base, which dispensed with or lacked an organized system of slavery. Supervision

of defense and the waterworks was reserved to the Imperial authority (1969a, pp. 26, 167–171, 206–208).

(3) A central feature of the *cultural ontologies* of China was the idea of *li,* the central idea shared by both the Confucians and Taoists, of harmony and inter-relatedness among all phenomena—mankind included—in the universe. The preference of the Taoists for withdrawal and "wordless" contemplation of Nature (discerning "The Way" and "the One") contrasted with the heavy emphases among the Confucians on literacy, scholarship and learning (1969a, pp. 34–35, 138–162, and chapter 8; 1971, 2, pp. 519, 528, 557). Here Needham would also include the way in which the calendar—very closely linked to religious and political concerns—played a central role due to the interconnectedness of celestial and human affairs.

(4) *Linguistic-symbolic technologies* of China refer to the complex structures of its language—as written by the literati and as spoken—the restraints built into its language by its pictographic cast and the early innovation of a genuine decimal number system (1954–71, 3, pp. 3–4, 21).

(5) *Community and associational structures and values* comprise the pieties, obligations, and norms of conduct deriving from traditional claims of family, clan, secret-society, and other claimant group structures (1969a, p. 207).

A few short pages will not suffice to show how these five complexes ground the explanation of so vast an amount of data as Needham has assembled. I would briefly indicate by way of illustration how their intersections help explain four phenomena which are equally critical for our purpose:

(1) The marked superiority of China over the West in technological inventiveness from the 2nd to the 16th centuries.

(2) The failure of empirical scientists and merchants to emerge in China as they did in Europe (Needham 1969a, pp. 38–41).

(3) The failure of an idea of "laws of nature" to emerge in China (Whitehead, Zilsel, and others; Needham 1969a, chapter 8, 1954–71, 2, chapter 18).

(4) China's failure to achieve a breakthrough to the experimental-mathematical mode of Galileo (1954–71, 2, pp. 540–43).

The evidence can be given in synoptic form as follows:

(1) The absence of any systematic exploitation of slave labor was in Needham's view a key factor in the superiority of China's technology

prior to the sixteenth century. The absence of slaves only secondarily gave incentive to the invention of labor-saving devices. Primarily, the absence of slave labor meant that the adoption of any such invention would pose little threat to the texture of society. Hence, China's socio-economic structure was more amenable to the propagation of inventions than, say, that of Rome (Needham 1969a, pp. 26, 167–171, 206–208).

China's cultural ontologies were also important factors in its early technological sciences. The Taoist practice of contemplation of Nature—to discern the "Tao," which would reveal the proper (anti-Confucian) "Way" for society (1969a, pp. 34–35, 158–162)—cleared the path for many advances in pharmacology, alchemy, medicine and some parts of physics. For example *wu wei,* the principle of nonintervention (the same principle which blocked the path to the notion of "Law of Nature") made the idea of action-at-a-distance (that is, of non-mechanical action) unproblematic, leading to the discoveries of the seismograph and of many magnetic phenomena unknown in the West until quite late (1969a, p. 210).

On the other side of the coin, the Confucians, in the service of the feudal bureaucracy, lent "orthodoxy," and thereby encouragement, to its projects—especially astronomy (for calendar-making) and hydrodynamics (1969a, p. 186). In the case of the latter two technologies, we see also the contribution of the environmental factors of monsoon and river systems. The agrarian-based feudal bureaucracy had to control seasonal floods which threatened both crops and safe passage of tax-barges carrying grain as in-kind payment (1969a, pp. 18, 32, 181, 195).

(2) As for the *non-emergence of merchants and experimental scientists in China:* here, as Needham tells the story, the feudal bureaucracy plays the leading role. Unlike the late medieval European city, where the mayor or burgomaster shared responsibility with the guilds, the Chinese city was governed by the bureaucratic representative of the Emperor (1969a, pp. 185–186). There was no gap of power in which the merchants could insert themselves. This situation was exacerbated by the fact that merchants held the lowest station in the society, after scholars, farmers, and artisans in that order. As Needham puts it, every merchant's son desired to become a bureaucrat. Furthermore, continual interventions (1969a, pp. 39, 196) by the bureaucracy in the name of the Emperor—e.g., sumptuary laws—in the economic affairs of merchants inhibited their attainment of power and influence.

Just as the merchants, so also the artisans—who were next to the bottom of the ladder in the society, and according to Needham were the richest source of experiment and invention—were inhibited from coming to the fore. Like the merchants, artisans could achieve little status (1969a, pp. 28, 39). Consequently, bright young men were drawn away from the crafts to the bureaucracy, applying their intelligences to the literary studies necessary to pass qualifying examinations, rather than to invention. Furthermore, the lack of training in the language of the literati made it virtually impossible for artisans to make even their discoveries understandable or respectable among the powerful Confucian scholars, the advisors of the emperor (1969a, p. 31).

The connection Needham finds between the failure of the merchants to achieve power and the non-emergence of experimentalists becomes clear when he puts his critical question in a new way: What was there peculiar to the Western development which pushed it beyond the situation which had crystallized in China? Raising the question in this form allows Needham to bracket the importance of theoretical traditions from the Greeks until the end of the 14th century (1969a, pp. 190–191, 217). It also compels him to stress changes in the economic and social realms. He writes:

> It may well be that concurrent social and economic changes supervening only in Europe formed the milieu in which natural science could rise at last above the level of the higher artisanate; the semi-mathematical technicians. The reduction of all quality to quantities, the affirmation of a mathematical reality behind all appearances, the proclaiming of a space and time uniform throughout all the universe; was it not analogous to the merchant's standard of value? (1954–71, 3, p. 166)

Needham also asked whether the lack of Chinese interest in exploiting technology in certain areas—and hence the failure to break through to Galilean science—could be attributed to the lack of a profit-motive due to the feudal bureaucracy. His answer on this issue is striking in its simplicity:

> Put in another way, there came no vivifying demand from the side of natural science. Interest in Nature was not enough, controlled experimentation was not enough, empirical induction was not enough, eclipse—prediction and calendar calculation were not enough—all of these the Chinese had. Apparently a mercantile culture alone was able to do what agrarian bureaucratic civilisation could not—bring to fusion point the formerly separated disciplines of mathematics and nature-knowledge (p. 168).

(3) The very idea of *"law"* was radically different in China and the West. Whereas in Rome an abstract system of legal concepts and procedures had to be developed to encompass adjudication of the discrepant interests within and among Roman groups and the rela-

tions of Romans to the many "peoples" of the empire (each having its own "law") coupled with the relative lack of internal differentiation, China's geographical isolation made this unnecessary (1954–71, 2, pp. 518–519). As a result, the notion of "laws" promulgated by a divine lawgiver was wholly foreign to the Chinese understanding of law. The latter started from the assumption that the entire universe of man and entities was a "pattern," a "ceaseless regularity," and it was not a commanded regularity; it was a spontaneous co-operation of all things according to their natures (1954–71, 2, pp. 557, 562, 579). (4) An issue which Needham himself describes as "a focal point" in his grand design is put most powerfully in his restatement of his challenges in the context of the relations of mathematics to science. Needham writes:

> If the foregoing pages have been numerous, we must reflect that from many points of view mathematics has always been a discipline of its own of equal rank with the whole of the natural sciences. The conclusion of the account of Chinese mathematics brings us to what might be described as a focal point in the plan of the present work. What exactly were the relations of mathematics to science in ancient and medieval China? What was it that happened in Renaissance Europe when mathematics and science joined in a combination qualitively new and destined to transform the world? And why did this not happen in any other part of the world? (1954–71, 3, p. 150)

As we presently see, this form of the question is very close to Weber's form of the question.

Looking more carefully into the inner mechanism of this development, Needham writes:

> . . . It has often been said that whereas previously algebra and geometry had evolved separately, the former among the Indians and the Chinese, and the latter among the Greeks and their successors, now the marriage of the two, the application of algebraic methods to the geometric field, was the greatest single step ever made in the progress of the exact sciences. It is important to note, however, that this geometry was not just geometry as such, but the logical deductive geometry of Greece. The Chinese had always considered geometrical problems algebraically, but that was not the same thing. (p. 156)

Needham does acknowledge that certain of the formative ideas which were to find fruit in Galileo appeared as early as the 13th century among the philosophers at Oxford (pp. 161–162). He hastens from this to explain that

> these ideas came later to be associated with the University at Padua, where Averroism was strong and logic was studied as a preliminary to medicine, not law or theology. Discussions there, between the + 14th and the + 16th

centuries, led to *a methodological theory which, except for the important element of mathematisation, showed some similarity to the eventual practice of Galileo.* (Emphasis added)

Needham comes close to anticipating our own view toward the end of his discussion where he declares:

Thus while the practice of the higher artisanate was akin to the second or experimental part of the Galilean method, the theorising of the scholastics foreshadowed the first or speculative part. But how widely they were aware that agreement with empirical fact was the ultimate test of hypotheses seems doubtful, nor is it clear that they always understood the importance of examining new phenomena . . . which had not already been used as the source of the hypothesis under test. Moreover, they rarely succeeded in advancing beyond the primitive style in their hypotheses. Robert Grosseteste of Lincoln (+ 1168 to + 1253) has been selected as the key figure in this natural philosophy, but the dual process of induction and deduction goes back to Galen and the Greek geometers, probably reaching Grosseteste through Arabic sources, such as the encyclopaedist Abū Yūsuf Yaqūb ibn-Ishāq al-Kindī (d. + 873) and the medical commentator 'Alī ibn Ridwān (+ 998 to + 1061). Through Grosseteste may have believed that organised experimentation beyond mere further experience should be used to verify or disprove hypotheses, it is not claimed that he himself was an experimentalist. He does seem however to have influenced the + 13th-century group of practical scientific workers which included the Englishmen Roger Bacon (1214 to 1292) and Thomas Bradwardine (1290 to 1349) in physics, the Frenchman Petrus Peregrinus (fl. 1260 to 1270) in magnetism, the Pole Witelo (c. 1230 to 1280) in optics, and the German Theodoric of Freiburg (d. 1311) with his admirable theory of the rainbow. It is curious that during the period in which these men were working, China was the scene of a scientific movement quite comparable. But after the early years of the 14th century there was a marked regression, and verbal argument again dominated in Europe until the time of Galileo himself. . . . There was thus a continuous line of experimentalists in Europe from Roger Bacon to Galileo, but after about + 1310 the contribution of scholastic philosophy ceased, and for three centuries practical technology was the order of the day. (pp. 161–162; emphasis added)

Needham proceeds to a careful dissection of the phases of the so-called Galilean method proceeding on the assumption that the new or experimental philosophy was characterized by the search for measurable elements in phenomena, and the application of mathematical methods to these quantitative regularities. He is obliged to acknowledge that the conscious experimental test of precise hypotheses which formed the essence of the Galilean method differed from instigative experimentation of technologists and craftsmen. He adds:

In such empirical ways it was possible to accumulate great stores of practical knowledge, though the lack of rationale necessitated a handing down of technical skill from one generation to the next, through personal contact and training. With due regard to different times and places there was not much to choose between China and Europe regarding the heights of mastery achieved; no westerners surpassed the bronze-founders of the Shang and Chou, or equalled the ceramists of the Thang and Sung. The preparations for Gilbert's definitive study of magnetism had all taken place at the other end of the Old World. And it could not be said that these technological operations were non-quantitative, for the ceramists could never have reproduced their effects in glaze and body and colour without some kind of temperature control, and the discovery of magnetic declination could not have occurred if the geomancers had not been attending with some care to their azimuth degrees. (p. 159)

VI

If, as I have suggested above, we look into the pages of a number of comparative historical sociologists of social and cultural process whom Needham hardly mentions by name; if, that is, we look into Weber, Durkheim, Marcel Mauss and, last but not least, Sir Henry Sumner Maine, we will be rewarded by finding critical new clues in answer to Needham's questions. The most rewarding of these bear upon two related clusters of concepts which are variously represented in the works of these men. The clusters and concepts may be described as follows:

(1) *rationalism, rationality, rationalization* and what I elsewhere describe as *rationale-structures* in their several and joint relations to the prospect of promoting the fullest rationalization of intelligence (Nelson 1968b, 1974c [and chapter 4 above]). (On the many issues involved here Weber serves as our main, but not our only, guide. A limiting aspect of this fact is that Weber's teachings on these heads are not entirely free of ambiguity.)
(2) the sources and outcomes of different passages to *universalism, universalities* and *universalizations* in the multiple contexts of thought and action in the varied spheres of social relations and symbolic cultures (Nelson 1969a). (None of our authors speaks about this second complex of themes expressly in any detail; the main contributors are Weber, Durkheim and Maine. Durkheim offers clues on the relations between changes in social structure to changes in the size and density of communities of discourse and persuasion [Durkheim 1915, 1933]. Weber's hints can be elaborated to yield an

important body of notions expressing the pattern of *"double dialectic"* of universalization processes in the movements and changes in the social and cultural spheres. Of this we will speak at greater length.)

To return to what I have called *the fullest possible or maximum rationalization of intelligence:* If decisive movement in this direction is to occur, it is requisite that great numbers of men and women be free to transcend the particularistic restraints of family, kin, class, caste and to allow their minds to wander freely in zones which are acknowledged to be immune to intervention. These zones are neutral zones protected against the insurgence of the political or theological censor and the group dictate (cf. above, chapter 6, pp. 111f).

The promotion of a full rationalization of intelligence demands that substantial numbers of persons be legally empowered and psychologically disposed to carry on mental production at the highest level of operation without being called to a halt by disabling private or public inhibitions or barriers.

We proceed now to set down some of the main answers to Needham's challenges.

VII

Although Marco Polo was astonished by what he saw on his visits to China, it is now evident that had he been oriented to a genuinely comparative analysis of cultural and social process he might have needed to report the giant strides which Europe had achieved during the twelfth and thirteenth centuries. The years between the first crusade and the Council of Vienne of 1311 were extraordinarily eventful for Western Europe (cf. chapter 12 below).

On the surface, China was a vast land Empire with extraordinary evidences of complex technologies of transport, communications, exchange, regulative legal and instructional systems, and so on. During the European High Middle Ages we are describing here, however, Europe underwent a succession of shocks which were to leave it stronger rather than weaker. The following developments will be recalled:

(1) The overcoming of Muslim power in the Mediterranean (Pirenne 1956).

(2) The development of self-governing cities by no means lacking in legitimate authority to rule on their own behalf, as a cryptic phrase of Weber mistakenly allows us to believe. (For reasons which remain to be discussed in detail on another occasion, Weber placed cities under the rubric, "non-legitimate domination [nicht-legitime Herrschaft]"; Weber 1968, p. 1212.)

(3) The spread of liberties in the several senses of both liberties *from* and liberties *to*. In the present instance the critical liberties are the liberties *from* the control of territorial powers and liberties *to* employ one's own abilities in the "public's" or one's own behalf (Weber 1968, pp. 1249–1256).

(4) The spread of many centers of learning, especially universities, which were dedicated to expanding the scope and exploring the roots of knowledge as then conceived. Of critical importance here was the liberty to philosophize on matters not dogmatically established (Sutton 1953).

(5) The systematic concordance of knowledge and learning. Most important here were the summing up of "scientific" theology, natural philosophy, political philosophy, moral philosophy, and law—Roman, canon, common, municipal.

From the sociological side, the 12th and 13th centuries were the seedbed of the modern European society. It was exactly the differentiation into kingdoms, principalities, cities, estates (*stände*), professions, universities and so on which helps us understand the extraordinary pulse-beat of the developments of the 12th and 13th centuries (Weber 1968, chapter 16, pp. 1212ff, chapter 15, 1928, chapters 28, 29).

It was the West which was to explore the vast new uses of numerous techniques and implements which were originated by the Chinese, Hindus, and Muslims. An especially striking case is the development in the West of the so-called Hindu-Arabic numeral system and the zero (Sombart 1915; Weber 1928; de Roover 1937). It was the West which was to give new force and meaning to the mathematical approach to natural philosophy and to move dramatically toward mathematical physics with the help of newly recovered Greek texts. In these centuries, the West was entirely to restructure the rationales of thought and action, knowledge, opinion, and conscience (Chenu 1968; Nelson 1969a, and chapter 12 below).

Only because we have been disinclined to study the evidence and traces of the comparative development of the symbolic technologies in East and West have we failed to see the most important elements in the development of the West.

VIII

In other writings I have emphasized the advance which Needham makes over many other investigators in his presentation of some of the key issues involved in his "challenges." I have, indeed, proposed that Needham needs to be included among the ranking pioneers in the comparative historical sociology of science and sciencing in civilizational perspective (chapter 5 and 6 above, pp. 87f and 111f).

The point has now come to enter some necessary demurrers. For all his expertise in science and sinology, Needham falls short of Weber, whom he hardly appears to mention, when it comes to the analysis of the interplays of science, civilizational setting, cultural orientation and lifeways of peoples—the Chinese and China being no exception in this regard.

Weber became interested in studying variations and variabilities in the cultural patterns of the great civilizational complexes of West and East early in his career. The entire thrust of his research led him to become increasingly committed to what I am calling the *comparative historical (differential) sociology of social-cultural processes and civilizational patterns.*

Everyone of Weber's works offers clues to his answers to Needham's questions. Weber's clues are easy to trace through the course of his writings from the essays of 1904–5 carrying the title, *The Protestant Ethic and the Spirit of Capitalism* to his possibly last essay, "The Author's Introduction" to the *Collected Essays in the Sociology of Religion*. The turning point in this sequence came with his initiation of the studies which went under the name, *Wirtschaftsethik der Weltreligionen* (the collective title given to the studies in the two last volumes of the *Gesammelte Aufsätze zur Religionssoziologie* [Weber 1920–1]). As one might expect, some of the best formulations appear in his work on Confucianism and Taoism which was translated under the title, *The Religion of China* (Weber 1951).

Here in the *Religion of China,* Weber strongly elaborates the immense supports to archaic traditionalism given by the clan and family structure, the absence of a *polis* and city of the Occidental type, the importance of the mandarin elite and the Confucian ethic, the prevalence of the ideal of the scholar-gentleman and the denigration of trade and systematic disciplined work (Weber 1951, chapters 5, 16).

Weber gives due stress to the fact that China far surpassed the West in many sectors of economy and polity but this does not lead him afield; he does not infer that the priority in developments of economy and technology will necessarily be paralleled by leadership in respect to pace and pattern in the passages to new phases of rationalized orientations and schemes of action. In the case of China, it was precisely the enormous authority and hold of practical prudence and this-worldly rationalism which barred a full rationalization at appropriate levels of individual conduct, social organization, and cultural framework (Weber 1951, pp. 151, 152). In the West, it was the thrusts to new transcendental expressions of the charismatic which broke down the old supports of traditionalism in the spheres of knowledge.

Deeply sensitive to the great role played by transformative breakthroughs in the West, Weber writes:

> If this development took place only in the Occident the reason is to be found in the special features of its general cultural evolution which are peculiar to it. Only the Occident knows the state in the modern sense, with a professional administration, specialized officialdom, and law based on the concept of citizenship. Beginnings of this institution in antiquity and in the Orient were never able to develop. Only the Occident knows rational law, made by jurists and rationally interpreted and applied, and only in the Occident is found the concept of citizen *(civis Romanus, citoyen, bourgeois)* because only in the Occident again are there cities in the specific sense. Furthermore, only the Occident possesses science, in the present-day sense of the world. Theology, philosophy, reflection on the ultimate problems of life, were known to the Chinese and the Hindu, perhaps even of a depth unreached by the European; but a rational science and in connection with it a rational technology remained unknown to those civilizations. Finally, western civilization is further distinguished from every other by the presence of men with a rational ethic for the conduct of life. Magic and religion are found everywhere; but a religious basis for the ordering of life which consistently followed out must lead to explicit rationalism is again peculiar to western civilization alone. (1928, pp. 232–33)

Weber was deeply interested in the characteristics of Chinese science and magic. He was aware of the fact that Chinese religion did not set itself in opposition to new theological and scientific outlooks; he also perceived that passages to new science often arise from new cosmological and ontological visions and needs. The very absence in China of stress on a transmundane God had a limiting effect on Chinese ontology (1951, pp. 155–159).

Weber's understanding of the strange turnings of rationalism in China and the inhibiting effects of the absence of transcendant points of reference are very great insights into the failure of the Chinese to proceed toward transformative breakthroughs. In this same connection Weber does appreciate the critical importance for the scientific development of the West of the Greek heritage, especially the importance of Euclid. Weber is not as helpful as one might wish in an analysis of the issues of universalism and universalization, but he does help us in this terrain. Few discussions of Weber on the Chinese are as revealing as chapter 6 in his *Religion of China,* where, among other issues, he discusses the absence of natural sciences in China. Here he anticipates many of the points which reappear in his "Author's Introduction" of 1920. At one point in the chapter Weber emphasizes the importance of the absence of powerful metaphysical interest in the Confucian tradition. He adds:

The development of mathematics had progressed to trigonometry—but this soon decayed because it was not used. Confucius evidently had no knowledge of the precession of the equinoxes which had been known in the Middle East for a long time. The office of the court astronomer, that is the calendar maker, must be distinguished from the court astrologer who was both an annalist and an influential adviser. The former was a carrier of secret knowledge and his office was hereditarily transmitted. But relevant knowledge can hardly have developed, witness the great success of the Jesuits' European instruments. Natural science as a whole remained purely empirical. Only quotations seem to have been preserved from the old botanical, that is pharmacological work, allegedly the work of an emperor. (1951, p. 154)

He puts his twofold general question in his usual way:

Why did the distinctively *modern* science arise in the West, and in the West alone? And why did China "*not* enter upon that path of rationalization?" What factors may have operated to cause it to fail to achieve full height and crystallization in the East until its adoption in the West? What factors in the sociocultural environment, what factors in the political and familial structures worked to check the full rationalization of science and intelligence in the Oriental World? (Weber 1958a, pp. 25ff; 1951, *passim*).

The reader wishing a full presentation of Weber's answers has no recourse but to read the entire corpus of the *Collected Essays in the Sociology of Religion*. At the very least he must examine the "Author's Introduction" to these *Collected Essays* and chapter 6 of the *Religion of China*. The reader will quickly recognize how strongly Weber stresses the enormous authority in the West as contrasted with the East of the universalizing and universalistic modes of thought and sensibility. Weber appreciated the prominent place accorded to systematic structure and other abstract formalisms in the West.

Weber allows us to see into his deepest intentions toward the close of his "Introduction." The *origin* of Western science, he explains, cannot be attributed to capitalistic interests:

Calculation, even with decimals, and algebra have been carried on in India, where the decimal system was invented. But it was only made use of by developing capitalism in the West, while in India it leads to no modern arithmetic or book-keeping. Neither was the origin of mathematics and mechanics determined by capitalistic interests. But the *technical* utilization of scientific knowledge, so important for the living conditions of the mass of people, was certainly encouraged by economic considerations, which were extremely favourable to it in the Occident. But this encouragement was derived from the peculiarities of the social structure of the Occident. We must hence ask, from *what* parts of that structure was it derived, since not all of them have been of equal importance? (1958a [1920], pp. 24–25).

IX

There is no hope of achieving a truly reliable comparative sociology of cultural change without first having a thoroughly grounded comparative history. To help resolve the issues, therefore, which divide Needham and Weber and other comparative historical sociologists of science and civilization, we need to turn our attention to the main lines of sociocultural development in Western Europe.

The first theme on which we must fix our attention is the extraordinary character of the era of the Crusades in World history.

As grateful as we must be to those who pioneered in the study of the 12th and 13th centuries—to Haskins, Gilson, Sarton and others—we are now aware that we have come a long way and that new reviews, perspectives and interpretations are needed. In our own day, R. W. Southern (1953, 1970), Marshall Clagget (1959b, 1961, 1968), Dom David Knowles (1963b, 1964), M. D. Chenu (1968, 1969), Pierre Michaud-Quantin (1970a, 1970b), and others have helped us to see the 12th and 13th centuries in a new light.

Briefly, the more we come to know the European cultural and social transformations of the crusading era the more we find ourselves having to conclude that they represent a watershed in the international history of civilizations. Without wishing to be understood as implying that the breakthroughs effected in the 12th and 13th centuries were without precedent or parallel, we need to say that anything short of a comprehensive survey or analysis of other civilizational complexes and civilizational encounters would not suffice to allow us to suggest the massive innovations which occurred during that epoch. Indeed, it needs to be realized that a large part of the significance of those days is based upon the extraordinary responsiveness—negative as well as positive—of men of that era to strands of influence coming from other eras and civilizational centers (Haskins 1957, 1961).

Eager appropriations and sifting of critical traditions and texts are the hallmark of the time. Not only did the men of this era recover neglected Greek and Roman texts—most notably Greek philosophical and scientific writings and the Roman law—they also labored to translate Arabic, Hebrew and classical works which had become part of the Islamic and Hebraic corpus. Thanks to Joseph Needham, Lynn White, Jr. (1971a, b), and others, each passing day brings new evidence of borrowings from China, India, and other lands of the East.

A prime responsibility here must be to delineate certain central changes of existence, experience, and sensibility which underlie the extraordinary productions of that time. To put my point briefly, the twelfth century is the era of the crystallization of new structures of consciousness which would have to be described as *rationalizing-and-rationalized structures* of *consciousness* which gain increasing dominance over *faith-structures* of consciousness and sacro-magical structures of consciousness (chapter 5 above, pp. 93ff). It is the time when monastic theology gives ground to scholastic theology, the time when new images and horizons of conscience, self, person, society, the cosmos, action, justice, forms of rule, institutions of law and learning take on a cast that have ever since been distinctive and primary features of the Western European world (see chapter 12 below; cf. Chenu 1969).

The key changes effected in this process are marked by the following emphases: in place of a sacro-magical, albeit sacramentalized, sense of the creation and creator, there appears a stress upon the need and ability of men to know and prove one's faith, to know and explain natural phenomena by the principles of natural philosophy, to offer rational justification of their acts and opinions.

X

It remains to review these comparative historical efforts in wider terms. Only by so doing can we identify the differences in the axes of social and cultural development: the differences in the axioms which ultimately are expressed in the unfolding projective sociocultural geometries of the orientations and institutions of the two worlds. Needham and Weber need to be brought into direct contact with one another if we are to describe these different geometries correctly. Let us consider the differences in notions assumed to be the ground of institutional development.

The hallmark of the West in the twelfth and thirteenth centuries is the crystallization of orientations and institutions which simultaneously rest upon the two-fold commitment to the *concrete individual person* and *objective universal*. The main notion on which the institutions are reared is that individuated persons are the bearers of rights and rationalized universals become the focal points of governing norms.

Such widenings of outlook imply new sanctions and new turnings to the future. Universalities of this scope presuppose the passages beyond sacro-magical and faith-structures of consciousness to rationalized structures of consciousness which place high premium on universal rationales and rationalities. Faith pursuing understanding of itself and its new contents, elaborates structures of rationality (cf. below, pp. 219f). The widened communication and communions take shape in the appearance of new forensic contexts in which discrepant views on the nature of rationality, the proofs of claims, need to be correlated in a manner which invites and requires rational assent. Illustrations of this process abound in Anselm of Bec, Abelard, St. Thomas, and other scholars (cf. Chenu 1968; Jolivet 1969; Gerber 1970).

The avenues by which the multiplicity of discordant perspectives and claims are brought into some harmony or concordance are dialogue and dialectic. Where alternative pre-rational methods exist for the establishment of the norms of action and belief, there seems to be limited recourse to these logical elaborations. Under certain sets of conditions, the new forensic contexts may be kept confined to limited areas and restricted domains.

History affords many illustrations of the limited institutionalization of universalism or rationality (or dialectic) within restricted spheres. We must not be surprised, therefore, if we find institutions apparently devoted to science or logic in many traditionalist societies. Here we need a distinction between a universalism of limited applications which is the base for an intellectual elite or a restricted meritocracy, and a societal commitment to universalization, that is, a commitment to encourage the expansion of the boundaries of participation to increase the avenues of communication and to develop skills in communication among those who have so far not truly entered into dialogue because of self-inhibition or the restraints imposed by others (cf. above pp. 111f and 178).

XI

The times we are describing, the 12th and 13th centuries, were eras of explosive innovations and abrasive conflicts in all spheres, and also of unrelenting efforts to find ways of balancing and uniting opposed positions and perspectives. Clearly there could be no going forward until—and unless—the unsolved conflicts of earlier days and incipient schisms of our own time were brought into new harmony and the contraries of *Yes* and *No* were in some way reconciled by a higher synthesis. The paradigms for these days were Abelard, Gratian, St. Thomas, and other summists.

A related and equally significant fact about the 12th and 13th centuries was the general conviction throughout the European Christian world that it was possible, with the exercise of dedicated will and unremitting intellectual labor, to discover an intelligible cosmos and to forge a rational world united in all its spheres, concordant in bringing complex harmony where appearance only presented dissonance.

Chenu and others who have perceived the critical importance of ideas of universality in the 12th and 13th centuries are correct in implying that there was a thrust toward universalistic perspectives at every level of action, reference, and significance. Everywhere men sought to take hold of the universal to depict a universe of order in nature, in law and everywhere else.

Thanks to the help afforded us by special studies and researches in semantics such as those of Michaud-Quantin, we are able to avoid falling into traps. The seemingly higher concentration of references to universality in the field of natural philosophy should not distract our attention from the very great evidence that universality was pursued everywhere else, in jurisprudence, in moral philosophy, and so on. The approaches to concordances of discordant canons were efforts to achieve universality. The treatises on human action on the cases of conscience were movements in the same direction. Why did they devote themselves so arduously to the writing of Summae and Encyclopedias? All the thinkers of the 12th and 13th centuries were inspired by the hope that they might rear a great edifice which would house all the confusing and vexing particularities of circumstance.

Our own research makes clear a main locus of this effort was the pregnant notion of conscience (see chapters 3 and 4 above, pp. 43f, 73f, chapter 12 below, pp. 223f), destined to be the spur and setting of an immense development of special interest to those interested in the comparative study of the shapes of diverse cultural patterns.

XII

Needham has clearly established that the Chinese attained very high levels of achievements in technology and protosciences from the 2nd to the 16th centuries.

The situation was quite different in the spheres of high-level science. Never having known the Greek heritage, the Chinese lacked an adequate base for the independent invention of Galilean science. The Chinese were always inhibited by restraints of various sorts—familial, political, social, and intellectual.

Needham notwithstanding, Western science did not have to wait until the days of Leonardo and Galileo to have its first flowering. From time to time Western science did undergo a loss of animation and fell into a sort of limbo. At no time, however, were the grounds for it totally gone. The Greek heritage lacking in China persisted in the West (Haskins 1960; Sarton 1968, vols. 2 and 3). Its revival and reinforcement in the 12th and 13th centuries were indispensable elements in the 14th century surge forward—especially at Merton College, Oxford, and Paris—of mathematics and mathematical-physical and logical modes of orientation (Crombie 1953, pp. 178–188; Dijksterhuis 1961, pp. 188–193; and see Crosby 1955, on Bradwardine; C. Wilson 1966, on Wm. Heytesbury).

Charles Haskins, Johann Nordström, and a number of more recent writers (Lynn White, Jr., William and Martha Kneale) have been right to ask whether 12th century developments in thought, art, and life might not have been in many ways more critical for the West than the later and better known Renaissance of the 14th and 15th centuries. It was the 12th century which saw the critical breakthroughs in many avenues of the sciences and logic, natural philosophy, the arts, letters, style, sentiment, the scholastic rationales of conscience and opinion. The present writer is pleased to find himself concurring with the views of the Kneales (1962) and Dom Knowles (1964) that Galileo had harbingers in the 12th and 13th centuries. One of these was surely Peter Abelard (1971); another was surely Adelard of Bath (1934).

How account for the extraordinary expansion of these novel interests? How explain the effort to find all one might know about the realm of the intelligible? Through what series of sociological and cultural changes would men come to develop such profound curiosities, skills and techniques? In my view there is need to grasp certain neglected facts about the changes in the structures of existence, experience, and expression.

Burckhardt and others apart, the 12th century witnessed the flowering of a wide variety of individuated selves and persons who were convinced that they had the ability to create arts, sciences, machines, laws, and even law codes, and were prepared to ameliorate their lot by deliberate innovation. The era sees an extraordinary widening of the horizons of consciousness and new strivings to absorb and explain new experiences and to map the worlds in which they are involved.

It is no wonder that the 12th and 13th centuries place such strong stress on critical notions and terms not before conceived as they then came to be. Presently, I shall explore neglected facets of these notions and terms in order to make clear what they imply in the way of the expansions and differentiations of the structures of consciousness.[4] Thanks to M.-D. Chenu and P. Michaud-Quantin, it is now possible to recover the aura and scope of certain recurring metaphors and terms, notably *universitas, civitas, communitas, persona, libertas, conscientia, aequitas, liber, machina,* about which I shall speak again later.

With Maine's help we are able to grasp the fact that the twelfth and thirteenth centuries saw surges toward the expansion of the number of those qualified to be persons in their own right (Maine 1861, chapter 2; see Mauss 1938, for some interesting facets of the early history of the notion "person"). Europe in the 12th and 13th centuries underwent an accelerated passage from status to contract, from what Maine called a *stationary* society to a *progressive* society (1861, chapter 5, pp. 164–165). We see the breaking down of invidious ascriptive solidarities and the release of "individuals" for independent action. Although the full expression of many of these developments comes in subsequent centuries, it may be noted that already in the 12th and 13th centuries we have the deliverance of domestic bondsmen from the yoke of their masters, the enfranchisement of sons from the power of their fathers, and the increased freedom of women to act as persons in their own right (pp. 163–164). It would be folly to say that this triple juridical revolution occurred everywhere in Europe at the same pace and to the same effect. Clearly the greatest advances in the spread of liberties and personhood occurred in the new cities, where freedom was open to all (Pirenne 1956).

We are in debt to Max Weber for a number of notions which are most helpful in understanding sociocultural process in the 12th and 13th centuries. I will speak now of only two central ideas: the first is expressed most acutely in the "Author's Introduction" (1920) to the posthumously published *Collected Essays in the Sociology of Religion,* which may have been written just before his death in 1920, where Weber offers clues of extraordinary value by way of explaining the differences between the path and tempo of Western development as opposed to the development of China, India, and the East generally. What Weber recognized here was that at an early date the West became committed to scientific and juridical rationalism which was marked by the highest degree of striving for system, abstract formalism, and above all, universality in the terms of reference (cf. Weber 1951, chapters 5 and 6, especially pp. 125–128 for suggestive observations on these issues).[5] The universalities of Greek philosophy and science, the universalities of Roman law and political theory, were an enduring heritage which recurrently entered into new fusions in the Western world. The 12th and 13th centuries built upon these universalities, both in Greek science and Roman law and were thus committed to structures of consciousness different in critical respects from those which had prevailed in China, in India, in Islam, among the Hebrews, etc. (cf. chapter 5 above, and Grunebaum 1961).

In two of his writings, notably in his "Types of Religious Rejections of the World and their Directions" (Weber 1946c, pp. 323–359), and in his special monograph on *The City* (1968, 3, pp. 1241–1260), Weber perceived another phenomenon that had an unique history in the West. This phenomenon was called by the name of *fraternization*, which occurred at different times and under different stimuli. In proclaiming universal brotherhood for all mankind, Christianity bore the promise of being an universal solvent of caste, clan, tribe, kindred, family and of other structures incorporating elements of invidious dualism (Weber 1928, pp. 262–263; and cf. Nelson 1969a). Also of grave importance were the fraternizations and confraternizations which Weber identified as the clue to the unique character of the medieval Western city. Nowhere else had so many cities come into being as a result of sworn brotherhood which proceeded against prescriptive rules of dominion created by ecclesiastical and temporal powers. This singular fact led to the appearance of new institutions which were havens for free men. Nowhere else would there develop so variegated a pattern of free workers, traders, professional men, notaries, patricians, clergy, etc. The maxim of the city was "city air makes free." Anyone found in the city or living in the city was presumed to enjoy freedom (Pirenne 1956; Weber 1968, 3, p. 1239).

It would be hard to understand the main features of philosophy, theology, and science during our era without making reference to the flows of experience and action in new urban settings marked by the great prominence of novel types of forensic forums. The cities and universities were places where men of varied and discrepant experience and viewpoints engaged in unending dialogue and dispute concerning all possible occasions, actions and events. It is this predominance of friction and dialogue that helps us grasp the paramount role of dialectic and scholasticism (McKeon 1935; Chenu 1969; LeGoff 1964, pp. 422–428).

The many new disputes and movements which have their source in the medieval city derive their strength from the fact that new groups and individuals continued to contend one against another in defining and shaping the ways in which they would carry on their lives. Panofsky (1967) is not wrong when he links Gothic art and scholasticism; the two are comparable and conjoint efforts to build new fabrics which harmonize discordant points of view and perspectives.

In their different ways, Greece, Rome, Israel, Christianity, the Medieval Church were well oriented toward universalism.

Weber does not enter into detail about the intervening processes, the dialectical conversions and changes in the social organization and in the cultural structure, but here, too, it is apparent that he perceives the critical importance of movements of fraternization and confraternization, all the conversions of brotherhood, including the passages from brotherhood to otherhood, from tribal brotherhoods to every form of universal otherhood (Nelson 1969a, pp. xix-xxv; Weber 1928, pp. 262–263). Elsewhere in his writings he perceives the vast distinction between the Western city and the city of the Orient and especially of the Far East (Weber 1958a, pp. 13–31).

Our main answer would then be that we cannot hope to cope with the challenges put to us by Needham without applying for aid to the cultural sociology of Max Weber and Henry Sumner Maine.

It is now apparent that many philosophers and theologians of the twelfth and thirteenth centuries had a deep need to make sense of the whole universe, to discover its patterns and structures, and phases of its coming into being, the connection among all of its parts, the measure of its motions. It is no accident that new thinkers of the time proceeded on the assumption that the world, being in a profound sense the work of God's hand, was necessarily a *machina mundi,* which could be understood as a *machina* if only the explorations were deep enough (see chapter 9 above, pp. 160f; Gieben 1964, p. 144; and Sternagel 1966 for Grosseteste, *De machina universitas;* as well as White 1969). Numerous treatises of the time are devoted to exploring the proofs and traces of the divine artificer in the world of nature. Many were no longer content to deduce knowledge of the natural world from the Book of Scriptures. Thus it is no surprise that from the beginning of our era, we encounter thinkers who draw heavily upon Muslim philosophers, scientists and commentators to understand the natural world (Sarton 1968, 2, *passim,* 3, pp. 109–152, 279–330, 485–542, and *passim*).

For many, everything that was made by the divine architect was necessarily intelligible and was itself a revelation. The usual suggestion was that there were *three* books—the *Book of Creation (or Creatures),* the *Book of Nature,* and the *Book of Conscience* [see above, pp. 156ff]. Knowledge concerning the world was deemed to be available in each of these books, and already in the 13th century we find authors who look to nature and natural philosophy, especially as made available through the writings of Plato and Aristotle and the Muslim thinkers for truth concerning nature.

The motives of the search were many, but the principal intention was to establish nature's workings. Particular stress on explorations of this sort occurred among the English Franciscans, and those closely associated with them. We need only mention in this connection the work of Robert Grosseteste, Roger Bacon, John Peckham, Peter Maricourt and others (Crombie 1953, *passim*). The flowering of this effort comes in the 14th century, especially among the logicians and philosophers associated with Merton College at Oxford (Murdoch 1969; Clagett 1967).

XIII

We may summarize the results of our survey thus far in the following ways:

(1) Although Needham evidently had a much greater knowledge of the sciences and of sinology than did Weber, the analysis he provided of the historical sociology of Chinese science does not match that of Weber. For reasons which need not be detailed here, Needham strongly tends to favor a hypothesis which inevitably simplifies a complex issue. Needham insists too much on the crucial importance for the development of modern science of the economic and cultural dominance of the so-called mercantile "bourgeoisie" and "merchant capitalism" (Needham 1969a, pp. 182–184). A more carefully woven net than this is evidently needed to capture the data now accessible to us.

Needham seems not to have been aware of a point made most forcefully by Weber, namely, that the breakaway developments which are connected with modern science cannot be described as one or another form of rationalism—whether prudential, calculative, ethical, philosophical. Indeed, it was precisely against a mistaken thesis of this sort over the all important role of rationalism in helping to explain the character of the disenchanted cosmos of the 19th century which constitutes the point of departure for Weber in his *Protestant Ethic* (cf. Nelson 1973b, pp. 72–74).

If I may be allowed to cite from one of my previous writings,

> The central object of Weber's interest was the uncovering of the cultural, so-called spiritual, foundations of the distinctive bureaucratic enterprise, organization, and outlooks of the modern Western world. Prudent "rationalism," he insisted, was not the view which spurred the spread of the vocational, innerworldly asceticism, as the dominant *ethos* of the modern industrial era. Far from being smooth and straight, the roads to modernity were paved with "charismatic" breakthroughs of traditional structures. These breakthroughs forged in an atmosphere of religious and social effervescence, had their issue in the overcoming of the invidious dualisms which had hitherto inhibited new rationalizations (and new rationales) affecting all patterns of action and conviction relating to work, wealth, welfare, regulation of self and society, political order, the sense of ultimate worth, the experience of justification, and so on.
>
> As the years passed, Weber was to strengthen his desire to discover whether and why the rationalization process had been more intensive and extensive in the West than anywhere else. Already, indeed in his first edition, however, the *civilizational* differences in the central orientations to "religion" and "world" proved to be of critical importance for his way of thinking. (Nelson 1973b, p. 78)

Needham so far seems to have missed the extraordinary importance for transformative development of the breakdown of traditionalist rationalism. His recurring identification of middle class membership with antitraditionalism is again and again put in doubt by Chinese and non-Chinese evidence (Needham 1969a, pp. 174, 184, 197, 211). In China, as in the West, many of the most important encouragements to breakthrough came from movements beyond ascriptive solidarities toward universalizing and confraternizing associational, communal, and cultural structures (Weber 1946a, pp. 323–359; 1951, pp. 142–145).

(2) In the West one of the greatest forces in the direction of universality was the Christian message as it extended the Jewish idea of tribal brotherhood of all to universal brotherhood of mankind. Another very powerful source of universalism was Greek philosophy and logic and Roman law. Wherever these had influence, movements to universality achieved great importance.

We may put our proposition in a partly paradoxical form. In many traditionalist societies the roads to universality were barred by the strength of the ethos of family and tribe, kindred and clan. At some moments in some societies there is a passing beyond these nuclear structures to new unions and associations. The passage originates in movements of wider social, civic, and religious confraternization and in aspirations to universality in law and science.

(3) In brief, the argument maintains that for a society to make advances in the direction of universality it is necessary that there be some levers spurring movements in that direction. In the cultural structures of the West, Christianity, Greek and Roman law were levers working to that end. In the Far East these did not serve as goads to change. Indeed, for this reason changes of a universalizing character in the modern East had to wait upon the influences of the cultural structures of the Western world (Bellah 1964; Eisenstadt 1968, pp. 243–251; cf. Nelson 1972e, pp. 125–126). Westernization needed to become universalization in order for secure advance on the road to modernity to be achieved. It seems highly probable that Weber was in a better position than Needham to appreciate the character and significance of these wider cultural and social movements. The relations of Weber and Needham to some aspects of the issues posed in Needham's challenges may be put as follows:

(a) For Weber the priority of China in technology by no means establishes China's superiority in science and social organization during the era up to Galileo (Weber 1951, chapters 5–6, especially pp. 151–152).

(b) It appears to have so far escaped Needham's notice that the 12th and 13th centuries in particular were eras of extraordinary social, political, and cultural innovation in European and international history. Needham would need to enrich his theoretical resources to do justice to the effervescences and renewals of the "medieval Renaissance" [as discussed earlier in sections VII and IX-XII].

(c) Needham neglects to pay enough attention to the extraordinary societal differentiations which distinguish the Western world from China. Where but in Western Europe could one find its incredibly diverse patterns of cities, parishes, countries, states, territorial sovereignities, universalities? Where but in the West could one expect to find the patterns of citizenship, the corporate structure of cities, the political philosophies?

(d) Needham also understates the evidence for the fact that the separation of church and state in the West gave vast opportunity for widened inquiry into fresh horizons. Neither the Pope nor the King nor the Emperor were ever in a position wholly to dictate the thrust of reflection and criticism. The case was different in China (cf. Weber 1951, especially pp. 13–32, 1968, 3, chapters 15-16).

(e) Needham has so far failed to give due stress to the fact that religious and ethical *universalism* in Christian thought and the surge forward of Christian sentiment and expression were of the greatest importance in providing a climate for the development of distinctively Western thought with its surge toward *generality, rationality,* and *universality.*

(f) Needham sometimes seems to assume that religion and theology could never have been spurs to scientific advance in the Western world. Religion did a great deal more than breed inquisition, as Needham in one place seems to imply. As a matter of fact, all of the major theologians of the Western world in the 13th and 14th centuries were curious about the natural world and trusted to be able to develop notions useful for exploring, describing, and explaining the natural world and natural motions. The advances in the physics and mathematics of the West developed by medieval and early modern theologian-philosophers, especially in the 14th century, are now carefully documented, as we have noted, in the works of Clagett, Crombie, Grant, Alexandre Koyré, Annaliese Maier, John Murdoch, and others.

(g) Weber had a stronger awareness than Needham has so far shown that the very Western image of human action and world changed dramatically in the 12th and 13th centuries. Collective consciousness gave ground to the individual conscience around which there evolved comprehensive logics of opinion, belief, and action. Interestingly, the treatises on human action were treatises on conscience and its cases, all of which provided a setting for the proliferation of highly developed structures and decision matrices of the moralities and logics of thought and action (Chenu 1969; Michaud-Quantin 1962; Nelson 1969a, pp. 236–246).

(h) For my own part, I must add that these new outlooks and sensibilities found expression in an ever more intensive pursuit of disciplined experience and trustworthy methods for assuring certain knowledge of the books of creatures (or creation), conscience, and nature. We need not be surprised that it was a thirteenth-century theologian-philospher-scientist—Roger Bacon—who was the first to sound the call for *scientia experimentalis* and to speak explicitly of "laws of nature."[6]

True, neither the men of the 13th century nor the hardy thinkers of the 14th century Oxford and Paris attained a Galilean level of scientific thinking and experiment. Weber and Needham are agreed on this point. Happily, scholars of the stamp of Koyré, Annaliese Maier, Beaujouan (1957a, b, 1963), Clagett, Murdoch, and others have taken us beyond the extreme claims of Duhem on this score in studies based on scrupulous control of the texts. And here I must stop this installment of our story.

The limits of the present essay do not allow me to go beyond the 14th century for fuller exploration of Needham's claims. Quite deliberately I have concentrated here on the moments of the civilizational turning points, the 12th and 13th centuries. In doing this I do not mean to suggest that everything of value in Leonardo and Galileo came from the Middle Ages through any simple or direct derivation. Clearly Galileo was very powerfully influenced by the new currents of science and sensibility which emerged in the era of the so-called Renaissance. The new editions of Ptolemy and Archimedes almost certainly were a very powerful inspiration for his work (Sarton 1955). However, from the perspective here adopted these constitute steps forward within a milieu and ambience which had achieved the new crystallizations in the 12th and 13th centuries (Kneale 1962; Knowles 1964).

Elsewhere in this series of volumes edited in the name of this Colloquium [i.e., *Boston Studies in the Philosophy of Science*] I have described (chapter 8 above), some of the central currents of thought and sensibility which achieve their culmination in the work of the early modern pioneers in science and philosophy. There is no need to repeat this material here. The substance of the present discussion can be communicated in a single sentence: Without an Abelard, a Grosseteste, a Roger Bacon, there might have been no Galileo.[7] A most critical spur to the breakthrough of the era from Copernicus to Newton was the great leap forward during the course of the High Middle Ages of *concrete universality* in all the spheres of existence and thought, in all the domains of life, public and private. The momentous—even revolutionary—achievements of the *medieval* Renaissance were vibrant elements in the sweeping surges of science and philosophy in the 16th and 17th centuries.

NOTES

1. It needs to be noted that a number of writers have begun to present Osiander's initiative in a very favorable light; cf. Koestler (1959); Wrightsman (1970); Christianson (1973). This is in the sharpest contrast to the views earlier taken by Kepler and Galileo. Kepler (1600–01); Galileo (1615–16); cf. Clavelin (1968).
2. Writings newly recovered from oblivion and partially transcribed by E. Garin (1971)—the *De veritate S. Scripturae* and its attached *opuscula* by Giovanni Maria Tolosani, a Florentine Dominican of San Marco—puts in doubt a number of points of view which have long prevailed among historians of science. An especially arresting and influential expression of the main element in this perspective will be found in Koyré (1964, p. 69): "The Catholic Church seems to have been not only quite unperturbed by Copernicanism before the advent of Giordano . . . , but altogether unaware of its 'inherent challenge'."

   In sharp contrast to this statement we now learn from Tolosani that he undertook to complete a confutation of Copernicus's *De Revolutionibus,* which his friend, Bartolomeo Spina, "*magister Sacri et apostolici Palatii,*" had been prevented from finishing by his death in 1546. Tolosani's work is represented in the opusculum *De coelo supremo immobili et terra infima stabili* . . . , which dates from 1546-1547.

   Tolosani goes to considerable pains to argue the case against Copernicus from the point of view of a professional scholastic and theologian. To his mind, Copernicus failed to marshall the requisite disproofs of earlier positions and the necessary demonstrations of his largely Pythagorean view. Interestingly, Tolosani explains that Copernicus was learned in languages and knew a great deal about the astronomical and mathematical sciences but little about the "physical or dialectical sciences." Also, Copernicus gave offense by his intemperance and lack of learning in the divine sciences. To our surprise, Tolosani knew that Copernicus was not the author and did not share the view of the anonymous Preface (Osiander). Lastly, these materials were known to Tommaso Caccini, Galileo's scourge, who carried on an unrelenting campaign against the astronomer in Florence, Bologna, and elsewhere. I reserve fuller discussion of these materials for another occasion. I owe thanks to Edward Rosen for bringing Garin's paper to my attention. [*Addendum originally added in proof:*] The *full* text of the Anti-Copernican statement by Giovanni Maria Tolosani described above has just come to hand (13.3.1974). The new essay by Eugenio Garin of Florence is entitled, "Alle origini della polemica anticopernicana," and is to be found in *Studia Copernicana* 6, Colloquia Copernicana 2; [Wroclaw/Warszawa/Krakow/Gdansk: Polska Akademia

Nauk], 1973, pp. 31-42. The text of Tolosani's *Opusculum quartum* will be found on pp. 32–42 of Garin's essay.

Only a very few points need to be added here to the description of the work in the note above. It will be noted that Tolosani places strong emphasis from the very outset on the *Book of Genesis* and the description of the creation of "the whole world and its parts." Scriptural passages believed to describe the creation are set forth throughout the first chapter of the *Opusculum*. As we have already noted, Tolosani acknowledges Copernicus's learning and ability but takes him to task for not explicitly responding to St. Thomas Aquinas's arguments and objections against the Pythagorean view. Tolosani criticized Copernicus also for not truly knowing the "*rationes*" of Aristotle and Ptolomy and for not confuting them in his own work. Toward the close of his third chapter Tolosani takes direct exception to some of Copernicus's physical ideas. The *Opusculum* closes on an accusatory note with Tolosani insisting on Copernicus's "Intolerable errors against divine letters." He explains that the "*magister Sacri et apostolici Palatii*" had been prevented by death from doing his own confutation of the *De revolutionibus*.

3. A particularly helpful and strong discussion of the issues involved in the present essay will be found in Southern's striking title essay in his *Medieval Humanism* (1970, chapter 4, pp. 29–60).
4. Close studies of historical semantics are an indispensable component of a comprehensive research program for a comparative historical, *differential* sociology of sociocultural process. Such a program involves cross-civilizational investigations of the careers of clusters of key terms with a view to uncovering their workings, analogues, and locations in different settings. In this spirit I have been at work for some time on the cross-civilizational analyses of a selected list of axial terms for Western civilization in the 12th and 13th centuries.
5. Weber notes the failure of Chinese philosophy to give birth to scholasticism "because it was not professionally engaged in logic, as were the philosophies of the Occident and the Middle East." Discussing Confucius he notices the "absence of *speech* as a rational means for attaining political and forensic effects, speech as it was first cultivated in the hellenic *polis*." He continues:

> Such speech could not be developed in a bureaucratic patrimonial state which had no formalized justice. . . . The Chinese bureaucracy was interested in conventional propriety, and these bonds prevailed and worked in the same direction of obstructing forensic speech. The bureaucracy rejected the argument of "ultimate" speculative problems as practically sterile. The bureaucracy considered such arguments improper and rejected them as too delicate for one's own position because of the danger of innovations." (1951, pp. 127–128)

6. Few recent writers seem to give enough weight to the distinctions among the *several sources of the several notions* of the law and author of Nature: the Greco-Roman, the Hebrew, early Christian, patristic, scholastic and early modern. My own sense is that the more we intensify our research the less certain we become that medieval nominalism and Calvin were as decisive for the ideas of law and order of Nature as Whitehead (1925), Zilsel

(1942b), Collingwood (1945), Needham (1962; SCC, 2, p. 539ff.), Oakley (1961) and more recently, Oberman (1973), have contended.

The main ground for my conviction arises from the substantial evidence I have already gathered, and hope to discuss elsewhere, that critical images and ideas regarding divine guidance, the construction of the natural world and creation long antedate the 14th century. Some bits of proof may be offered from the history of the images of world-machines and the divine geometer-engineer-architect in the medieval and early modern period:

The notion of the world-machine occurs again and again in the 12th and 13th centuries (cf. Sternagel 1966). Thus Robert Grosseteste even wrote a treatise, *De Machina Universitatis,* which is briefly discussed in S. Gieben (1964, pp. 144–168; cf. esp. pp. 144, 151, 153 for Grosseteste's exemplarism.)

Other sources of images of God as the Divine Artificer shown directly engaged in the geometric construction of the world were set forth with the authority—and through the citation—of scriptural passages: *Wisdom* xi. 20, "Thou hast ordered all things according to measure, number, and weight"; *Proverbs* viii. 27, "He set a compass upon the face of the depth"; *The Revelation of John* xxi, ff. These images and texts have a long history in the early and high Middle Ages. References to the spread of the first of these images will be found in L. White, Jr. (1971b), where emphasis is placed on the Winchester Gospel (ca. 1000 A.D.); also see A. Blunt (1937–1938), who offers a spirited discussion of Blake's representation of Urizen setting compasses on the world.

It must also be noted that a goodly number of thinkers and writers applied ideas of God's omnipotence *to rule out notions of natural law.* Thus it became necessary for Galileo directly to controvert Urban VIII for his (the Pope's) arguments from Divine Omnipotence in the *Dialogue.* How checkered were the afterlives of these images and proof-texts may be seen in one especially revealing episode. G. B. Riccioli wrote his *Almagestum Novum* at the behest of the hierarchy to defend the inquistorial and papal rulings on Galileo. Nonetheless it carries a frontispiece calling attention to the hand of God ordering all things by measure, number and weight. (Bologna, 1651; cf. ill. in Stimson.) [Reproduced above, p. 159].

It is hard to bring this note to a close without calling attention to the provocative conclusion of Needham's last chapter on "Human Law and the Laws of Nature" (1969a). There Needham raises exceptionally challenging questions:

> . . . But historically the question remains whether natural science could ever have reached its present state of development without passing through a "theological" stage . . . The exact degree of subjectivity in the formulations of scientific law has been hotly debated during the whole period from Mach to Eddington, and such questions cannot be followed further here. The problem is whether the recognition of such statistical regularities and their mathematical expression could have been reached by any other road than that which Western science actually travelled. Was perhaps the state of mind in which an egg-laying cock could be prosecuted at law necessary in a culture which would later have the property of producing a Kepler? (p. 330)

On the first of Needham's questions above it is now urgent to read the recent essay of Helmut Koester (1968). Identifying Philo "as the crucial and most important contributor to the development of the theory of the natural law," Koester adds:

> Only a philosophical and theological setting in which the Greek concept of nature was fused with the belief in a divine legislator and with a doctrine of the most perfect (written!) law could produce such a theory, and only here could the Greek dichotomy of the two realms of law and nature be overcome. All these conditions are fulfilled in Philo, and the evidence for the development of their theory of the law of nature in Philo is impressive. (p. 540)

7. Just as this goes to press there comes to hand a suggestive essay by Joseph Levenson (1953) which appears to raise doubt on a major theme and conclusion of the present essay. Levenson finds that a number of Ch'ing philosophers remind one of Abelard, but none came so far as Francis Bacon. Levenson writes:

> In European history, divergence from idealism could take the forms of the pre-scientific nominalism of Peter Abelard . . . as well as the form of Sir Francis Bacon's . . . inductive empirical science; our Chinese thinkers' affinity was with Abelard, not with Bacon. (pp. 158–159)

Levenson's view seems to me insufficiently morphological. He misses the extent of Abelard's breakthrough in the domain of the structures of conscience. Also, he fails to note the importance of Abelard's thrust in the direction of a new logic of intentionality. Lastly, Levenson exaggerates the role of Francis Bacon in the development of high science. In the same spirit, the philosophers described by Levenson hardly seem to parallel Abelard. Comparisons are, indeed, a prime necessity of comparative historical differential sociology; they also, however, regularly involve serious risks.

On the present issue, there is an excellent passage by the Kneales:

> . . . On the other hand, it is arguable that the exercises of the medieval universities prepared the way for modern science by sharpening men's wits and leading them to think about the methods of acquiring knowledge. For it is certainly a mistake to suppose that all the philosophers of the Middle Ages believed in systems of deductive metaphysics, and that experimental science began quite suddenly when Galileo or some other Renaissance worthy made an observation for the purpose of refuting a generalization of Aristotle or Galen, just as it is wrong to suppose that Luther was the first to suggest reform of the Church. (Kneale 1962, p. 226)

PART IV

Civilizational Analysis and the Study of Existences, Experiences and Expressions

ELEVEN

On the Structure of Consciousness and the Omnipresence of the Grotesque

My remarks grow out of a sense of frustration which came over me recurrently in the course of a session at the Whitney Museum. I found everything said by all participants to be of great relevance and interest, but I did not quite see how I could apply the prevailing categories of the disciplines familiar to me—sociology, history, anthropology, psychoanalysis, more generally, the social sciences and humanities—in such a way as to throw new light on the questions under discussion. I therefore decided that it was urgent that I proceed at once to sort out the thoughts I have been having since we last met. The key results of my efforts during the interim lead me now, frankly, to avow my conviction that such discussions as we had last time—and will doubtless have again today—evidently call for a fresh look at certain familiar paradigms in the professional study of human culture and action.

I

Showing over 150 slides of every sort and description, Mr. Sowers revealed the grotesque to range across the entire universe of art and, for that matter, over almost every epoch. The great depth of illustrations were a perfect foil for the multifariousness of his categories—and both together served to goad me into strenuous efforts to ascend to new ground.

The Grotesque, Mr. Sowers said, is the "*investment of anything with a far* more *or far* less *complex, vital, animate, conscious, genial or menacing*

Originally published as "The Omnipresence of the Grotesque," in *The Psychoanalytic Review* 57, 3 (1970): 505–518. The substance of the paper was read to the Riverdale meeting of the Society for the Arts, Religion and Contemporary Culture on April 25, 1970. Reprinted with permission.

order of existence than we ordinarily attribute to it." Mr. Sowers added that when all of his illustrations were taken together, it might be found that there were three essential forms, *archetypes,* if one preferred, which insistently recurred in them with multiple variations and idiomatic difference.

1. *The Monster.* The first form of the Grotesque to which Mr. Sowers referred he described as the *Monster,* by which he meant "some either humanoid or superhuman, *but not human,* incredibly strong, animal with a diabolical, semihuman countenance."

2. *The Orifice.* The second of Mr. Sowers' family of illustrations established the recurrence of the images of orifice: jaws, caves, and so forth. The variety and range of these illustrations was truly extraordinary.

3. *The Maelstrom.* Mr. Sowers reached perhaps his peak of analytic insight when he remarked that the images comprised under his third head essentially represented to us a "condition or a situation, full of chaos, claustrophobic jamming, nightmarish atmosphere, flashbulb lighting and dark background, all sorts of contrary, extremely unpleasant, disordered conditions which suggest *a world of depravity and disorder.*"

Mr. Sowers' next moves carried him into another sphere: he went ahead to develop a series of categories of formal devices most frequently discovered in the illustrations of the grotesque.

The first device he described under the name of "*aggressive* symmetry"; another as "overanimation"; a third, as the "obvious disintegration of form"; a fourth, as "attenuation." Mr. Sowers then spoke of "aggressive protrusions" and "highly overcharged *over*-articulation"; and then there was the reduction of all kinds of living forms of seeming *mechanism* or, as one might say, *under*-articulation. Then there was the Grotesque of "obsessions"; the Grotesque of "distinctions"; the Grotesque of "misplaced ability"—and, at this point, it seemed to me the speaker slipped into largely descriptive perspective and used a largely aesthetic vocabulary: "visual *double-entendre,*" "pure camp," the "Grotesque of Squalor," and so forth. Coming to his close, Mr. Sowers issued an important caution:

> The Grotesque is bound to be a culturally relative thing, so that what will strike a people at one time as being grotesque will not necessarily do so for others elsewhere, and so forth.

II

During the entire course of this presentation, I continued to ask myself a single set of questions: "To what sets of experiences did these images, these forms, and these devices refer? To what sets of experiences . . . could they be referred? Was there any kind of general view that one might evolve as to the sources, roots, meanings of the experience of the Grotesque? Where could one look for keys as to the issues, problems, dilemmas, displacements or whatever, which may have been roots and sources of the Grotesque? What elements or settings of sociocultural process, what aspects of group cultural experience, or the existences of individuals help explain the varieties of the grotesque?" Thus, as I must now confess, Mr. Sowers' presentation had the effect of deepening a suspicion I had from the start, namely, that far from constituting a single expression or naming a distinct species of events, the word "Grotesque" related to a mass of very different contexts and ranges of reference which spread over a wide expanse of densely tangled terrain.

And so having early lost my way on a dimly lit stretch of the road, I started on a new journey, whose results follow:

From one point of view, there is nothing more grotesque in the world than what we call "normal, everyday reality." This awareness has been illustrated with particular force by the contributors to the so-called "Theatre of the Absurd." Ionesco has been unambiguously clear on this score; he has repeatedly said that he has no intent in conjuring up works of the imagination in the conventional aesthetic sense. Nor will he allow that he considers himself a pessimist. Rather, in the idiom of our land, he has insisted on "telling it exactly as it is." Ionesco knows that when the ordinary is *re-presented,* it has a sense of total uncanniness; he feels obligated, he explains, to tell the story without adornment "so that people who feel a need for solutions can at least have something to depart from."

Are there circumstances when "normal reality" takes a more than normally grotesque appearance? The answer is yes—when times are unusually stressful; when sensitivities are exceptionally heightened; when an extraordinary shift in *consciousness* has resulted in a widened sense of a Great Awakening. Changes of this scope cause the familiar inconsistencies of the "taken-for-granted" world to appear void of reason, *absurd.*

I must admit that I am aware that strict warrant is lacking in the social sciences for some of the terms I have used in the last paragraph and some others I shall need to use from this point forward in my essay. For this reason I turn abruptly to a cluster of ticklish semantical problems associated with terms expelled or lost from the languages of the social sciences and humanities too soon, before their older meanings and possible newer uses were fully explored. I speak first of three older terms which have long struck me as critical coordinates of a renovated theory of sociocultural process, the linked terms *Existence-Experience-Expression*. Only a few remarks about the so-called interfaces among these notions can be allowed here.

The world "existence" has long been a technical term in philosophy which has received new and vastly extended use in the "Existentialisms" of our day. Despite its roots in the classical past and its great range in present-day thought, it is not now a phrase in the working vocabulary of any of the cultural or social disciplines. As a frame of reference comprising every sort of factual situation, it can have many new uses in the study of sociocultural process. To understand the structures of cultural "experience" and expression we must surely know the effects of existential determinations.

The word "experience," dear to many philosophers since at least the times of Kant and Hegel—the Americans are especially numerous in this lot—oddly happens to be a word not in use or allowable in sociology, anthropology or psychology. There is no estimating the price paid by social scientists for the lack of this concept. I can allow myself only a few sentences to suggest some of the uses to which the term "experience" shall be put here. "Experience" in the present context refers to any trace which serves as an input in an ongoing process of response and making. Standing at the crossroads, experience is thus both a confluence and a source of change in existence and expression alike. The structure of events associated with the occasions of experience emerge at the crossroads of our journeys across life's way.

The third in my series—"expression"—seems to provoke least opposition today, at least in the generic sense in which I intend it, namely as any utterance, sign, or symbol or enactment to which a meaning can be imputed.

Our triangular perspective offers us many advantages over competing models. With its help, the social scientist and the humanist scholar are reasonably protected from falling into theoretical traps in the mapping of the links and directions of influence in sociocultural process. Our paradigm reminds us that experience cannot be described as a simple function of existence: all experience is also subject to the influence of intertemporal frames of expression. Conversely, cultural expression is never an immediate outcome of existence, individual or social; all cultural expression goes through the filter of experience. In short: if we omit the notion of experience, we omit the tides of human action and reaction; we omit the sources of innovation.

Grotesque images are the forms gestated in all the settings of our experience, against the horizon of the antecedent traditions of expression, on the basis of the existences we are fated to have. As I have written elsewhere, we all move in orbits of one sort or another. We all have the experiences of our worlds and we all, in one way or another, do actually give vent to or create expression of one or another sort. There is nothing in what I have said which would suggest that our *existences* are simple functions of our social locations, that our *experiences* are no more than the stimulus-sets of this time and place, or that our *expressions* are direct derivations of current patterns of existence or experience.

There are three other terms to which I find myself needing to have fresh access for the sake of helping to evolve an enriched vocabulary of the social sciences and humanities. I shall detail them in turn.

The word "consciousness" has no place at all in the vocabulary of current psychology, sociology, or anthropology despite the fact that it is in very frequent use, usually over on the Left, among those who want to talk about the *new consciousness* or *working-class consciousness*, the *myth of the objective consciousness*, or the relation of *consciousness* to society. Few of those who use the term normally undertake to say very much about it that is precise. (The situation is not much better for the critical word "unconscious"!)

"Civilization" is yet another critical word not now in systematic professional use at this time. There are hardly a handful of sociological colleagues across our country who would allow the words "civilization" or "civilizational" as technical terms of a responsible professional sociological language. (The latest edition of the *International Encyclopedia of the Social Sciences* does not have an entry under "Civilization"; instead, the reader is advised, "*See* Urbanization." Should he proceed to "Urbanization," he would find nothing on Civilization, nor, for that matter, would he find a great deal on Urbanization either from critical points of view that have the civilizational dimension.) In short, the sorts of distinctions that one would want to make in the area that are critical for the present purpose are not available.

How can we explain these gaps? I may not do more now than speak to this issue in a prefatory way. If we would make sense of our present maps of knowledge, we need to realize that each of our disciplines arrived at its present state in the course of a prolonged process of fission of older and more comprehensive units and unities. More recently than most historians suppose, particular sets of questions were disengaged from a primary matrix and were encouraged to claim an independent existence outside the matrix in which the questions had previously been imbedded. I have elsewhere documented the story of a sort of paradigm instance of this kind of atomization. Related processes occurred in all the social sciences.

Today, the questions which men put to themselves can only be put in terms of the specialized languages, the semantics of the individual disciplines. It is no wonder that I am taken to wander in so many areas in the effort to ask the sorts of questions central for the explorations of the grotesque.

Luckily, men committed to their specialties often do better than they *know*. This is attested by many splendid studies, especially in anthropology, literary criticism, and so-called humanistic or literary sociology. Yet the fact is that so long as roads do not readily open out for new concepts; so long as fresh accesses to imperfectly explored old terms are lacking, it is not possible to ask questions in the proper way. I come now to a more decisive point.

III

The very large number of illustrations offered by Mr. Sowers ranged across very diverse historical periods and cultural settings, and yet they assumed a small number of forms. Shall we say that this fact proves the existence in the "collective unconscious" of a set of archetypes which are carried forth in the racial inheritance? Or shall we say that possibly there are structures of consciousness which, when triggered by some set of wider experiences in a given social or historical or cultural context, are likely to come up in almost the same fashion each time for other reasons?

Contemporary psychologists, sociologists, and anthropologists tend, in the main, to think they have ways of accounting for the symbols we are discussing without necessarily referring to any of the wider sociocultural historical frames.

All of us must recall innumerable cases of claims by men distinguished in the above mentioned disciplines, that some very fundamental and far-reaching collective alteration in structures of consciousness and expression are simple derivations of experiences *in the primary environment of family.*

This pattern of explanation is the one offered in most of the current discussions dealing with the "generation gap." It hardly matters whether authors we read are strongly committed to a strict Freudian or neo-Freudian or a social-psychological socialization model. The fact remains that for the majority of these writers the explanation which seems to have worth or relevance is the one which claims the primacy of some set of experiences within the original setting or socialization of family interaction.

Mr. Sowers appears to have felt under no special obligation to propose explanations of his examples of *Grotesque,* whether *monster, orifice,* or *maelstrom.* Even though I cannot promise to explore genetic issues in detail here, I find myself needing and wanting to make some distinctions in this sphere. Images of *monster* and *orifice* may—as a matter of fact they *do*—have a nearness to primary associations which are likely to recur in almost all peoples everywhere as a result of intense and searing experiences in the primary environments of coming into being, growth, and development. Renderings of these phenomena in the arts, especially the plastic arts, prove again and again to exhibit the workings and influence of primary process.

We are at a different horizon, however, when we seek to understand the illustrations of *maelstrom*. Here we see a much wider range of reference, and it is harder to connect with any single environment—it calls for a much wider setting of experience than either *monster* or *orifice*. The complex structures of societal or civilizational experience or, for that matter, even subsocietal, social group experience, are often of first instance in this case.

The fact is: there is no making sense of the total range of our experience without including reference to horizons and to vectors far beyond those of small groups and families. I see no reason to extend unqualified acceptance to the notion largely favored by American sociologists and psychologists, and said to derive from the fundamental work of Georg Simmel. For me sociology is a good deal more than the study or analysis of social (in the sense of *interpersonal*) interaction.[1]

Psychologists and sociologists omit these wider frames of reference from their horizons at great loss. Philosophies of methodological individualism have worked great confusion here by casting doubt on both the reality and relevance of societal, international, civilizational and, indeed, even intercivilizational frames. Societies *do* run courses, but these courses are surely not necessarily linear; societies even exhibit *malaises*, distempers, and even functional disorders.

I have elsewhere contended that critical changes are now occurring dyschronically in central institutions and the structures of consciousness. Our society and our world are taking on the appearance of the Grotesque precisely because of the nightmarish jammings of these dyschronicities. The deep fissures traced by these jammings are leading great numbers to speak of desiring "to destroy *the system* ONCE AND FOR ALL," to cry out for a great deal more than a change of command or procedure in the way of running the economy; they demand the substitution of wholly new values for the ones now believed to be in the saddle. From every corner these days, calls for new heights of counter-civilizational and civilizational fulfillment are coming to us from the societal depths. Civilizational, societal, subsocietal, and social group-processes continue to fuse into powerful images of maelstrom and Great Awakening.

Today the Grotesque is our everyday way of being and living, appearing and seeming, dressing and wandering forth into a demented world. Many of the most outstanding works of mdern literary and visual art prove to re-present experience to us in terms of carnival or circus. A related sensibility is attested to by the use everywhere today of such terms as "scene," "put down," "put on," "camp," "laugh-in"!

The Grotesque is now revealed as the supreme vehicle for *"putting down* (and off) the old" and *"putting on* the new." *Annihilation* of the "gloomy actual," reads the scenario, is succeeded at once by total *re-creation ex nihilo*; ever pregnant Chaos engenders the Cosmos of the Everlasting Now—the Perfected Future!

IV

I allow myself a few related illustrations from the everyday life of our time. As odd as the manifestations of our time may seem, they are surely not unique.

Indeed, the Grotesque is once again an aspect of every single person's experience, just as it was in the days of Diogenes the Cynic; just as it was centuries later in the days of Dante, whose haunting visions of the Grotesque call out to us from the cantos of the *Inferno*; just as it was in the era of the Black Death and the Hundred Years' War when great numbers of men and women disported themselves too strangely for many contemporaries to understand. Everyone of us would do well to reread the chapter on "The Dance of Death," in Johan Huizinga's masterly *Waning of the Middle Ages*. Also: how forget the Ranters, Diggers, Levellers, Seekers, Families of Love in 17th-century England? The Lycanthropists and Nihilists of 19th-century France and Russia? Again and again, people—old and young alike—are found fashioning odd costumes and sometimes making their own flesh their only dress, ranging across the world.

There may, however, be one more aspect of our situation. The apparent anomalies of our time are general, universal; they are also deliberate. Everything undergoes the kinds of shifting that one would expect in a society which, as Kierkegaard saw, had this inexhaustible capacity to absorb any sort of dissenting innovation and turn it into some kind of form of impersonation, new life-style, new stage-setting, new points of departure for spectacle, business, or busy work.

When I walk out the door of my classroom into New York City, I always look very carefully around me and insist on trying to identify the cast of characters, *dramatis personae*, striding on the twin stages of Fifth Avenue and Fourteenth Street. Inevitably, the "actors" range all the way from the Cynics or early Christians to tomorrow's Spacemen;

the hair style, garb, walk, facial and other gestures, proclaim the man. A little while ago, the 14th century was IN (*fashion*), but now it seems OUT *(of fashion)*. Only yesterday one of my favorite students was very much IN—wearing the Rubens' beret. But that was *yesterday!* As for resurrections of the 19th century, they occur too often to make news. Simultaneity has not yet conquered all. Despite the apparent depth references of our acts, our perceptions remain traditional. (The collective "our" here refers to all of us, including today's artists, scholars, actors.) If we had a depth-historical relation to time-space, we would have much less trouble understanding the resurrection among us of groups of people reliving the lives of the Diggers or the Families of Love and Charlie Mansons.

Nothing that has once become history ever disappears. Anything that once was can always be. History always—and never—repeats itself. Yet it is not necessary to say, as Marx insisted we must say, that if history repeated itself it would manifest itself the first time as Tragedy and the second time as Farce. This is a notable cultural epigram but not an unexceptionable rule of history.

V

Images of the Grotesque, especially images of the maelstrom, regularly seem to multiply when large numbers of people find it impossible to function, much less thrive, in their everyday worlds. The hopes for finding a modicum of joy and meaning in ordinary life—what I prefer to call *generic succession existence*—are now felt to be choked.

Then, each one is affected in a measure by the experience of askewness. We must recall that a large number of Mr. Sowers's pungent illustrations last time attested to the passage of persons and groups through heightened states, including some of a so-called "pathological" or "psychopathological" sort. The feelings and fantasies at work in these episodes are very frequently beyond the power of even gifted men and artists to control or moderate; when these states occur they reveal a terrible insistence and sameness. This is why the symbols are so often indiscriminately ascribed to experience assumed to be prior in the biogenetic sequence: there is a blunder here, sociological no less than logical, which results from the failure to distinguish between an originative source and a proximate expression. The role played by symbols usually ascribed to primary process does not constitute proof that the originating issues root in the biological or familial frames. Much of today's evidence of primary process traces to the perturbations in societal, political, and civilizational settings. We turn now to these wider settings.[2]

Every man is ever in the midst of predicaments from which neither culture nor civilization can wholly shield him. So long, however, as human agents are able to feel that the tides of change and the streams of cues are congruent with their needs and expectations, they are generally able to achieve some meaning which will sustain them through their lives. In times of heightened strains and sensitivities, ever larger segments of experience seem empty of rhyme or reason—*Grotesque*. The various kinds of discontinuities envelop increasing numbers in a sense of maelstrom.

The fact of change is not of itself the decisive element in the grotesque. We are all able to tolerate inconsistencies, but we cannot tolerate inconsistencies that are so severe that we fail to find anywhere around us any sense of vital center or purpose or meaning or any rationale to which we can in fact offer our allegiance or loyalty. Few of us can live at ease in maelstroms.

Lastly—within our now enlarged frame of reference—there is another disjunction which implies especially deep crises in civilizational values. This occurs when the social organization seems to be working at a very high level of "productive efficiency"; when the cultural system seems to be maintaining itself with the help of vast numbers of persons who are specialists—yet many people going through their rounds in the phases of life cycle feel let down by the incongruous evidence that the system works to provide them largely negative psychic and moral incomes. Their senses of fitness are outraged. In the language of the Existentialists, the outcomes seem arbitrary, "nauseous," absolutely *grotesque*.

Whether we be ready or not, junctures of this sort call upon us to offer more than merely reactive response to the flux of experience in the spirit of the Grotesque. We find ourselves challenged to evolve new societal structures, new structures of new consciousness and to find new patterns of meaning. We are desperate to work our way out of maelstrom.

The achievement of new structures of this scope is always difficult. Today the difficulties are doubly great because of the concurrences of confusing conjunctions and discontinuities. Extending a fertile suggestion made by an art historian in the pages of Karl Mannheim, I find myself inclined to describe these movements under the names of *"contemporaneity of the non-contemporaneous"* and *"non-contemporaneity of the contemporaneous."*

Neither of these situations is purely of the present moment, or purely modern, or purely Western. Yet there have been few eras when the sense of commingling of all events and time scales have been so powerful or when, on the other hand, the feelings of discontinuities in the senses of reality have been so rife among clashing generational groups.

Indeed, the mixed experiences of *simultaneity* and *non-contemporaneity* are as general and as encompassing as they have ever been in the history of civilizations. A hundred factors conspire to extend imagination in global and cosmic directions. Arts, sciences, technologies work together to reinforce the sense that the *entire* past and present are caught up in ecstatic embrace. All history has now become material for a living theatre of ongoing spectacle. The past is felt to be wholly alive and the insurgent future is waiting in the wings for immediate appearance.

The "grotesque" new ways do not inevitably have to be mere "put-ons," or totalistic aggressions; some are experimental designs for living in a world that is only now being forged through new structures of consciousness. After new journeys—"trips"—there will come new maps. It hardly matters whether we welcome the results or not; many of the new ways of being now in the making will constitute themselves as the prevailing modes and the new social metaphysics.

The "contemporaneity of the non-contemporaneous" is a less taxing frame than the more familiar *"non-contemporaneity of the contemporaneous."* It is the latter which is cited nowadays as the *central* element in the gap of the generations. The structures of our varying sensibilities are not so much mistaken in their objects as they are differently related to the myriad changes going on in our societal experience and in our ways of life. Those who have become hardened crystals of now unlivable pasts are opposed by those who have yet to forge viable ways. Odd conjunctions and disjunctions occur at every turn. The warring forms of consciousness embroil all in a state of civil war.

We have come full circle and are back at the center from which we have never truly departed. *The omnipresence of the grotesque* is the very hallmark of our time.

NOTES

1. I am pleased to report that my long-standing conviction on this matter is now supported in a newly discovered, as yet unpublished, manuscript of Max Weber's, one of our foremost sociologists and historians of culture and civilization, who declares that Georg Simmel made a serious mistake in this definition of sociology [now see Weber 1972].
2. A reader of the foregoing (Marie C. Nelson) suggests that I add the following to lines 5–8 of p. 515:

What *does* root is the need—when social chaos offers no adequate protective structure—to take refuge in disconcerting acts and alarming self-adornment as aggressive self-protection to actively ward off evil. And when this is sufficiently widespread, such grotesquerie of act and appearance assumes also the passive function of self-protective coloration. An instance of this might be the wearing of the Safari hat which proclaims at one and the same time: "I am a White Hunter *and* a harmless tourist."

In this connection, attention is called to Mrs. Nelson's previous paper, published under the name of Marie L. Coleman, "An Integrative Approach to Individual and Group Psychology" (1949).

TWELVE

Eros, Logos, Nomos, Polis: Shifting Balances of the Structures of Existence

Earlier I sought to distinguish several principal types of "structures of consciousness" illustrated in the histories of the areas which have been the cradles and homes of the world religions (chapter 5). Here I hope to throw light on the shifting blends and mixes in the crucibles of men's myriad histories of *eros, logos, nomos, polis*—central elements of sociocultural process (Nelson, 1962d, and chapter 11). I continue to be spurred forward in these efforts by the sense that once again today in both the so-called "East" and the so-called "West" decisive shifts are occurring in the shapes of *civilizational contents* and the balances of *intercivilizational ascendencies*.

A key notion in this context is the changing fortunes of societies and of men's sense of their fruitions; both prove to be in their most equable situations when hopes for the realization of vital balances of *eros, logos, nomos* and *polis* are not (wholly) unavailing (cf. Merton 1970 [1938]; and chapter 2 above). In the absence of such options, received structures of rule and assurance as often as not break down. The needs for fresh expression in the ways of *eros, caritas,* resentment, defiance, alienation, or withdrawal course through society (Marcuse, 1955, 1964). Where apathy does not come to prevail there are thrusts to fresh actualizations of *eros,* of *diké,* of *charisma.*

Unhappily, many who have written on these themes have been insufficiently attentive to modalities and gradations. Simple polarizations such as those offered by Herbert Marcuse (1955, 1964) and Norman O. Brown (1959, 1966) will not suffice (cf. Nelson, 1961c).

Originally published as "*Eros, Logos, Nomos, Polis:* Their Changing Balances and the Vicissitudes of Communities and Civilizations," in Allan W. Eister (ed.), *Changing Perspectives in the Scientific Study of Religion* (New York: John Wiley & Sons, 1974), pp. 85–111. Reprinted by permission of John Wiley Sons, Inc.

Throughout this paper, I shall be citing many illustrations from the twelfth and thirteenth centuries. My reasons for choosing that era are two—both weighty ones. The more one ponders these two centuries, the more one becomes convinced that *they constituted prime seedbeds of the institutional and cultural developments of the Western world* (Clagett et al., 1961; Haskins 1927; Nordström 1933). Indeed, I would take the next step and say that when these centuries are considered in inter-civilizational perspective, they prove to have been *a watershed in the international history of the world* (see Bozeman 1966; Barker 1948; ICHSCDM 1963–70).

To make this point clear, I adopt to my purpose an incisive observation by Whitehead. From the wider cross-cultural point of view, Whitehead explained, the great turning point in the history of Western civilization was the Scientific Revolution of the seventeenth century rather than the Protestant Reformation. The latter, he remarked, was in many critical respects an episode in the *domestic history* of the European people (Whitehead, 1968). The position taken in these pages rests upon the notion which Whitehead shared with others that there is no way of truly understanding the seventeenth century without appreciating the roots of those developments in the twelfth and thirteenth centuries.

The twelfth and thirteenth centuries are the decisive era in which the initiatives in all spheres shifted from the Middle East to the West. It is the era in which there occurred the reception of the new Aristotle, the reception of Roman law, the development of the canon law, the renewal of Greek political philosophy, the emergence of the universities, the development of scholastic philosophy, natural theology, natural science, the crystallization of critical logics associated with the "court of conscience" in all spheres of act and thought and so on (Haskins 1927; Clagett 1961; Vinogradoff 1967; Nelson 1968b; Chenu 1969).

My second reason for selecting the twelfth and thirteenth centuries is that within this era, more clearly than in any time before our own, we are able to witness the struggles both to fuse and to defuse blends of *eros, logos, nomos,* and *polis.* This was the era in which the decisive break occurred in the emergence of incorporated cities (Weber 1958c [1921]; Pirenne 1956), an institution with few parallels or precedents in non-Western history; a free self-governing citizenry; differentiation of occupations in handicrafts, manufacture, trade, shipping, with organized groups of artisans, merchants, moneylenders, financiers, professional philosophers, lawyers, doctors, notaries, mathematicians and so on.

The changes which occurred in all spheres during that era—in economy, polity, society, culture—were truly extraordinary. It was an era of decisive thrust of the so-called "West" to recover the Mediterranean as a European lake and the route to the East—Far and Near East alike; it was the era of the change of fortunes in the struggle with Islam (Davison 1926; Thode 1904); it was an era of the establishment of the cities, new institutions of communes of various sorts, intentional communities of almost every type, surges toward hierarchical political structures and countervailing surges against every type of nomistic control restraining the free flow of *eros* (Douie 1932; Morrison 1969).

A history of European social and religious developments from the beginning of the tenth to the middle of the fourteenth century offers supreme illustrations of the sorts of struggles we are again witnessing today. I refer to decisive struggles over the way in which the social fabric should be threaded, struggles over the ways in which the structures of consciousness should be formed. At no time so much as in that era and in our own can we see the convulsions which derived from the civil wars in the structures of consciousness. The battle ranges between the existential structure of "faith-consciousness" and the more objective "rationalized-structures" of consciousness (see above, pp. 95ff).

In the main, the twelfth and thirteenth centuries witnessed the spread of the rationalized-structures at the expense of the faith-structures. Today we are seeing the reversal of that trend. In short, from the point of view of a comparative historical sociology informed by a strong interest in the shapes of civilizational complexes and intercivilizational encounters, our twentieth-century days since the outbreak of the First World War need to be viewed against the backgrounds of wider horizons which include reference to the twelfth and thirteenth centuries.

Before entering into the main highway of my present story, I first need to make two brief detours in the hope of clarifying its comparative historical and "anthroposociological" horizons. The particular relevance of some suggestions of Feuerbach will be considered in the second of these prefatory turns.

I

This essay is written in the conviction that *sociology* and history—and one must add, *anthropology* and history—have a relation to one another like that which Kant ascribes to concepts and percepts. Anthropology and sociology—one wishes one would have available a broadened expression, *anthroposociology*—prove to be empty without history just as concepts are empty without percepts. And history without anthroposociology is blind just as percepts are blind without concepts. It is on these assumptions that I will feel free to refer here to matters historical as well as matters anthropological and sociological.

Nor do I suppose that so-called past events have occurred once and for all. I am convinced that many pasts live again today; also, that many of the extraordinary features of the tumultuous sociocultural processes of the twentieth century are best understood as the re-presentation of what had been assumed to be extinct.

Thus, in my view, anthropology and sociology are strongly rooted in depth-historical understandings of structures of existence, experience, and expression. We are hardly likely to range beyond *schematic* histories if we do not come into close contact with the actualities of existence; the myriad strivings to realize hopes, dreams, myths, in social and symbolic forms of various sorts; if we do not somehow see the meshes and mixes of *eros, logos, polis, nomos,* in wider civilizational and even intercivilizational perspectives. *Schematic* histories are exceedingly weak bases from which to play creative new moves into an as yet unknown future.

II

I must now speak about Feuerbach, whose insights prove to have wider implications for our theme than have yet been clearly set down (Feuerbach 1957, 1967).

In his *Essence of Christianity* (1841), which Marx and many others called a masterwork of their century—and which in my view remains a critical source for all interested in the study of sociocultural process—Feuerbach sought to explain the relations between the structures of men's existences and experiences, and the structures of their religious consciousness and theologies; and, as some readers may recall, after a rather systematic analysis of the doctrines of Christianity, he came to the view that "the key to theology was *anthropology*"—that is, the understanding of man's "nature," man's "essence," man's *"experience"*—was the key to the understanding of theology.

If Feuerbach's theses be true in some sense, if it be the case that man has indeed secreted his "essence" into his theology, if men have painted themselves and their histories into their faith, then it behooves us to look very carefully at men's theologies. Apparently in order to recover men's histories and the understandings of their natures and their characters as persons, we must go by way of their theologies. This way of looking at the matter has not received the attention which is its due.

Before all else, it has to be perceived that men do not naturally know their own natures, any more than they naturally know their own histories. Men's natures and histories are in no sense immediately accessible, tangible, palpable, available; men have to continue as it were, to seek to discover their natures through tenacious efforts, and they have to find some sort of terms in which to express the structures of their own *existences,* the characteristics of their *experiences.* They have got to look at their *expressions.* And since their theologies are a foremost form of their expressions, they must look intently at their theologies as well.

For this reason among others, as I have already indicated, I shall be placing strong stress here on the vicissitudes of men, institutions, cultural forms and theologies in the twelfth and thirteenth centuries. It is these very centuries which witnessed the systematic codification of great theologies—*scholastic* as well as *monastic* (Leclerq 1945; Chenu 1968; Bertola 1970), *natural* as well as *moral* and *mystical;* it is these very centuries which now have to be designated as the seedbeds of the central structures of Western social and cultural organization. In fact, these centuries witness the constitution of what may be called the groundplan of Western civilization in a sense more technical than familiar.

If we be allowed to extend Feuerbach's thesis with the help of Durkheim and Mauss, we must be prepared for a surprising outcome: the existences and experiences which have found their way into the theology will *re*-present themselves to use in ways we never expected. Irresistibly as we assay the civilizational process of that era from the point of view of the shifting blends of *eros, logos, nomos, polis,* we will find strong evidence that men did, in fact, discover themselves in their theologies during the twelfth century. In making these discoveries, we also discover that the twelfth and thirteenth centuries are the era in which the notions—your notions and my motions—of Western man were recovered (Gilson 1940a; Knowles 1964; Peter Abelard 1971; Murray 1967). The distinctive characteristics which define the image of man, of society, and of civilization were uncovered or discovered in the twelfth century. If anyone would suppose that in the eighth or ninth centuries there was accessible to any great number of persons images of man, society, community, civilization, that were comparable to such images after the twelfth century, he would simply be mistaken. The history of the structures of consciousness or conscience is neither continuous nor consecutive, nor is there any consecutive or continuous understanding or awareness of man as an agent in all modalities of his personhood and existence. How, then, did men of the twelfth and thirteenth centuries find the ways in which to express their new experience of themselves?

I shall have to ask readers to join me in looking closely at the facts: if the data attest to anything, they attest to very critical transformations in the structures of consciousness and conscience. Why so? Where can this best be seen?

III

No setting presents the stirring changes of the twelfth century so strikingly as the confrontations of St. Bernard of Clairvaux (Knowles 1964; Kneale 1962) and Abelard (Williams 1935)—confrontations which, in my view, were to have extraordinary political, sociological, cultural, and civilizational and intercivilizational consequences. These encounters offer us a paradigmatic instance of the passage from what I have described as the faith-structures of consciousness to the rationalized-structures of consciousness (above, pp. 94ff).

I will talk first of Bernard, then of Abelard.

Bernard felt himself called to devote his life to the recovery for man of the meaning and experience of love, God's love for man in Christ, man's love of God through Christ.

Hoping to enjoy a state of mystical union with the Godhead, wishing to be suffused with the word of God, eager to be possessed by the structures of faith-consciousness, Bernard meditated upon the Trinity (Williams 1935). He meditated upon the mysteries which were involved in the relations and separate distinctions of persons in a single essence; he meditated the distinctions and yet sameness of the divine nature; and he also continued to meditate on the actual human existence of Christ.[1] In the process, Bernard discovered something that had not been fully disclosed or fully acknowledged before in Western theology. He discovered the Christ-*Man* as Everyman's Perfect Friend.[2]

It was to be anticipated that one who meditated on the Trinity and Christ with the stark intensity and conviction of a Bernard—such a one would discover the man-God. The reasoning runs as follows: man had been put into the Trinity, into Christ, when the doctrines of the Trinity and the hypostatic union of two natures was developed. Whoever troubles to study the relations between Christian doctrine and the shapes and images of man and society and community and relations, and power and purpose and sequences of causality, and grace and spirit and all of these other notions—whoever pursues this path must not be surprised to discover that when the time for reappropriation occurs, what proves preminently available for recovery is that which had been secreted into theology.

Now, what Bernard strove to reappropriate was the incarnate son within the divine-human exemplar, the imitation of whom was the way of light, life, and truth. Bernard's discovery was one of the ultimate points in the history of Western culture. All images of *eros, logos, nomos, polis* which were Bernardine in inspiration represent a fundamental form of the option again and again taken in the West in the form of faith-consciousness.

At the very moment Bernard was giving ultimate shape to the *eros-as-logos* structure, the notions of person and truth were being recovered in other ways and were actually being elaborated with other intentions and other purposes. I refer to the work of Abelard and the scholastic—as distinguished from the monastic—theologians. Only a word may be allowed here about a critical step in the passage from the monastic to the scholastic thinkers.

The early scholastic writers, including Anselm, said that they believed in order to know. *Credo ut intelligam; fides quaerens intellectum:* such were their telling expressions. Those who apply these expressions often enough, those whose faith insistently seeks understanding, find themselves imperceptively shifting on their axes; they discover themselves passing from *wanting to know in order to believe* to *wanting to know for the sake of knowing* (Barth 1960; Knowles 1964; Kneale 1962).

The passage from the first stance to the second stance occurred quickly and with immense effect; it occurred in a single generation, and, when it occurred, those that had believed in order that they might know were discovering that there were others who were intent upon knowing in order that they might believe, and they were both performing the same job. What led Bernard to oppose Abelard, to pursue him so relentlessly? Some find it hard to understand. Before facing this question directly, we must talk of the institutional expressions of love and friendship.

IV

Eros and friendship are to be found everywhere in the life and work of Bernard. He reveals a comprehensive theology of love; it hardly needs saying that all monks who were in any way influenced by him, all communities that were in any way inspired by him were grounded in a notion of love. The notions of love and friendship were ruling conceptions in the twelfth and thirteenth centuries, as one sees quickly by looking at the sociocultural structures of those centuries (Egenter 1928; Bloch 1964).

Different institutions need to be regarded from the point of view of the characters of the mixes and the strains which developed within each of these structures. *Eros*—in the form of *philia*—*logos, nomos, polis* were fused, blended in various ways in all of them.

In some settings, a pro-structural stress on eros was greater than the pro-structural stress on the expression of *logos,* (see, e.g., Jocelin of Brakelonde [Butler 1949]; Christina of Markyate [Talbot 1959]).

To understand the nature of communities bound in one or another form of *eros,* or in sublimated *eros,* that is to say, in *amicitia* (friendship), especially *spiritual friendship,*[3] we would do well to look closely at two institutions, feudalism and monasticism. The first may not instantly strike us as an outstanding illustration of the notion of friendship, but it is the case that the *pro-structural* stress on friendship undergirded the entire structure of the relations of "Religion" and the "World," at almost every level and function in the Middle Ages. Thus we must look for prime crystallizations of *friendship* structures in the feudal world as well as in the cloister.

Feudal relationships preeminently involved relations of trust and loyalties—individuals who were tied to one another were obliged to undergo risks of life and limb for one another, to stand hostage and surety for one another, to come to one another's aid in continuing conflicts over prerogative and place. The seriousness of this pledge is not reduced by the fact that there were limits to the amount of time in which the fullness of these services could be demanded; the absolutely ruling ethic was grounded in the idea of the friend-foe relation (Bloch 1964; Nelson 1949a, pp. 141–164).

To miss the fact that invidious friendship was the base of the feudal ethic is to miss the passion as well as the pride of the medieval world, indeed, of the entire premodern era. All relationships, vertical as well as horizontal, rested on the same hinges; the name of these hinges were friend and foe.[4]

Among friends everything was common, especially enemies. All members of the nobility without exception were conceived and conceived themselves to be responsible to act in the spirit of the one example.

It is from the cloister that we receive perhaps the most critical underlying principle of the dualism of "Religion" and "World," that is, the image of the Perfect Friend. In this era, Christ becomes the perfect friend. "Greater love hath no man than he lay down his life for a friend" (Gospel acc. to John xv:13). Christ laid down his life for all mankind. As the true and perfect friend he therefore represents a sort of exemplar—both for the nobility and for the clergy, notably the monastic clergy. And, indeed, Christ is presented to all as the model for all. It was in this spirit that it was possible for the Church to preach the crusades, and to draw calculating nobles into acts of supererogation and acts of sacrifice beyond the line of duty.

Too few seem to know that the medieval era produced vast numbers of treatises on the subject of friendship, especially "spiritual friendship." The notion of friendship was, of course, not invented in the Middle Ages; *amicitia* is a classical ideal and it has its counterparts in many other societies and civilizations (Butler 1922), which are not distinctly Greek or Roman. It is a very critical conception, but it does have notable variations in the institutional and cultural forms that were in crystallization in the medieval era.

It is odd that so little has been made of the fact that there is no other way by which great numbers, pledged to the triple vow of chastity, poverty, and obedience, could retain threads of communion which bound them in spiritual affections. As brothers and sisters in monastic houses and convents dare not love one another in the flesh, they nonetheless are pledged to love one another in the spirit and to express their love in the conjoint imitation of the example of the Perfect Friend (Nelson 1939–44).

Each has one paradigm—Christ—who gave his life and his love for all mankind. Each is expected to make his or her own life a perfect imitation of the divine example. Each is expected to keep his or her own bridal chamber stainless and ready. For, as Meister Eckhart and so many others reminded their contemporaries, "Behold the Bridegroom cometh!" (Eckhart in Blakney 1957). None will have such occasion to lament as the foolish virgins unprepared for the holy matrimony.

Those who have talked about the Protestant ethic existing in the Middle Ages entirely miss the immense authority of these structures of consciousness and sensibility. One must remember that each of the monasteries and convents and each of the new religious groups were the centers of circles from which there radiated the structures of *eros* and sublimated *eros*. Whoever wishes to understand the development of the Church must see that it is a community bound together in love and by love, a love that is bound together beyond a secular expression of desire and affection. Unless this ethic had persisted the Church could not have continued at all.

The cracks of the medieval structure came from two sides—the routinization of the charismatic and the sacralization of the profane—which recur in different measures at various times. One of the most challenging tasks for anyone who has an historical interest in sociocultural process is to study the vicissitudes of these blends and relationships of *eros*, *logos*, and civilization. No set of circumstances so powerfully illustrates these changes of phase as the main changes that occurred between the Cluniac Reform and the beginning of the so-called Babylonian Captivity of the Church.

I allow myself some comments at this point, which perhaps may not be as familiar as they might be, in reference to the notion of friendship as a structural principle for modern society.

Too few seem to perceive that in the medieval world and in the early modern world—prior, actually, to the Puritans—a full religious sacralization of the family or of family property did not exist. There did, indeed, not occur the sacralization of what might be called the special friendship with one's own wife. In the "world," the notion of temporal friendship was regularly conceived as friendship between two who had an almost blood-brotherhood relation, where they were altogether possessed of a single soul while they had bodies twain (Mills 1937).

Now, how did the change occur? That is, how did the premodern—the classical, medieval, Renaissance notion of friendship—become the modern ethic of impersonal service on behalf of an impersonal goal? If there was anything at all about the earlier medieval ethic, it was stubbornly interpersonal. Now, how did we get our contemporary view?

Throughout antiquity and the Middle Ages friendship was conceived as the union that transcended all calculation and egotism whether of family or of person. From at least the time of Plato forward, the moralists and novelists insisted on the preeminence of friendship, going so far as to deny that one's wife or member of one's own family could truly be friends in the highest sense. The stress on Christ as the perfect and true friend continued into the Renaissance, and the Elizabethan period in England. Interestingly, the first powerful assault on the idea of friendship was the Puritan attack on Elizabeth and the courtly style. Puritans correctly grasped that so long as friendship and friendship circles were held in the highest respect, there was no possibility of achieving sanctification of the special love within the family (Schücking 1966).

It was the Puritans, above all, who mounted the attack on the ethos of friendship which prevailed in England until their day. The very idea that charity begins *at home* involves the sanctification of the home. It constitutes a very extraordinary extension of the notion *caritas incipit in se,* and it is, indeed, of course, something of an extension and elaboration of the notion that one is to love one's neighbor as oneself. Too few have noticed that such a sanctification goes far beyond the idea that we have to love ourselves as we love our neighbors. The new maxim, "charity begins *at home,*" is the sacralization of a collective egoism of the family and its property (Nelson 1949a, pp. 141–164). It is the indispensible base of the newer structure which came into being with the Protestant ethic. Having elsewhere discussed some of the other facets of this story, I hasten to return to the vibrant twelfth century.

V

To come back now to our briefly suspended confrontation. The second actor in our dialogue—Abelard—now demands to be heard.

I will address myself first to Abelard's own work and teaching; I will then turn to the activities of a number of notable scholastics and jurists, including some who were Abelard's students.

Throughout this discussion I will put the emphases on the changes in the structures of rationales which came to be elaborated as a result of the distinctively Abelardian notion that all activity had to be conceived as the action of individuals who were possessed *of consciences* which had a power to opt between alternatives of relative worth or of varying credibilities.

Doubtless, the idea of *conscientia* is to be found in antiquity; we can find it in Seneca, Cicero, and other authors, but the wider story which follows here has many surprises. The notion of conscience *(conscientia)* only comes to the fore with the breaking down of the structures of collective consciousness (Durkheim 1955: Burckhardt 1860; above pp. 43ff) and the growing strength of the need to establish assessments of individual liability and individual responsibility, or individual blame. The more carefully we survey the historical developments, the more likely we are to find that it was not until the eleventh, twelfth, and thirteenth centuries that the term conscience came to undergo an extraordinary dialectical, logical, philosophical, religious, theological unfolding (Lottin 1942–60). And it was set forth in very great detail, starting with Abelard's necessity to understand the contents of faith (Abelard 1971).

The more closely we scrutinize the documentary remains, the more we perceive the critical importance of notions of responsibility and guilt in the working out of the problems of intentionality.[5] We must not be surprised if we find the following among the questions most insistently put forward:

Who may be said to have been responsible in the case of the crucifixion (Abelard 1971; Nelson 1947)? Are consent and knowledge pre-requisites of sin? How shall the sinner make proper restitution? How respond to the dictates of "conscience"?

It is with Abelard that there begins the true unfolding of a variety of perspectives and logics which are connected with the analysis of the possibilities of conscience—conscience seeking to realize itself in the world here and now; seeking to make itself viable and operative, meaningful and fruitful in myriad ways.

There is, in fact, no possible activity which does not fall under the governance of conscience in this era. We are mistaken if we identify the notion of conscience exclusively with the sense of guilt or with retrospective remorse; the medieval conscience is mainly prospective.[6]

An added word may be needed to explain how *opinions* fell under the governance of conscience. Abelard and others were convinced that every opinion had relative degrees of value as a truth-function or truth-claim. As time passed, the scholastic writers began to talk about the varying degrees of probability of opinion (Nelson 1963b and chapter 7 above) relative to some particular logical object or state of affairs. It would not matter what the probability was, it was a relative probability of opinion in respect to the meaning of a moral norm or an opinion on the state of nature or anything else.

In short, Abelard and his followers present us with a twofold logic operating from the same hinge.

Two additional facts must be understood if we are to understand what came to be called the Court of Conscience. The complexities that must arise in the effort to administer so comprehensive a set of structures relating to so varied a range of predicaments are impossible to exaggerate.

A second point which needs to be recalled is that Abelard did a great deal more than evolve a philosophical schema. The institutions of the conscience were all predicated on the assumptions that all who were Christians were answerable. The notion of answerability was a paramount idea. There is no one, no pope, no king, no one who is not answerable to the tribunal of conscience—the king has his chancellor, who is the keeper of the king's conscience and from whose activity the entire law of equity is crystallized (Vinogradoff 1967). And the Pope has his confessor.

All are answerable; there must therefore be an institution which reviews the cases of the world in the light of conscience. The circumstances of the world being myriad, there is no single or direct move from any set of principles to a decision in all cases.

Now when everyone in Christendom was made responsible under the Fourth Lateran Council of 1215 to confess at least once a year, there was a need for an highly elaborated moral theology in which all the cases of conscience were considered as they arose in the practical life of the times (Thompson 1957; Michaud-Quantin 1962). There was a period in which Christian moral theology and moral theologians made bold to consider every conceivable public issue in concrete terms. And they had the temerity to name the people who were engaged in the questionable activities at the very highest level of functions, and judge them publicly for all to read about.[7]

The high point of this morality of prophetic witness covers a period roughly from 1170 to 1230. Stephen Langon was such a witness, (Nelson 1933; Baldwin 1970). Another man of the highest importance in this development was Cardinal Robert de Curzon, who is often known as Robert de Courcon, the Cardinal-Legate of Pope Innocent III in France and author of an extraordinary *Summa* with an almost unique treatise on usury. I did not know at the time I was writing my master's thesis on the theme many years ago, that the tradition that Robert represented in such an extraordinary way was to have an early end as many good traditions do.

Even before the first half of the century had ended, reference to great public questions passed out of fashion in the *summae* of cases of conscience; instead, a sort of split occurred. The moral theologians moved ever closer to the very profound and yet more rarified atmosphere of advanced moral metatheology, or meta-moral-theology, and they did not deal with any of the cases except in a very general sense. And at the lower level, there were those who became experts in developing a kind of highly routinized, rationalized analysis of the predicaments and circumstances and worked them all out alphabetically. By the end of the thirteenth century, those deputed to serve official roles as judicial officers and physicians of the soul in the Courts Christian, in the confessional, could just do it "by the book." The mighty effort of society of the High Middle Ages to achieve a public moral governance of itself ground to a near halt. So far, at least as this horizon was concerned, "prophecy" gave way to "routine."

VI

Abelard and Bernard were not at infinite removes from one another. Both men put their faith in the virtuous imitation of a divine exemplar. Both placed profound stress on intentionality. But, in the words of a recent writer, Abelard was preeminently devoted to eludicating the *logic* of love—divine and human alike—the logic of willing, consenting, sinning, indeed, of existing and acting in all the dimensions of experience.

Abelard needed *to know* what he believed. He needed to know who shared in the guilt of Christ's crucifixion; how to speak of the nature and persons of the Trinity; how to interpret the Faith to make it invulnerable to attack from whatever quarter; how best to ground moral obligation and argument.

By pursuing these issues relentlessly, Abelard became the foremost twelfth-century architect of structures of the rational consciousness.[8] Abelard worked out a logic of intentions. Without him there is no possibility of understanding the separation which constitutes the logical crux of the foundations of Western civilization.

In brief: I am saying that all the axial structures changed dramatically in the twelfth to the thirteenth centuries. Collective consciousness apparently gave ground to individual consciousness, and around these there developed comprehensive logics of opinion, belief, action, and so on; there treatises on human action, conscience, and its cases provide a setting for the proliferation of highly developed structures and decision-matrices of all of the moralities of thought and action.

To oppose *eros* and *logos* as absolutely antagonistic principles is to miss the point; this view has turned the whole question of the future into *a one-dimensional prospect* which seems to characterize the projections of some of the millenarian thinkers named above. We have no recourse against one-dimensionality as prospect if we see one-dimensionality as constituting the total retrospect.

IN SUM

Rarely has so much social innovation and new culture come with such dramatic impact as in the period we have been discussing. The twelfth century witnesses extraordinarily dramatic fusion and institutionalizations of *eros, logos, nomos,* and *polis.* Vast funds of new passion pour into new polities. Prior structures suited for an earlier age now give ground to new collectivities—cities, guilds, universities, monastic communities, popular religious movements (see Pirenne 1956; Le Goff 1957; and the articles by A. H. Thompson, H. D. Hazeltine, E. A. Armstrong, G. Mollat, C. Roth, E. Power and others in the *Cambridge Medieval History,* vols. 5–8 [and the related discussions earlier in sections VII, IX–XI of chapter 10]).

It is no wonder that the great confrontations between the faith-structures of consciousness and the rationalized-structures of consciousness occurred at that time. As we have observed above, transformative thrusts toward the advanced structures of the rationalized consciousness and rationales in the canon law, theology, and the public life gathered great momentum throughout the twelfth century. It is against these settings that it seemed best for our purposes to explore the historic encounter between St. Bernard of Clairvaux and Abelard.

Already by the third quarter of the twelfth century there are signs of tension within the newly rationalized sectors and structures. Insurgent mass movements, coming from below the surface, declare their will to reform the Church and restore it to its mission as the exemplar and protector of the poor. The prime symbol of the new time is Peter Waldo—in his rejection of the profits of trade for the life of *a*cosmic love (for Waldo see Wakefield and Evans 1969, pp.

200–210; Davison 1926; and Thode 1904). From this point forward there is an insistent recurrence of movements to blend the commitment to poverty and reform.

Peter Waldo is the harbinger of St. Francis of Assisi. It is not by chance that St. Francis sought to rededicate his movement to poverty and the life of simple faith. His followers were to own nothing either collectively or individually, were not to pursue learning, were not to mingle in governance.

The most decisive struggles of the thirteenth century were those which raged among the Spiritual and Conventual Franciscans—struggles which were deeply influenced by the eschatological visions of Joachim of Fiore. We must not be surprised that the crisis in the Franciscan order and the struggles over Joachism came to a peak in the Pontificates of Innocent III and Gregory IX.

Although many view the thirteenth century as the era of the full crystalization of Gothic Art and Scholasticism, there is as much warrant for another perspective, in which the Christian rationalized civilization that was developed in the twelfth century is seen to be undergoing mighty hammer blows from which it is never fully to recover. By the end of the century there had occurred the ultimate attempt of spiritualism to effect charismatic renewal of the Church through the person of Celestine V. The rulers of territorial states had entered upon a civil war against the popes. Boniface VIII had undergone humiliation at Anagni and the Babylonian captivity of the Church had begun.

Only one group failed to triumph in the midst of these debacles, the Spiritual Franciscans. For a time, it did appear as though the Spirituals would attain the heights of their hopes. With such leaders and champions as Peter John Olivi and William of Occam they hoped to humble the papacy. They did not succeed. In this very critical hour the popes were able to count upon the support of the lay estate and the patricians in the Italian cities. The Spiritual Franciscans were hunted down and persecuted. The middle of the fourteenth century initiates a new era in the life of Christian civilization, an era which was not ended until the Reformation.

We offer these evidences as proofs that developments of European society and Western civilization involved hectic changes in the mixes and blends of *eros, logos, nomos, polis*. Now, as then, issues of the fusions and conflicts of these elements are felt to be of peak religious and civilizational significance.[9]

Few readers who have accompanied us thus far in our journey are likely to escape the feeling that the profound turnings of institutions and sensibilities we have just reported for the twelfth and thirteenth centuries seem to have a decidedly contemporary ring.

Then, as now, religious, theological, and political antagonists insisted on pitting *eros* and *logos* against one another in the conviction that lasting truth, justice, and happiness could be realized only by the conquest of one by the other (e.g. Walter of St. Victor, post 1179 [Glorieux 1953]; Marcuse, 1964, 1966; Brown 1966, 1970; Roszak 1969). Our effort to be faithful to the vicissitudes, changing balances, and realizable hopes of the cities, communities, and civilizations of this world has here led us to project a fourfold (actually, a 4 + *n*) "dialectic" of the mixes in socio-cultural process. We have been unable to suppress the memory that a dialectic of two opposed terms all too often ends by forgetting the "social-reality principle" (Nelson 1962c, and chapter 2; cf. Turner 1968).

NOTES

1. An especially helpful statement of these issues and links will be found in Mother Adele Fiske's two-part essay on St. Bernard's exploration of the idea of friendship. Mother Adele writes:

> To St. Bernard, all love is rooted in the love of God and therefore is an "affection," not a contract: "Affectus est, non contractus." This affection of love is, if not the equivalent, at least analogous to the vision of bodies. God is "sensible" to the heart that loves; love then is a vision. For Bernard, knowledge is based entirely on a likeness of the subject knowing to the object known. For man to know himself is to see the image of God; he is miserable, for he is nothing of himself, and the "likeness" has been lost by sin, but he is also great, for the image remains though defaced *curva,* bowed to the earth. The image for Bernard is in the will, in freedom. This is the bedrock of Cistercian mysticism, and it may be perceived behind all that St. Bernard thought and felt about friendship. (1970, pp. 1–2, and see the exceptionally interesting formulation on pp. 3–4, 15)

2. Many readers will detect the relation of the argument of these lines to the work of Henry Thode on St. Francis. It is therefore necessary to make clear that, so far as the structure of sensibility and image associated with St. Francis in the powerful pages of Thode are concerned, Bernard is here presented as a critical precursor rather than a full crystallization. Particularly interesting insights into this will be found in an essay by J. M. Déchanet (1955) entitled "On the Christology of St. Bernard." Déchanet makes two observations central to our theme. He notes how profoundly Bernard prepares the way for Francis. Thus he writes:

> On a beaucoup insisté—beaucoup trop peut-être—sur ce côté sensible et quelque peu sentimental de la Christologie bernadine. On n'a pas assez remarqué et fait remarquer que les choses n'en restent pas là et qu'après avoir orienté—un des premiers semble-t-il—la dévotion chrétienne vers une attitude qui trouvera dans un saint François d'Assise son expression la plus aiguë, et son orientation définitive dans la Dévotion Moderne, saint Bernard est revenu à l'antique tradition de l'Eglise.

However, one must not overlook the fact that Bernard's Christology does allow for a sharp distinction in a double approach to Christ through a double love and a double knowledge "selon le chair et selon l'esprit-libre" in accordance with the text (as Déchanet reminds us) of St. Paul.

The full context of this passage is of central importance for the argument of this essay. See Déchanet (1955, p. 65).

3. Mother Adele M. Fiske, RSCJ, has published many learned and valuable essays on this theme which are now published in the CIDOC Series at Cuernavaca, Mexico. The most interesting of these essays from the point of view of the present paper are listed in the References.
4. Certain insight into this theme, albeit from a special point of view, will be found in the writings of Carl Schmitt (1927, 1934). Cf. the work by Schmitz (1965); also cf. Fijalkowski (1968).
5. A fuller understanding of the varying experiences of guilt and systems of imputation of liability is a major desideratum of present-day comparative historical sociology. Nietzsche and Freud have not spoken the last word on the matters above. A great deal remains to be derived from Durkheim and the Durkheimians; cf., e.g., Glotz (1904), Fauconnet (1928), Harrison (1912), Murray (1925), Conford (1934, 1912).
6. Scarcely any authors have given due weight to this critical fact. Studies and texts on the developments of the idea of conscience will be found in the following: Kirk (1927), Lottin (1942–60, esp. vol. 3, Pt. 2, 103–468), Hofmann (1941), Walter of Bruges (ed. Longpré, 1928), [and Section V in chapter 3 above].
7. I made this point many years ago in my master's thesis at Columbia (1933), and my *Idea of Usury* (1949a). The point is not given due stress in recent studies of moral theology in the twelfth and thirteenth centuries. See, e.g., Baldwin (1970); Michaud-Quantin (1962).
8. For appreciation of Abelard's role here, see Knowles (1964), Kneale (1962), and Jolivet (1969).
9. I am bound to report that as I was completing the last draft of this paper, there came to hand an eloquent essay by Father M. D. Chenu (O.P.) which paralleled the emphasis of the present essay on the critical importance of the twelfth-century developments for the history of the transformations of conscience and consciousness. See Chenu (1969). A glimmering of a related idea will be found in A. V. Murray (1967).

THIRTEEN

Sociology and Civilizational Analysis: A Reply to Friends and Critics

It is a pleasure to have the chance to offer some comments on the essays which are presented as a discussion of my paper "Civilizational Complexes and Intercivilizational Encounters." The fact that such a symposium is now published here is itself a strong sign that the perspectives and problems I have been seeking to set forth are already becoming part of the agenda of sociologists who wish to go beyond more familiar frames of sociology today. . . .

I

Some have suggested[1] that I have not always stopped to define certain notions as precisely as one might wish. By way of defense I may say that it appeared to be necessary first of all to offer documented empirical illustrations of the particular advantages and uses of the perspective I was eager to set forth. That is why I have written papers which deal in considerable detail with particular civilizational complexes and encounters of "East" and "West" in the eras of the Crusades and the "early modern revolution in science and philosophy." I have been intent upon making sense of the structures of consciousness and conscience exhibited in the histories of the great civilizational complexes which have been vehicles of world religions. Here my strategy has been to enlarge the horizons offered by Weber through the infusion into his main structure of the critical suggestions which are found in the work of Maine, Durkheim, Mauss, the Durkheimians and neo-Durkheimians, and most recently Joseph Needham. I have also been strongly influenced by particular studies of the histories and sociologies of great civilizational complexes, the world religions and intercivilizational encounters. I have been selec-

Originally published as "De Profundis . . .: Response to Friends and Critics" in *Sociological Analysis* 35, 2 (1974): 129–142. Reprinted with permission.

tive in using noted contributors to our field because I have not always felt that they could be readily integrated in the systematics and problematics I was intent on carrying forward.

I would agree that there are many typologies of structures of consciousness—including the so-called *unconscious*—which might be useful for a wide variety of purposes. *I am not engaged in studying the elaborations of the structures of consciousness from the perspectives of a genetic epistemology or an ontogeny of the imagination.* Indeed, I have reservations about the reliability of such efforts when they are not precisely linked to the verified histories of particular peoples in specified places undergoing the urgencies, passions, and predicaments of their existences.

Let me clarify here my sense of the sociological. I begin with what might be called a "predicamental vision," that is, the sense that peoples everywhere, whatever the form and structure of their arrangements, inexorably find themselves in the midst of predicaments which call forth urgent responses in the way of passions, actions, efforts to achieve mastery and control through multiple forms of affiliation, organization, imposition and imputation. I see no way of doing sociology without clearly relating to the structures of *existence* of peoples, the structures of their *experience,* and the structures of their *expressions.* Unlike those who conceive expressions to be simple precipitates of interest or epiphenomenal outcomes of their economic arrangements, I see expressions as efforts to give structure to the character of their experiences and histories, and to define the central coordinates which map all the ecologies of their natural, social, and cultural environments and experiences.

No matter how "primitive" or "complex" the nature of their societies, peoples cannot live together continuously, share common life ways, and speak common languages without creating expressive structures of meaning and symbolic designs—creating cultural ontologies and mappings of their historically situated life-worlds—which constitute the immanent textures and media of their coexistence over time. These cultural productions and ontologies are not peculiar to complex civilizations, although they are differentially constituted in different societies and historical milieus. Although no particular cultural ontology is *a priori* necessary, there are no societies, whatever their levels of complexity, without cultural ontologies; for no people can coexist continuously over time and constitute a "society" without creating persisting communities of cultural identification, meaning, and reference, i.e., structurally persisting phenomenologies. Hence the stress in my work upon the intertemporal character of the cultural productions or the structures of expression. . . .

Where I differ with many others is in my primary commitment to make sense of histories. And I deliberately use the plural because it is premature to talk of a single history of Man, whether as species, as individual or as creator. I do not see how a comparative historical and systematic sociology can currently take generic-species-essence Man to be the determinate unit whose fortunes and directionalities are to be analyzed. . . . It is thus a central matter for me to understand how actual peoples in actual places, in their predicatments and with diverse interests come to be linked in structures with relative degrees of commitment to the regulative and symbolic designs which issue from their activity.

In an essay of a decade ago (chapter 2 above), I urged that there was a distinct advantage to seeing the various questions in terms of the contents, profiles and interconnections of six sets of "cues": the *percipienda, sentienda, agenda, credenda, miranda, emulanda.* I have supposed that the *percipienda* constitute the most comprehensive and determinate frame of reference. . . . *By percipienda* I mean all the ways in which elements in the cosmos, including so-called social, political, historical environments are to be perceived and defined. These cultural ontologies include reference to all of the categories set down for systematic study by Durkheim, Mauss and their followers and are incorporated in the *percipienda.* But a great deal remains to be done in studying what links can be found between the *percipienda* in an enlarged sociocultural sense and the perceptual types which are discovered through inspection of artifacts and works of art. As I see it, it is not helpful to treat the latter structures as being more comprehensive and more determinative than the categorial structures.

Some have charged me with grounding consciousness too narrowly in issues of moral and intellectual responsibility. Some explanation seems warranted here. One of my primary interests has been to ascertain how groups variously located and disposed seek through their forensic and other disputes to make sense to themselves and others. Within this framework I have a second reason for putting stress on responsibility. I suspect that across the world great numbers of peoples have been committed to the view that they had themselves performed or acquiesced in some sort of crime or wrong for which reparation needed to be made. Wherever we discover the order I have called *praxisms* as the central feature of the society, wherever, that is, great numbers are joined together in the performances of acts in some sense ritually prescribed, we have sacro-magical structures at work. To explain what is revealed to us by the complex odyssey and turbulent transformation of a people struggling to master and give meaning to themselves and to their inner and outer worlds I have so far placed more stress on the developments, institutions, and frameworks associated with legal, political, social, religious, economic spheres than with the distinctively artistic orders. The reason is a simple one. I needed to understand what explains how people conduct and comport themselves; what explains the suppositions they apply in defining their patterns of interaction; what explains the ways in which they affiliate and organize; the way they map their worlds and constitute their cosmos. When this wider view is taken then the structures that Kavolis (1974) illustrates from Gebser (1966) seem difficult to establish as the ground for central sociocultural processes and patterns.

In the same spirit, I do not agree with the claim that I conceive the rationalized structures of consciousness to represent the ultimate consummation of human effort and to be relevant in the same measure in each instant of time. The warrant for my view can be readily verified in a number of papers I have written which explicitly deal with affectivities and communities (Nelson 1969a, 1972e, 1973g and chapter 12 above). I do not hold that a narrow rationalism or scientism constitutes the consummation for which we all devoutly pray. What I do believe, however, is that a mindless romanticism which would gainsay the most critical structures reared in complex societies, those resting upon the widest universalities of reference, would represent regressions. The issue is not one between classicism and romanticism, whether in the arts or in politics; the issue is one between some kind of consensual order including reciprocal answerabilities and mere a-nomism. Would the matter become clearer if I were to assert that I have long held our against the assumption that the muddy parables of "scientism" offer wider access to the "realities" of our existences and experience than the historic masterworks of art, religion and myth?

As to Durkheim and Foucault, I hasten to explain that when I speak of the structures of consciousness embodied in the cultural logics and epistemologies, I refer to the way in which aspects and elements of the world are mapped by those who have been charged or charge themselves with the responsibility to define and mediate the central structures of meaning. I do not make the mistake of supposing that these structures are always fully accessible to all members of the society, or that, for that matter, they are all available to all members of even the most sophisticated educated elites. I find, however, that the structures that constitute the prime civilizational values are the ones to which there is usually considerable commitment on the part of social actors. The consensual understandings of society are by no means fully understood by all. They do, however, represent the values which are held in esteem because they have appeared to prove themselves in the conduct of the common life. In this connection, I would urge readers to have a look at Rudolph von Jhering's interesting work, the *Struggle for Law* (1915 [1879]). Throughout my paper (chapter 5 above) I have indicated that the various structures exist in varying preponderances and saliences in groups and individuals across time. I do not wish to use the term strata to describe the form of this compresence or co-variance. So far, I have never encountered a social group which does not appear to embody to some degree the various structures of consciousness I have described. In this connection, I never claimed in this essay to have offered a full account of possible types or their co-variance. It would, however, be an error to suppose that my paper proposes a schema of universally linear developmental sequences of mutually exclusive stages of consciousness ot monochromatic horizons of cultural awareness.

Anyone who has looked at the papers I have recently done on Weber's *Protestant Ethic* (Nelson 1973a, 1974c) must know that I make, as Weber did, many distinctions in the family of terms connected with rationalism, rationales, rationalities, and rationalizations. Rationalism of certain sorts can be found in even primitive societies. Indeed, Weber never supposed that the mark of the modern world was rationalism pure and simple, or that calculative or prudential rationalism was the spur to the transformative breakthroughs of the central structures of polity, consciousness, law, economic organization, normative regulations of conduct, pathways to perfection, systems of "cure of souls," collective identifications, and so on. . . .

Finally, I may say that only rarely have I referred in my own work to Mead because of an unhappiness over the fact that he appears to have been excessively yoked to sociological perspectives unduly remote from his forgotten wider interests in communities, communication, and communions. Indeed, a great deal of the work currently done by sociologists in a microsociological vein is deliberately, rather than accidentally, intent upon denying the validity of the approach through institutions, societal and civilizational patterns, to say nothing of civilizational complexes and intercivilizational encounters.

As for the appearance of "idealism" in my essay, let me suggest that sustained effort by anyone to establish the wider frames we avow will involve him, too, in the semblance of "idealism." Everyone risks the appearance of "idealism" who wishes to do justice to the immense authority and influence of inherited cultural traditions, symbolic designs and technologies, directive and regulative systems, and so on, in the systematic study of the comparative and historical sociology of civilizational complexes and intercivilizational encounters.

It should be no surprise that I have had to forge my way through the welter of contemporary sociology in order to be able to recover points of departure for widened efforts. Essays that I have written on Parsons and Merton make clear my sense of great indebtedness to their work but also my sense that not everything they have done on the shoulders of Weber and Durkheim appears to represent advance in the spirit of comparative historical, differential sociology. In any case, my own effort is one of relating to our new times, marked as they are by synchronicities and dyschronicities of over at least a half dozen revolutions whose outcomes are crossing one another in the trajectories of our own existences, and whose fallout for the future remains as yet unknown to us. A fact which needs to be said in this connection is that I do not so readily, as do other colleagues, accept the assignment to economics and political science and anthropology all responsibility for dealing with areas outside the boundaries of the United States. I share the convictions of Durkheim, Mauss, and Weber that a sociology of—and for—our time must be comparative and comprehensive, even global, to be true to the challenges all of us, including sociologists, are now called upon to confront. In this sense I consider that all sociologists have to acquire greater mastery than many now have in the history of at least two civilizations: non-Western as well as Western. In another connection I have contended that without such awareness of the histories of variable mixes of social and cultural elements we cannot make sense out of our own day. I am sincerely convinced that a re-reading by critics of some of the essays I have written will show that I do not make the mistake of equating civilization and rationalism or rationalization.

II

The foregoing answers to my friendly critics would fail to reach the mark if I did not make certain points unambiguously clear. I will be forgiven if in this effort I restate selected passages of the Symposium essay ("Civilizational Complexes and Intercivilizational Encounters"), and draw upon other papers, one jointly written by Vytautas Kavolis and myself and another ["Scope and Method in the Comparative Sociological Study of Civilizations"] to be presented at the VIIIth World Congress of the International Sociological Association [Toronto, August 1974].

A

I begin with a re-statement of the "Setting and Perspectives" of the Symposium essay:

My efforts grow out of a number of deep convictions and commitments: Sociology is at a turning point in respect to the horizons it is obliged to confront and the perspectives and methods it is obliged to adopt in order to make sense of the perplexing and tumultuous sociocultural processes of our time. We dare no longer suppose that these processes can effectively be gotten at by confining ourselves to settings which are local, parochial, or instantial in terms of the level of interaction (for example, small groups); the social-system setting (the firm*), the geographic setting (the local community); the temporal horizon (the current moment, the present).

We are obliged to see that many of the most important phenomena of processes and productions of our time are occurring everywhere across the world and they are occurring most intensely in those levels and in those settings which have been least systematically studied by sociologists or anthropologists. I refer to the *societal level,* the *civilizational level* and the *intercivilizational settings* and *encounters.* In truth, these last-mentioned settings have always been of enormous importance and it would be a mistake to suggest that there have been eras when great complex polities have escaped being in the throes of processes at these levels.

The comments that have been made by my friendly critics in respect to the exact boundaries of civilizations and the relations between "cultures," "civilizations," "nations," "states" lead me to observe that I took pains in my essay to indicate that "I reject the widespread habit of using the word *civilization* to refer to the sociocultural processes of selected great nations." Instead, as I continue:

I prefer to reserve the term "civilizational complex" for a segment of the paradigmatic cultural patterns in the *sphere of the expressive* and *instrumental productions of societies* or *societal complexes.* In speaking of complex societies in this way I refer to *political societies,* which contain numerous internal differentiations, including nations, classes, institutions, and varieties of cultural experience.

Also, by way of clarification I admitted to "sharing a preference" for some frames of reference and styles of analysis initiated by Weber, Durkheim and Mauss.

By the *civilizations* of peoples I wish to refer to the governing cultural heritages that constitute the accepted milieus of 2 + n societies, territories, areas which generally enjoy or have enjoyed a certain proximity. These strongly-based acceptances will normally be discovered to constitute configurations of the following elements: identities of language, the highest level of the technology of the group, which I would call the "prime material facilitations and skills" of the group; the central patterns of reciprocities including juridical rules; the fundamental canons governing the decision-matrices in

*I have altered this word to make my point more clear. (BN)

the spheres of opinion and act . . . the taken-for-granted structures of consciousness, comprising cultural world-views, logics, images of experience, self, time, the beginning and the end, the extraterrestrial powers. I have previously described some of the last-mentioned schemas under the headings of dramatic designs, directive systems, systems of spiritual direction, symbolic economies. . . .

The comments of several symposiasts lead me to point out that I was at great pains to make clear the fact that there are intra-civilizational as well as inter-civilizational conflicts throughout the course of civilizational histories. Thus, I wrote:

To speak of a single civilizational atmosphere in the history of any large societal environment is to miss the fact that conflict over civilizational ascendencies is endemic in large societies.

In the same vein I continued:

Civilization can, indeed, be based on diverse geometries. However, insofar as civilizations come to serve as cultural prerequisites for the relatively enduring organization of different sorts, they will generally be found to involve patterning of related elements. These patternings are found to comprise arrangements of coordinates defining cultural ontologies, epistemologies, and logics; directive systems, dramatic designs, and sociopolitical frameworks; and technologies of different sorts—symbolic as well as material.

Also by way of responding to queries by my commentators concerning the contemporary civil wars in the cultures of consciousness and conscience, I would note that I remarked:

From a current perspective, it seems that the half-century since 1920 has been one of especially intense passages towards collective conscience and collective consciousness. Those who call us to the collective reappropriation of redeeming symbols through the collective unconscious may indeed prove to be more popular prophets than the defenders of rationality. The tides are now turning strongly against the very idea of *rationale*, and, indeed, every image of *nomos* and of *nomism*. On every side we are seeing efforts to achieve new forms of corporate embodiment, new fusions of identity, and new patterns of non-contractual consensus.

The shifts we have witnessed and are now experiencing in the spheres of collective sensibilities and in the structures, *rationales* and streams of conscience and consciousness are no less critical for the future of culture and civilization than the changes in the economic and political spheres which have followed in the wake of heightened industrialization.

B

Several times already in this essay I have referred to a paper, of which Vytautas Kavolis is co-author, which deals with "Comparative and Civilizational Perspectives in The Social Sciences and Humanities. . ." (Nelson and Kavolis 1973e). In this paper it is contended that the main distinctions in the so-called comparative approaches refer to eight "Horizon-Approaches" and divide into two main groups: A) those directed to the study of "societies" and "cultures"; B) those directed more properly to the study of "civilizations."

Speaking first of *societies* and *cultures*, we distinguished:

(1) the comparative study of total societies as social systems;
(2) the comparative study of *analogous institutions of total societies in the social-system perspective*;
(3) the study of *cultural patterns* or "culture-personality" profiles as *wholes;*
(4) the transcultural study of specific cultural processes and effects.
The paper turns then to the four horizons which related to "civilizations" and described them as follows:
(5) the *directly* comparative study of so-called civilizations as *wholes;*
(6) the *"comparative" survey* of all civilizations for the purpose of establishing *irreversible directionalities* of development;
(7) the comparative study of *institutions* and *"symbolic designs"* against the backgrounds of determinate civilizational settings;
(8) the comparative study of the histories, sociologies, psychologies of civilizational complexes and processes as these are perceived and work in the settings of intercivilizational relations and encounters.

On careful review of the literature it is contended that Horizon-Approaches 7 and 8 are now proving most valuable in the design of current research. Celebrated examples of *Horizon-Approaches* 5 and 6 while often distinguished by extremely suggestive insights are regularly marred by serious defect in method and result. They (5 and 6) offer us almost no opportunity to do rigorous empirical work of the sort we now require nor do they permit us to move toward a more systematic understanding and nomenclature. *Horizon-Approach* 6 too often takes on the character of a theodicy. From our point of view, it does not matter greatly that nowadays the theodicy is often expressed in the language of cosmic evolution or of general systems theory. In this spirit the paper asserts:

It is civilizational *complexes*—alternative major patterns of diversity at each processual level—that constitute the main focus of concern for civilizational theory as distinguished from traditional evolutionary theory, which is concerned with general directionalities of development. We speak of these last two *Horizon-Approaches* (7,8) together because they share similar methods and purposes. It must be admitted, however, that the accents of these approaches will differ from time to time.

Horizon-Approach 7 is marked by the following features:

(1) comparison of institutional (political, family, etc.) or cultural (science, religion, art, etc.) spheres or of problem areas (modernization, rationalization, militarization, universalization, etc.) *of the same general type* from several largest-conceivable units of civilizational pattern;

(2) particular situations, institutions, or problem areas are studied with explicit attention to and respect for the uniqueness and specificities of distinguishable phenomena, yet, simultaneously, as tests of generalized theoretical formulations.

Horizon-Approach 7 is the one most favored by contemporary master historians, historical sociologists, anthropologists and others who have wished to have firm grounding in a determinate civilizational setting while they are carrying on comparative analysis. Until now the most impressive studies in this mode have tended to be holistic and configurational rather than oriented to specific comparisons of institutional or symbolic frameworks. Major forerunners in this field have been Maine, Weber, and Mauss. In our own time there are the recent studies by A. Bozeman on comparative politics and law, M. Lapidus on Moslem cities, monographs by Clifford Geertz, Robert Bellah, Melford Spiro and others on Oriental religion and society. Papers in this genre have been appearing for some years in the pages of *Comparative Studies in Society and History*, edited by Sylvia Thrupp of the University of Michigan.

Horizon-Approach 8 offers us particularly valuable horizons for relating to actual cases of *intercivilizational relations* and encounters. Exceptionally powerful results have developed in the course of exploring questions arising in the study of civilizational encounters in the era of the Crusades, Hellenistic world, the 16th and 17th centuries, and our own times of abrasive civilizational conjunctions. But there are a host of special questions under this rubric which can only be gotten at through asking questions which challenge us at the very roots, questions such as those put by Max Weber in his "Author's Introduction" (1920) and by Needham in his *Science and Civilization in China* and *The Grand Titration*.

The essay goes forward to describe the strategies of *civilization-analytic perspective* as a distinctive approach to the comparative study of civilizational patterns and encounters:

The operational characteristics of a distinctive civilizational (or *civilization-analytic*) perspective seem to be as follows:

(1) All socio-humanistic patterns and processes are studied in historical-sociological psychological depth.

(2) Equal emphasis is given in principle to: (a) *social structures* in all institutional spheres; (b) *structures of consciousness* at every level of differentiation; (c) *symbolic designs* on various levels of generality of formulation and depth of influence (both artificial and natural languages); (d) *changes* in the manner of such interaction and shifts in the balances and ascendancies over long periods of time.

(3) The relative weight of social structures, structures of consciousness and symbolic designs is presumed to be a part of the problem to be determined by empirical research.

(4) Sustained and detailed attention is paid to the variables and mechanisms of collective psychology (or psychohistory); and to the psychobiographies of particular individuals in whom factors of collective (or even individual) psychology clearly became determinants of social structures or symbolic designs.

(5) Smaller-scale situations are related through appropriate intermediate links to the largest conceivable empirical units of sociocultural analysis. These are: so-called *evolutionary "phase state" levels,* defined by the direction, or main alternative directions, of socio-technological development, and *civilizational designs,* defined as relatively enduring symbolic configurations, shared either by a number of independent societies or by a variety of ethnic or national groups encompassed within an empire, and responsible for the major sociocultural differences existing at each processual level.

Horizon-Approaches 7 and 8 describe ways of studying all sociocultural processes and issues within a distinctive *civilization-analytic perspective* which proves preferable on general intellectual grounds (comprehensiveness, depth, compactness) to existing horizons and approaches in history, anthropology, sociology. The comparative depth-historical study of process and pattern in *civilizational perspective* offers great advantages over current varieties of structuralism, structural functionalism, schematic Marxism, phenomenology, inductivistic empiricism, and so on.

III

As I approach the end of the present essay, I am persuaded that the critical issues calling for clarification in this field have to do with the specification of the frames of reference, tacit assumptions and premises which lie below the level of rival and alternative perspectives. The only hope of making headway among social scientists and humanists committed to different viewpoints is to establish strong proof that they will not be able to do justice to the critical significance today and, indeed, throughout recorded histories, of civilizational diversities, encounters and conflicts—intra-civilizational as well as inter-civilizational—when they attempt to understand historically-embedded structures of social existence, sociocultural experience, and sociocultural expression. Now, more than ever, there is an urgent need to perceive the historical underpinnings of the variable and varying mixes of civilizational elements in comparative, *historical differential,* sociological perspectives. Everything about our times requires us to have a planetary sense of civilizational patterns and conflicts of civilizational complexes. What, then, stands in the way of the wider acceptance of such changes in perspective?

My answer at this juncture would be that adoption of widened frames is barred at the start by the overwhelming and often taken-for granted commitment of great numbers to what I am compelled to call "*uniformitarianism.*"[2] The grounds for this inclination are by no means all derived from the justified claims of science. Uniformitarian systems are inexorably more prescriptive than descriptive. They inexorably involve the nullification or the post-hoc construction of the past as well as the homogenization of the present and the uniformitarian stereotypification of the possible future. We cannot stop now to explain why it appears to be the case that uniformitarianism seems to have the character these days of a *Weltgeist,* but it is quite clearly the case that all of our current patterns of thought and sensibility seem to be locked into uniformitarian schematisms of one or another sort. From my point of view it makes little difference whether the uniformitarianisms are put forward in the name of culture, civilization, existentialism, phenomenology, psychoanalysis, Marxism, structuralism, general-systems theory or whatever. Even mixes of these orientations regularly take on a uniformitarian cast.

All uniformitarianisms tend to rule out as irrelevant the actual histories of actual peoples in actual places over time and space. In erasing the past, uniformitarianism aborts the capacities of groups, individuals, and peoples to relate themselves to meaningful traditions, compels them to wander without moorings by vitiating the contact with those memories and symbols which have become deeply submerged in the structures of the histories in which their collectivities and plural communities of reference are imbedded. Wherever uniformitarianism triumphs, there occurs the robotization of vast numbers who are thereby rendered ready accessories to thrusts to anomisms, primitivisms, scientisms of one or another sort, to millennialist politics and so on.

Born in desperation out of a need to schematize histories, actualities and diversities, uniformitarianism readily lends itself to systems of control, whether of consciences or actions or institutions or cultures. This proves to be the case whatever the actual motives of the authors of uniformitarian views. The insistence upon actual histories, processual patterns, diverse futures, is the only defense against the totalitarian advance of uniformitarianisms.

As I now see the matter, the critical decisions which face sociologists are not between Marxism and Hegelianism, communism and capitalism, Islam and Israel, Christianity and Buddhism, China or Russia or whatever. The decision which has the most momentous applications for all of us is that between simple uniformitarianism, whether in the guise of advanced science, advanced anti-science, advanced rationalism or advanced mysticism, *and* alternate perspectives that cling to histories, that demand truth be told about diversities and possibilities, that allow for innovations grounded in new mixings of experience and histories. What, one might ask, are the possibilities for innovation if histories are erased, presents are homogenized and futures are aborted, by the schematization of histories? All uniformitarianisms prove in the last analysis to work in the same way. All readily lend themselves to totalizations, to systems engineering, system-designed methods of surveillance, totalistic control and domination. All allow themselves to be developed as a base for bio-medical and genetic engineering.

It is perhaps now clear that my effort will not be fully understood if it is only perceived as a desire to achieve an answer to a particular historical or sociological question such as the resolution of what I have called Needham's "Challenge" or the sorting out of a comparable challenge which has been powerfully set forth in recent papers on the alleged "decadence" of Islamic science. If I study comparative histories and sociologies as intensively as I do, it is because I deliberately wish to strengthen the case *for* the *Civilization-Analytic Perspective* and to strengthen the case *against* uniformitarianism. The struggle against uniformitarianism is the struggle for actual histories, for actual presents and for such futures as we will have and can make in the welter of our multiple commitments, constraints, and actions. In struggling to this end, we put ourselves on guard against uniformitarian schematisms of every shape.

A *civilization-analytic* approach now seems to be the only one which will keep us from treating the specious present as the universal rule or the paradigm for all future culture, organization, and order. It is, therefore, my hope that a goodly number of sociologists will come to perceive that they cannot escape triviality and error if they do not carefully inspect whatever units of investigation they choose to make their own—*in civilizational perspective.* The most important blunder into which we are thrust by narrow perspectives is that we fail to recognize the variabilities in the mixes of economic, political, social, cultural elements in the different civilizational settings of the world. Who will today deny that the disregard of a *civilization-analytic* perspective has trivialized the study of ethnic communities and intergroup relations now so critical everywhere in the world?

As we prepare to celebrate the 200th anniversary of the Declaration of Independence does it not behoove us to recognize that our own so-called "first new nation" is unintelligible without being studied against the background of the civilizational structures with which it has been in contact, the civilizational commitments of those who are its citizens and denizens, the relations it has been having with non-Western civilizations, the societies and civilizations of the Far

East and Near East, of Eastern Europe, of Africa, of Latin America and so on?

I end on a note of hope: the appearance of this Symposium in this Review is notable proof that uniformitarianism is not everywhere in power. Time remains to reverse the drift toward uniformitarianism in favor of the view so powerfully represented by Maine, Durkheim, Mauss, Weber and other pioneers committed to the comparative, *historical differential,* sociology of civilizational patterns, complexes and encounters.

NOTES

1. The present paper, as announced in the opening paragraph, is a response to the issues raised by various contributors to a symposium published in 1974 which discussed and criticized Nelson's paper, "Civilizational Complexes and Intercivilizational Encounters" (chapter 5 of this volume). The participants included Burkart Holzner, Edmund Leites, Vytautas Kavolis, Donald Nielsen, Edward Tiryakian and the present editor. The titles of their papers are listed in the references. Professor Nelson apparently liked the results of his response which clearly attends to many larger issues which have since been raised by other writers, including Jürgen Habermas and Niklas Luhmann. I have therefore removed nearly all references to particular participants in the original Symposium under the assumption that the general issues are of greater interest in the present context. Secondly, in the light of Nelson's conscious formulation—in the last year of his life—of what he called "The *New* Science of Civilizational Analysis," reflected in both a six week seminar given at the University of Minnesota in the Spring of 1977 and the opening chapter of this volume, the present statement is his last sustained discussion of these issues regarding the new mode of civilizational analysis. Readers wishing to have access to Nelson's complete remarks and the companion papers may consult volume 35 #2 of *Sociological Analysis.*—Ed.
2. The phrase is adapted for my purposes from A. O. Lovejoy, *The Great Chain of Being* (1960 [1936]).

Epilogue: Priests, Prophets, Machines, Futures: 1202, 1848, 1984, 2001

I. FROM JOACHIM OF FLORA TO THE "PROTESTANT ETHIC"

One could, if one wished, start the first installment of the story I have chosen to tell here now with the year 1202. Those who have strong memories will recollect that year for at least two reasons. That was the year in which there died the famed Abbot Joachim of Flora, a visionary exegete of Scripture, who was cited by Spengler as the spiritual source of the new image of time which came to prevail in the era of *Faustian* man and culture. In his own day, Joachim's commentaries echoed loudly in a new sense of Apocalypse and realizable eschatology. It was also the year Europe was to become acquainted for the first time with the Hindu-Arabic numeral system including the all-important zero. The source was the *Liber abaci* of Leonardo Fibonacci of Pisa. That event, claimed Werner Sombart, might be said to mark the birthday of modern capitalism. Given the zero, it was possible to innovate in spheres of major importance in the worlds of practical life and science. The zero came to play a large part in the rationalization of the economy of the household, the rationalization of the activity and exchanges of firms of every sort, the rationalization of the administrative and public statistics.

I do not claim that the year 1202 or any other date can be called the Archimedean point of world history; I am simply suggesting that many notable future developments of critical relevance to the present essay need to be traced back to the historic crossings—I include fusions and conflicts alike—of the cultural themes associated with the works of the Calabrian Abbot and the Pisan enterpriser.

Reprinted from *Religion and the Humanizing of Man*, edited by J. M. Robinson (Waterloo, Canada: The Council on the Study of Religion, 1973, revised 2nd ed.), pp. 37–57. Reprinted with permission.

Joachim sought to give expression to his conviction that the world was on the edge of a new era, a time of an everlasting Gospel, when men should see each other face to face and not as through a glass, darkly. He was convinced that we had already passed the age of the Father, the age of the Old Testament, when all men lived in a state of bondage to their Fathers, and when institutions had a repressive and authoritarian character. He was further convinced that the Age of the Son had been and was being represented by the monastic communities which dwelt in mutual regard, concord, and love, and where they were guided by the New Testament, which was the Book of the Age of the Son.

The Third Age was already upon us—the third age, the third time, the third kingdom—and in this third age, the age of the Holy Ghost, there was to be no longer any mine or thine, no superior or inferior; there was, indeed, to be the time of total undividedness and absolute freedom.

Joachim of Flora merits the accolades which Spengler and others since have heaped upon him. Joachim was preeminently a Trinitarian thinker who clearly had in mind to overcome the divisions among peoples, classes, religions, nations, and to work toward a higher unity. He is a medieval precursor of a thinker like Teilhard de Chardin, one who wishes to speed the advent of a perfect Unity of Mankind, mind, and spirit—a kind of noosphere.

Despite—perhaps because of—the questions raised about his Trinitarian views at the Fourth Lateran Council of 1215, his wider message quickly carried to all who were in favor of the apostolic renewal of the church. His influence was especially strong—indeed, he took on the appearance of a sort of John the Baptist—among all those in the Franciscan order who preached the doctrine of Apostolic poverty and sought to prepare for the New Time. Many of the foremost innovators among the Franciscans counted themselves adherents of the Spiritual Fraticelli, as they called themselves.

From the point of view of this essay, the most striking fact is that many noted leaders of the hunted Spiritual party combined the Joachite emphasis on Apostolic poverty with a philosophy of mystical enlightenment (illuminism) which at one and the same time celebrated the supreme worth of mathematics, the idea of an "experimental" science, and utopian hopes for a new technology.

It is no easy matter to say to what extent the thrust to the new technology may have been, indeed, rooted in a mystical enthusiasm for the workings of natural magic. Roger Bacon's writings certainly communicate this flavor. Profound religious impulses may well have played a much larger part than a bourgeois secular spirit in the thrusts toward science and technology in the High and Latter Middle Ages.

From this point of view, Roger Bacon, Leonardo da Vinci, Galileo, and Kepler have a great deal more in common than is generally realized.

The deeper subterranean urges toward the drive to know nature and expand human productive powers are eloquently evoked by the recent writings of Paolo Rossi of the University of Florence. The union of theology, science, and magic arts is asserted again and again in the Promethean image of Renaissance engineers and artists (Rossi 1970).

Although Henry Adams's nostalgic phrases clearly exaggerate the extent of unity of the various sectors of the life of Christian society, it is nonetheless the case that some of the most striking abbeys and cathedrals could not have been built without the cooperation of all sectors and segments of the population. *Mont-Saint-Michel and Chartres* offers us the most eloquent account yet rendered of the power of the Virgin in winning the love and hearts, the faith and trust of men to proceed to transcendent achievement. In that work he writes:

> The Queen Mother was as majestic as you like; she was absolute; she could be stern; she was not above being angry; but she was still a woman, who loved grace, beauty, ornament—her toilette, robes, jewels; who considered the arrangements of her palace with attention, and liked both light and colour; who kept a keen eye on her Court, and exacted prompt and willing obedience from king and archbishops as well as from beggars and drunken priests. She protected her friends and punished her enemies. She required space, beyond what was known in the Courts of kings, because she was liable at all times to have ten thousand people begging her for favours—mostly inconsistent with law—and deaf to refusal. She was extremely sensitive to neglect, to disagreeable impressions, to want of intelligence in her surroundings. She was the greatest artist, as she was the greatest philosopher and musician and theologist, that ever lived on earth, except her Son, Who, at Chartres, is still an Infant under her guardianship. Her taste was infallible; her sentence eternally final. . .
>
> The palaces of earthly queens were hovels compared with these palaces of the Queen of Heaven at Chartres, Paris, Laon, Noyon, Rheims, Amiens, Rouen, Bayeux, Coutances—a list that might be stretched into a volume. . .

This extraordinary unity of conception and design—this marriage of all the arts, including the practical arts, in the celebration of the faith in the Trinity and in the Virgin—was not to go without rupture. Again, Adams needs to be quoted:

> The architects of the twelfth and thirteenth centuries took the Church and the universe for truths, and tried to express them in a structure which would be final . . . and this is true of Saint Thomas's Church as it is of Amiens Cathedral. The method was the same for both, and the result was an art

marked by singular unity, which endured and served its purpose until man changed his attitude toward the universe. The trouble was not in the art itself which presented different aspects as man moved. Granted a Church, Saint Thomas's Church was the most expressive that man has made, and the great Gothic cathedrals were its most complete expression.

The split between other-worldliness and inner-worldliness did less to inhibit the development of technology either in the monasteries or in the world itself than many have supposed. Although the prevailing dualism did function so as to place those who had chosen the triple vow of poverty, chastity, and obedience in a grade much above that of anyone who lived within the world—the so-called "status of perfection"—medieval. Christianity was not world-denying or acosmic. As Max Weber was to perceive, the monks of the Middle Ages were in a decisive sense spiritual progenitors of the ethic of "inner-worldly asceticism," the spiritual foundation of the vocational ethic of modern Occidental rationalism.

The monastic emphasis on a methodical life strictly lived in discipline under a rule—a life measured by a standardized community time—is the nucleus of the so-called "Protestant ethic" which has played so fateful a part in the life of the modern world.

II. THE RELIGIOUS-SCIENTIFIC-TECHNOCULTURAL REVOLUTIONS OF THE SIXTEENTH AND SEVENTEENTH CENTURIES: NEW PERSPECTIVES

The developments between the fifteenth and eighteenth centuries need to be seen from a number of related but distinct perspectives. The linkages among these have to be spelled out in some detail in order to make clear the tangled character of the development of the forms of mastery of nature.

Great leaps forward were made in the sciences and technology. There also occurred exceptional changes in the structures of orientation, notably in the spheres of religion and polity; great changes in the organization of action and administration.

I will speak of these now in turn, beginning with the so-called religious-scientific-technocultural revolutions of the sixteenth and seventeenth centuries. Some of the perspectives which I shall suggest on these matters may seem novel: they grow out of specialized research I have done in recent years.

The Protestant Ethic as described by Weber and others cannot be identified as the main spur and goad to the emergence of the modern scientific revolution. The key facts about the relation of Protestantism to the new science and technology are more complex than many suppose: (1) Very few of the major theologians and leaders of the Protestant churches looked with favor upon the spread of the new science; (2) the main concentration of Protestantism was on the elimination of the medieval casuistical theology—moral theology even more than natural theology—of the Roman Catholic Church; in the same way the Protestant Reformers protested the inherited concepts of church, faith, grace, God's sovereignty, the sacraments, and so on. The net effect of their instruction was to give support to the breaking down of invidious dualisms between religion and world, and the promotion of a wider rationalization of conduct in the form of the spread of self-reliance and self-control. The weakening of an all-powerful hierarchy did indeed make it possible for there to appear social and professional circles which regarded the pursuit of practical knowledge with greater favor than did the Catholic Church.

The thrust toward experiment, while not originating in Protestant lands, was stronger in Protestant culture areas where it encountered less resistance. It therefore becomes necessary to see the distinctions in the particular blends of religion, science, philosophy, and technology which occur in different lands and different areas.

The critical moments in the passage to the scientific revolution of the early modern era are associated with the attacks by the pioneers—mostly Catholic in background—of early modern science and philosophy upon the inherited logics and decision-procedures of the traditional sciences and disciplines. There was a new impatience with every form of probabilism and exegetical casuistry of opinion, a new hunger for certainty and certitude, a new urge to achieve mathematical precision in the rendering of the *Book of Nature.* Fictionalist views of scientific theory lost ground to emphases on the truth of natural laws discovered in observation and controlled experiment.

The extent to which experiment guided this process has been greatly exaggerated. The principal thrust was toward the uncovering of formal mathematical expressions for patterned regularities observed in the numbered *Book of Nature.* In this respect, as we have hinted above, the most critical advances had their origin in Catholic culture areas among men trained in Catholic settings. It is in truth in these areas that the formal sciences, especially analytic geometry and calculus, make very great headway in the sixteenth and seventeenth centuries. Only a very few writers have correctly assessed the extent to which the new technology and science depended on advances in mathematical analysis—on pure rather than applied mathematics. John U. Nef is one of the few scholars to appreciate the importance of developments in mathematics for the later developments of large-scale industry and precision engineering.

III. TOWARD THE ACTUARIAL WORLD-VIEW AND REGIME: *MATHESIS UNIVERSALIS* AND POLITICAL ARITHMETIC

I have laid stress upon a distinction between two central structures of orientations and rationales which have had long histories in Western civilization. Following a maxim of Piaget based on Durkheim, I have called the schemas "the moralities of thought" and "the logics of action." Since the Middle Ages, both of these structures have regularly rested upon a single hinge, namely, the notion of *conscientia* which had the combined senses in Latin and other languages of "conscience" and "consciousness." These were the grounds on which many central institutions with the Church, the university, law, and polity rested.

From this perspective, the central fact about the Protestant Reformation was the breakthrough it fostered in the inherited images of action. The Reformation stress on conscience was a momentous challenge to the established organization of the systems of spiritual direction, the so-called Forum and Court of Conscience of the medieval period. New openings could now develop for reconstitutions of structures of self, action, association, polity, law, liberty, and other horizons and milieux of life, public and private alike. . . .

Already in the later Middle Ages, however, we note strong thrusts coming from many quarters to substitute an outlook which was in the end to shatter the medieval rationales of conscience and their institutionalized expressions in moral casuistry, the dialectics of opinion and the cure of souls in large parts of the world. These thrusts gathered ever accelerating momentum in the Renaissance and the Reformation, and entered into new orbits in the seventeenth century.

The new world-view may be given the name of political arithmetic. This new view postulates that the only proper norms for legislation and policy have to be derived from the calculated estimations of the probable outcomes of alternative programs in the sense of relative profits or losses to the State. The casuistry of justice between persons gives grounds to the public-political-*collectivistic* point of view. "Reasons of State" come to prevail over dictates of conscience. Findings concerning the balances of power and trade come to predominate in the establishment of rationalized policies assumed to be in the wider public interest and to maximize general utilities.

We are not yet in a position to offer an integrated comprehensive account of the developments in these spheres. Clearly, as once before, we have to note that the newer developments do not begin with Protestant theology nor are they expressions of a Protestant Ethic in action. The thrust toward the new political arithmetic—the new moral arithmetic—occurs strongly in Italy in the fifteenth and sixteenth centuries. The evidence would seem to be that images of the balance of power arise everywhere in response to the new political and religious situations. Geopolitical outlooks come to take precedence over the previous modes of organizing group responses to moral and religious predicaments.

A remarkable and wholly forgotten pre-Soviet Russian study by E. Spektorsky has clearly established that the sixteenth and seventeenth centuries need to be viewed as times in which every subject, every perspective, and every major thinker moved toward metric expression. The two ruling conceptions were *mathesis universalis* and panmetric. Every branch of learning was compelled to take on the form of geometry. The efforts of Descartes, Kepler, Galileo, Spinoza, Hobbes, Leibniz, and others in this field are most memorable.

We do not need to go so far as Cassirer in making Machiavelli the predecessor of Galileo, but there can be little doubt that in his way Machiavelli was eager to substitute a form of policy science for the prior conceptions of the rule of princes.

By the middle of the seventeenth century the passage to political and moral arithmetic—and indeed to individual moral arithmetic—was very far advanced. This was a decisive step on the road to the full rationalization of activity and technique in the management of group and individual life. Probabilities of perplexed and doubtful consciences give way to statistics which have the appearance of precision and which may be offered as mathematically certain demonstrations of the need for one or another policy.

In another connection I have shown how critical this movement was in the legitimation of a notion of an *average rate* of interest—a notion utterly central to the full institutionalization of capitalism. These passages toward political arithmetic and statistical averages are the highway to what I have called the *actuarial society* of the more recent period. The passage from the prescriptive and principial structures to the actuarial ones now began on their fateful way. Where they have led us is clear for all to see on every night's news reports on television.

IV. THE NEW INDUSTRIALISM AND THE MIX OF RESPONSES

The emergence of the so-called "Industrial Revolution" is more than the introduction of a new technology. It involves an immensely complex series of changes in the conduct and the human organization (and disorganization) of human affairs in every dimension of both group and individual life. The enclosure of the common lands, the breakdown of the inherited structures of villages, local communities, and localities were indispensable moments in the passage toward modern industry.

A great number of sources make it plain that the working groups were not as disposed to engage in self-regulation under factory managers as is often supposed. Edgar S. Furniss, Jr., has told the story of the pressures directed against the working groups through the stimulation of inflationary advances of prices over wages. The introduction of labor-saving machinery was a basic element here.

The critical moment in the emergence of a new bio-sphere and eco-environment must be connected with the emergence of new cities which became centers of urban blight in the very hour they became the homes of modern industry. A host of witnesses testify to the outrages committed against nature and man in these new towns and these new factories. Blake, Carlyle, Morris, Ruskin, and many others maintained unrelenting attack on the ravaging of the landscape as a result of the march of modern industry. We may soon expect to have new intensive studies of the extent to which, in the very earliest new towns, problems of littering and pollution came to be critical.

The earliest important attacks on technology and the evil effects of new industrialism came not so much from theologians, as from workingmen, cultural critics, philosophers, and poets. The wider discontents of the actual wreckers of machines, the so-called "Luddites," who had their counterparts in lands other than England, are only now beginning to be studied closely.

The major cultural critics carried on a continuous inquiry into the very important dangers which the new industrialism threatened to the spiritual values of the past and the hopes of the future. Most important among the dangers explored are: the spread of forms of domination and new forms of dependence; the mechanization of spirit and robotism; the permeation of society by crassness and commercialism; the turning of life for many into sequences of boring and repetitive acts; the subordination of ends to means, the dissipation of religious substance, and the cheapening of religious symbols.

If we wish to recover the spirit of these early critiques, we have to recall especially the writings of Blake, Coleridge, and Wordsworth in England; Herder and Lessing in Germany; the great romantics and social philosophers in France.

Raymond Williams's fine book, which remains our best source for the developments of the attitude of British literary men toward industrialism, correctly stresses that from the time of Coleridge until our own day, every aspect of the relations between culture and economic-industrial function continues to be very problematic. The fullest detail on the plight of industrial workers in the new factories and towns of the first half of the nineteenth century will be found in the careful Reports of the Factory Inspectors and in Frederick Engels's *Condition of the Working Classes in England in 1844*. New strains were to be sounded on the Continent by political economists and publicists such as Sismonde de Sismondi, Fourier, Marx.

In France, the response to the machine was also strong. France was not without its Luddites, but France saw a concerted effort to discover a positive meaning and future worth in the development of scientific knowledge—it is not an accident that both Saint-Simon and Comte did seek to create new religions more congruent with the industrial order than had been the religions of the past. In France, a so-called positivistic position developed after the French Revolution which was, in fact, stronger than any which emerged in Germany.

"The New Christianity" sought to proclaim a kind of credo and even a pantheon for the new scientific-industrial technocratic faith. A similar effort was made by Comte, who claimed a place for a new science, a science which he christened *Sociology*, to which he assigned the responsibility for discovering scientific laws analogous to Newton's laws. The new science was to be the creed of a new church and a new Religion of Humanity.

For the Germans the most critical fact from the start was the risk which a new mechanism gave to the destruction of the hoped-for Faustian universality of man. It is for this reason that the animus against mechanism was especially strong. The antithesis between mechanism and spirit is everywhere to be found in German thought and, of course, continues in many guises in the writings of Hegel, Marx, Simmel, Weber, and others.

V. THE RATIONALIZATION REVOLUTION

The story of the rationalization movement has only so far been told in bits and snatches. In truth, engineers close to industrial production had been aware of the fact that the simple installation of machinery was no guarantee that there would be efficient use of the new Prometheus. It was not, apparently, until the consolidation of the ideas of Frederick Winslow Taylor and Frank Gilbreth in the United States of America that a way was seen toward the introduction of full rationalization. It was only as this occurs that we may speak of the rationalization movement.

Taylor was the poet and the prophet of the new era and it is, therefore, not strange that the phenomenon should be known as

Taylorism or Taylorization in many parts of the world. Even Lenin spoke of the need to introduce Taylorism into the new Soviet industry not long after the Revolution. In Germany, the ideas of Taylor came into already existing notions of coordination that had been elaborated in the course of the innovation of high science, high technology, the universities, and so forth. Oddly, Walther Rathenau was one of the most powerful opponents of the ongoing mechanization of spirit. He was a prophet of the total coordination of all resources. How ambiguous was the response to Taylorism may quickly be gathered by reading the profile of him which appears in *USA* by John Dos Passos.

Our attitudes to the perils and promises of technology and science in our own day have become totally transformed as a result of the awareness that everything now operates on a completely new scale. Nowhere can we be said to possess insulated chambers where new marvels or new horrors may be allowed to take place. The world has become indivisible as it never was before and all events have achieved the character of instantaneity.

We are not yet able to say exactly when the pace began to quicken in such a way as to threaten the present outcome. My own sense is that the future could already be perceived in the extraordinary spirits and integrations of technologies in the first World War. The American mobilization was probably the most extraordinary effort which industrial society had made up to that time. The German mobilization was an extraordinary premonition of it.

VI. THE VIRGIN AND THE DYNAMO

The twentieth century is now in the throes of a total revolt against the spirit and workings of technocultural civilization and the so-called myth of "objective consciousness" on which it is allegedly based.

The first critical expressions of this new mood occur soon after the emergence in Germany and elsewhere of the accelerated drive toward total rationalization.

Already in the first decade of the twentieth century, the new sensibility finds powerful utterance in outstanding writers of many lands and points of departure, including Henry Adams, Max Weber, and Eugene Zamiatin, who recognized the full meaning of the total coordination of all resources in accordance with the principles of instrumental rationality.

By the time Henry Adams had come to write his extraordinary autobiography, *The Education of Henry Adams,* the world seemed many years older than it did while he was writing *Mont-Saint-Michel.* The power of the dynamo now seemed to have attained full expression. In

The Education of Henry Adams, he tells us that the new world began with the coming of the railroads and the violation of nature's quiet and privacy. Indeed, it was Hawthorne's statement in one of his Notebooks which led Henry Adams to date the new world from the year 1844.

However, it was only the visit of Adams to London in 1900 in which he saw the great dynamos in action which led him to write on the power of the dynamo as being the modern form of the power of the Virgin. He writes:

> . . . Satisfied that the sequence of men led to nothing and that the sequence of their society could lead no further, while the mere sequence of time was artificial, and the sequence of thought was chaos, he turned at last to the sequence of force; and thus it happened that, after ten years' pursuit, he found himself lying in the Gallery of Machines at the Great Exposition of 1900, his historical neck broken by the sudden irruption of forces totally new . . .
>
> The historian was thus reduced to his last resources. Clearly if he was bound to reduce all these forces to a common value, this common value could have no measure but that of their attraction on his own mind. He must treat them as they had been felt; as convertible, reversible, interchangeable attractions on thought. He made up his mind to venture it; he would risk translating rays into faith. Such a reversible process would vastly amuse a chemist, but the chemist could not deny that he, or some of his fellow physicists, could feel the force of both. When Adams was a boy in Boston, the best chemist in the place had probably never heard of Venus except by way of scandal, or of the Virgin except as idolatry; neither had he heard of dynamos or automobiles or radium; yet his mind was ready to feel the force of all, though the rays were unborn and the women were dead.
>
> . . . Before this historical chasm, a mind like that of Adams felt itself helpless; he turned from the Virgin to the Dynamo as though he were a Branly coherer. On one side, at the Louvre and at Chartres, as he knew by the record of work actually done and still before his eyes, was the highest energy ever known to man, the creator of four-fifths of his noblest art, exercising vastly more attraction over the human mind than all the steam-engines and dynamos ever dreamed of; and yet this energy was unknown to the American mind . . .

Henry Adams is only one among many writers toward the close of the last century who gives expression to his deep conviction that the doom of Western civilization impended. He conceived that the multiplication in the fire power of guns might eventuate in the mutual destruction of all mankind. This prediction was to be reaffirmed again and again by others as man's power to destroy himself did indeed increase. There is no one who will, in this connection, forget the astonishing premonition of Freud as he closes his *Civilization and Its Discontents*.

There was yet another factor which led outstanding scientists and scholars to share premonitions of the running down and even the end of the world. Talcott Parsons, in one of his reflective insights, speaks of many thinkers as having been in the grip of a sort of "nightmare of entropy" at the end of the nineteenth century. Many of the ablest minds did, in truth, fear the early onset of irreversible tendencies to entropy.

VII. "INFINITE" UNIVERSES, "PROJECT TRINITY," "PROJECT APOLLO," AND GENESIS: SEVENTEENTH—TWENTIETH CENTURIES

How are the structures of our consciousness now affected by the new science, new technology, new technoculture, and the no less than half-dozen revolutions which have been making their way across the world during the present century? How are Priests and Prophets responding to the expanding proofs that we may be endangering the life of everyone across the world through the effects of the productions and distributions of warheads, wastes, and noxious pollutants on land, sea, and in the air? To questions of this kind comprehensive and secure answers are not now available. Neither social scientists nor humanists have gotten far enough in their studies of these matters to permit us to speak with any authority.

I can here only hazard selected comments which address several particular issues which have a continuing excitement for men of the Western world. The first addresses itself to the question of how man is responding to the continuing stress on such images as those of infinite universes and life on other planets.

We will recall that the late Arthur O. Lovejoy and Alexandre Koyré have helped us see that the vision of an infinite universe which contained an infinite number of worlds like our own was the key element in spurring the new sensibility of the seventeenth century. The discovery of hitherto unknown stellar events disclosed by the new telescope doubtless involved a struggle to reshape images of the system of the world, recalling to mind the drama involving Galileo, Cardinal Bellarmine, and the Holy Office; but it was Bruno's vision even more than Galileo's *Dialogue of the Two Chief Systems* which implied deep changes in the way Everyman experienced himself in the cosmos.

From the writings of Donne, Pascal, and others we are aware that there was a sense of anguish that all "coherence"—secular as well as sacred—might be irrevocably gone. Who can forget Pascal's great cry from the depths in his *Pensées* when he confesses to his anxiety over the fact that now man saw himself a speck of dust tossed on the oceans of two infinities, the infinitely vast unveiled by the new telescopes and the infinitely minute revealed by the new microscopes?

As far as our manifest behavior in our working states goes, issues of this sort rarely appear to trouble us today. Few are found who openly express dread over the meaning of the new technoculture for the drama of God's Grace and man's redemption. Few seem to concern themselves unduly about the uniqueness of man's place in the cosmos.

I would doubt that very many of the participants or onlookers in successive space efforts have felt that the case for the three-story universe was truly weakened by the achievements of Armstrong, Aldrin, and other astronauts. As far as manifest effects of the descent on the moon are concerned, the world, no less than the moon, is still "Tranquillity Base."

There is, however, another side to this matter.

The care taken to guarantee the homespun religiosity of the American astronauts surely implies a strong sensitivity to these issues. Indeed, if the testing of the atomic bomb at Almogordo was labelled *Project Trinity,* the execution of the space effort at Cape Kennedy could easily be described as *Project Genesis.* The apparent preference of Greek titles for the successive flights and missiles—Apollo, Gemini, and so forth—does not detract from the fact that the astronauts read to us directly from the *Book of Genesis.*

Indeed, the very questions which confront us now call us back in every way to the *Book of Genesis*—and the theme of possible new Beginnings. Actually we have been moving in the margins of Genesis and its images of creation, the world, man, and the dominion of nature from the start of our present essay.

VIII. TOMORROW'S DOOMSDAYS AND APOCALYPSES

The encounters of Faiths, Sciences, and Machines have reached new peaks of intensity in our own twentieth century. We are yet too close to the events to be able clearly to delineate cultural outcomes and to set up rank-orders of their significance. We must be content now to make the following observations.

As far as religious institutions and sensibilities are concerned, advanced science and technology have had mixed effects not easy to compress into linear equations. The three-story universe of traditional Christianity is playing an ever decreasing part in the explicit doctrines of churches and peoples, but we would err if we said that this older image of cosmos has entirely disappeared from the belief structures of everyday men. Few of the changes result from direct explications by astronomers or physicists of the structure of the new universe revealed by the new science.

As might have been expected, the ruling structures of men have proved mainly responsive to the critical horizons of their life-environments, their compelling experiences, and the immediacies—public and private alike—of their historical existences and times. The response of peoples to the new marvels wrought by the new science and technology has, however, been far from unitary or consistent. Those who have not been alienated by the overcommitment of the sciences to destructive purposes of warfare have come to regard our time as an Age of Miracle and Natural Magic. This is readily illustrated in the scenarios of science fiction which explore the mysteries of interplanetary-travel and the intercivilizational contacts with extraterrestrial populations and polities.

Not surprisingly, the activities associated with nuclear fission, the building of ultimate weapons, and the preparations for Doomsday, have evoked a sense of the cosmic, ever a seedbed of new religious images associated with the creation, origin, meaning, purpose, and destiny of the world and universe. So far, however, the awarenesses which were so frightening to Donne and Pascal—the feeling of man's significant lack of centrality in an infinite universe—have had little recurrence in our time. Not many ordinary men of the highly rationalized parts of the world seem to be distressed by images of the removal of the earth from its central position in the cosmos, or even the possibility that other peoples may now be inhabiting other planets. We cannot safely predict, however, what the future will bring in these areas.

The most critical encounters of Science, Faiths, and Machines have occurred in the societal and intercivilizational experience of men today. The sense that the opportunities for man are shrinking at a geometric rate, that his effective freedoms have been set at naught by the omnicompetence of military and political establishments, and the proof that the environments—the air, seas—of mankind are now undergoing pollution at alarming speeds, have contributed to the sense that a civilization believed to be reared on the myth of the indifferent consciousness must not be allowed to endure.

In this era of Doomsday and Apocalypse, new mystery religions, new superstitions, new magics, spread everywhere. Everywhere men seek relief from the ordeals of civilization through instant actualizations.

At the very moment that our Games of Life are taking on the shapes of Dances of Death, hopes multiply that bio-medical engineering and cryonic suspension will offer us a new immortality and even collective resurrection.

The more closely we study the evidence, the more it becomes apparent that, in the main, Faiths, Sciences, and Machines reinforced one another in the development of Western civilization. Theological rationalism and mysticism cooperating with a natural science and natural magic helped to produce the new cosmology and technology of the later Middle Ages and the early modern era.

The contemporary technocultural cosmos is the fruit of the fusion at white heats of science, technology, and organization, which occurred in the wake of successive restructurings of rationales of conscience and opinion, and the breakthroughs in the moralities and logics of thought and action since the end of the Middle Ages. Beyond Joachim of Flora and Leonardo of Pisa, the key figures in developing these algebras were: Roger Bacon and Francis Bacon; Luther and Calvin; Galileo and Descartes; Jeremy Bentham and Benjamin Franklin; Saint Simon and Marx; Frederick Winslow Taylor and Henry Ford; Lord Rutherford and Niels Bohr—and all their heirs and epigones in our modern and contemporary worlds.

Judaism and Christianity made their contribution to the building of Western civilization.

In their own ways, Prophets and Priests alike have contributed to its present impasses.

Can they not—will they not—contribute to the building of newer and stronger foundations for the newer and stronger world now waiting to be born?

APPENDIX

OUR ECOLOGIC CRISIS: ITS HISTORICAL ROOTS AND THE RECENT STOCKHOLM CONFERENCE

It would be impossible to conclude this paper without taking note of the now renowned essay by Lynn White, Jr., "The Historical Roots of our Ecologic Crisis" (*Science* [March 10, 1967], pp. 1203–1207). Much as I admire its learning and humaneness of spirit, I have to admit that I cannot accept its every turn of argument without some demurrer. I would readily agree with Professor White on his first point:

> . . . Viewed historically . . . modern technology is at least partly to be explained as an Occidental, voluntarist realization of the Christian Dogma of man's transcendence of, and rightful mastery over, nature. . .

His dramatic lines of explanation and assessment seem to me partly right, partly overstated:

> . . . as we now recognize, somewhat over a century ago science and technology—hitherto quite separate activities—joined to give mankind powers which, to judge by many of the ecologic effects, are out of control. If so, Christianity bears a huge burden of guilt. . .
> . . . the present increasing disruption of the global environment is the product of a dynamic technology and science which were originating in the Western medieval world against which Saint Francis was rebelling in so original a way. . .

I cannot agree, however, that the remedy he proposes speaks clearly to all the issues at stake:

> Since the roots of our trouble are so largely religious, the remedy must also be essentially religious, whether we call it that or not. We must rethink and refeel our nature and destiny. The profoundly religious, but heretical, sense of the primitive Franciscans for the spiritual autonomy of all parts of nature may point a direction. . .

It is hard to see how we can make great headway in realizing the ends we must pursue by our general acceptance of the principles of Zen Buddhism or adoption of the ways of the Beatniks, whom Professor White describes as the "basic revolutionaries of our time." It is of the utmost importance for the new world in the making that we get beyond "poetic" ways of conceiving the relations of East and West.

We must not be surprised that the contemporary crisis in respect to the so-called bio-sphere and eco-sphere is finding form in schematic versions of historical process which at times take on the hue of Manichean melodrama. Although there is no denying that an exceptional figure like St. Francis of Assisi might set an example of a new sense of cosmic harmonies and the integrity of all creation, we must agree with the sense communicated by Clarence Glacken in his interesting study of nature and culture in Western thought from ancient times to the end of the eighteenth century which he has entitled *Traces on the Rhodian Shore.* Glacken writes:

> It is often said that what distinguishes the modern from the medieval and classical periods is the modern sense of triumph in the control over nature in contrast with an earlier and unrelenting dependence. Such contrasts rest on an underevaluation of the extent of environmental change in classical and medieval times, on the belief that an advanced technology and sophisticated theoretical science are required for extensive and permanent change, and on a too sharp contrast between the so-called industrial revolution and the industry and technology of the past. One may wonder at the failure of the thinkers of the Middle Ages to create a theoretical science comparable with that of Galileo and Newton; fail they did, but they lacked neither an empirical knowledge of forestry, agriculture, drainage, nor a technology permitting them to induce sweeping and lasting changes in their environment. In fact, they made some of the most drastic changes in landscapes in human history up to that time.
>
> An ascetic ideal was the original stimulus in evolving a philosophy of man as a creator of new environments. The early saints purposefully retired from the world, and they fancied that by their clearings they were re-creating the earthly paradise, reasserting the complete dominion over all life that existed before the Fall. The attractive force of these retirements, both to other monks and to the laity, and organized efforts at conversion led to Christian activism, in which taming the wild was a part of the religious experience. One of the many great roles St. Bernard played was to increase the Church's potential for landscape change. Under his influence one can see the Cistercian order changing from remoteness and renunciation to a role of active Christianizing of new and old lands alike. The success of such undertakings depended on practical knowledge and sense like that expressed in the *Instituta capituli generalis of 1134*; "victus debet provenire de labore manuum, de cultura terrarum, de nutrimento pecorum"

In the age of the great *defrichements,* lay ambition and church ambition alike called for activity and change as a part of economic expansion and of conversion. The result was a yearning, to use a modern expression, for control over nature. In the later Middle Ages the interest in technology, in knowledge for its own sake whether to improve thinking or to better the human condition, in clearing, and in drainage and the like betrayed an eagerness to control nature. As in all epochs of human history, modification of the physical environment is linked with ideas, ideals, and practical needs. The period of great cathedral building embodied a religious ideal; it also meant vast quarrying; probably more stone was removed from the earth in this period than in any comparable period of the past. In the three centuries from 1050 to 1350 stone quarried in France built eighty cathedrals, five hundred large churches, and tens of thousands of small churches. The Christian duties of conversion and lay expansion and colonization meant firing, clearing, burning. The grain and the grape have their practical, their cultural, and their religious history.

The life of Albert the Great provides a clue. He shared with his contemporaries and the Christian thinkers of the past a belief in a designed earth, in nature as a book revealing the artisanry of God, in the need to know nature for religious and practical ends; he thought also of the role of the environment in cultural matters, and he saw the force of clearing, of burning, of domestication, of manuring. That is what it was, a chain from theology to manuring. (1967, pp. 349–51)

As these very lines are being written, spokesmen for 114 nations of the world are gathered in the hope of arriving at some ways of "preserving our small planet." If the intentions of this present essay are on the right track, solutions will not come easy. The tangles are too knotted, the motives too mixed, the ailments too complicated, the remedies too "costly." This is always the case when the interests are so mixed as the ones at stake in Stockholm.

If the world for whose birth we wait can come forth from its present anguish, "miracles" of every description—"social," "political," "religious"—shall have to occur. At the end of a statement such as this, one may be allowed to add without ado: Where "miracles" are awaited, are we not in the sphere of "faith"? Are we not also in need of Wisdom?

The achievement of improved relations of theologies, sciences, machines, and faiths will only come with continuous effort to use wisdom, however gathered, in the melioration of the patterns of human existence and the uses of human knowledge. The day is coming when all men who own and control machines shall have to offer proof that these do not involve social and cultural costs which are prohibitive. This has already begun in some lands, including our own.

There is no possibility of mitigating the problems of the world without effecting improved arrangements of the production and distribution of productive resources. There have to be better ways than we now have of creating wealth and sustenance, and better ways of reconciling differences and avoiding lethal conflict.

References

A. THE WRITINGS OF BENJAMIN NELSON

1933 "Robert de Curzon's Campaign Against Usury." Unpublished M.A. Thesis. Columbia University.

1939–44 With Joshua Starr. "The Legend of the Divine Surety and the Jewish Moneylender." *Annuaire de l'Institute de philologie et d'histoire orientales et slaves* 7:289–338.

1944 "The Restitution of Usury in Late Medieval Ecclesiastical Law." Unpublished Dissertation. Columbia University.

1947 "The Usurer and the Merchant Prince." *Journal of Economic History* 7:104–22.

1949a *The Idea of Usury: From Tribal Brotherhood to Universal Otherhood.* Princeton: Princeton University Press.

1949b "Blancard (the Jew?) of Genoa and the Restitution of Usury in Medieval Italy." *Studi In Onore Di Gino Luzzatto* I: 96–116

1951 "The Moralities of Thought and the Logics of Action." In *Value Conflicts, Moral Judgments, and Contemporary Philosophies of Education.* University of Minnesota: Lithograph.

1953 Co-editor with Arthur Naftalin, Mulford Q. Sibley, and Donald C. Calhoun. *An Introduction to Social Science: Personality, Work, Community.* New York: Lippincott. Second edition, 1957; third edition, 1961.

1954 "The Future of Illusions." *Psychoanalysis* 2, 4:16–37.

1955 Co-editor with John Mundy and Richard E. Emery. *Essays in Medieval Life and Thought, Presented in Honor of Austin Patterson Evans.* New York: Columbia University Press.

1955–56 "Adventure of Ideas." *Psychoanalysis* 4, 2:44–46.

1956 Reprint of (1954). Pp. 958–79 in *Man in Contemporary Society* II. New York. Columbia University Press.

1957

(a) With Marie L. Coleman. "Paradigmatic Psychotherapy in Borderline Treatment." *Psychoanalysis* 5, 3:28–44.

(b) Preface to *Freud and the Twentieth-Century.* Edited and selected by Benjamin Nelson. New York: Meridian, pp. 5–8.

(c) Foreword to *Psychoanalysis and the Future. A Centenary Commemoration of the Birth of Sigmund Freud.* Edited by B. Nelson and the board of editors of *Psychoanalysis.* New York: National Psychological Association for Psychoanalysis, pp. v–x.

(d) "On Dr. Walker's 'Five Theories'." *Psychoanalysis* 5, 3:26–27.

1958

(a) Introduction to *Sigmund Freud on Creativity and the Unconscious: Papers on the Psychology of Art, Literature, Love, Religion.* Selected and edited by Benjamin Nelson. New York: Harper Torchbook, pp. vii–x.

(b) "Social Science, Utopian Myth and the Oedipus Complex." *Psychoanalysis and the Psychoanalytic Review* 45, 1–2:120–126.

(c) "Questions on Existential Psychotherapy." *Psychoanalysis and the Psychoanalytic Review* 45, 4:77–78.

(d) With Charles Trinkaus. Introduction to *The Civilization of the Renaissance in Italy.* 2 volumes. New York: Harper Torchbooks, pp. 3–19.

1959

(a) "The Great Divide." *Psychoanalysis and the Psychoanalytic Review* 46, 2, 66–68.

(b) "Communities-Dreams and Realities." Pp. 135–151 in *Community. Nomos II,* edited by Carl Friedrich. New York: Liberal Arts Press. [A revision of (1954).]

(c) *O Século De Freud.* [Portuguese edition of *Freud and the Twentieth Century*]. Sao Paulo, Brazil.

1960 "Psychological Systems and Philosophical Paradoxes." *Psychoanalysis and the Psychoanalytic Review* 47, 3:43–51.

1961

(a) "Contemporary Politics and the Shadow of de Sade." *Psychoanalysis and the Psychoanalytic Review* 48, 4:30–32.

(b) Review of *Consciousness and Society: The Reorientation of European Thought, 1890-1930,* by H. Stuart Hughes. (New York: Alfred Knopf). *American Sociological Review* 26, 3:473–474.

(c) "Introductory Comment" to "Apocalypse: The Place of Mystery in the Life of the Mind," by Norman O. Brown. *Harper's Magazine* (May):46–47.

1962

(a) "Comments" (on Grant's "Hypotheses in Late Medieval and Early Modern Physics"). *Daedalus* 91, 3:613–616.

(b) Preface to *The Point of View for My Work as an Author* by Søren Kierkegaard. New York: Harper Torchbooks, pp. vii–xxi.

(c) "Sociology and Psychoanalysis on Trial: An Epilogue." *Psychoanalysis and the Psychoanalytic Review* 49, 2:144–160.

(d) "Phenomenological Psychiatry, Daseinanalyse and American Existential Analysis: A 'Progress' Report." *Psychoanalysis and the Psychoanalytic Review* 48, 4:3–23.

(e) "Faces of Twentieth Century Analysis." *The American Behavioral Behavioral Scientist* (February):16–18.

(f) Review of *Religion and Economic Action,* by Kurt Samuelsson (New York: Basic Books). *American Sociological Review* 27, 6:856.

(g) Editor. *Psychoanalysis and the Social-Cultural Sciences: Contemporary*

Perspectives. Special issue of *Psychoanalysis and the Psychoanalytic Review* 49, 2.

(h) *Freud e il XX Secolo.* [Italian translation of *Freud and the Twentieth Century*]. Verona: Arnoldo Mondadori.

1963

(a) "*The Balcony* and Parisian Existentialism." *Tulane Drama Review* 7, 3:66–79.

(b) "Casuistry." *Encyclopaedia Britannica,* 5. Reprinted vol. 5, pp. 51–52 in the 1968 edition. Chicago.

(c) "Sartre, Genet, Freud." *Psychoanalysis and the Psychoanalytic Review* 50, 3:156–171.

(d) "Über den Wucher." *Kölner Zeitschrift für Soziologie und Soziolpsychologie* 15, 7:407–47. [A translation of passages from *The Idea of Usury.*]

(e) "Hesse and Freud: Two Newly Recovered Letters." *Psychoanalysis and the Psychoanalytic Review* 50, 3:11–16.

(f) Editor. *Psychoanalysis and Literature.* Special issue of *Psychoanalysis and the Psychoanalytic Review* 50, 3.

(g) "A Professor's Avowal of Personal Faith." *Newsday* (March 26): 37.

1964

(a) *"Actors, Directors, Roles, Cues, Meanings, Identities: Further Thoughts on 'Anomie'." *Psychoanalytic Review* 51, 1:135–160.

(b) "Religion and Development." Pp. 67–68 in *Proceedings of the Sixth World Conference,* Society for International Development. Edited by T. Geiger and L. Solomon. Washington, D.C.

(c) "In Defense of Max Weber. A Reply to Herbert Luethy." *Encounter* 23, 2:94–5.

(d) "Max Weber's *The Protestant Ethic: 1904–1964.*" *Abstracts* of Papers presented at the 59th Annual Meeting of the American Sociological Association. Montreal, Canada, pp. 22–3.

1965

(a) *"Self-Images and Systems of Spiritual Direction in the History of European Civilization." Pp. 49–103 in *The Quest for Self-Control: Classical Philosophies and Scientific Research,* edited by S. Z. Klausner. New York: Free Press.

(b) *" 'Probabilists,' 'Anti-Probabilists' and the Quest for Certitude in the 16th and 17th Centuries." *Actes du Xme congrès internationale d'histoire des sciences* (Proceedings of the Tenth International Congress for the History of Science) 1:269–273.

(c) "The Psychoanalyst as Mediator and Double-Agent." *The Psychoanalytic Review* 52, 3:45–60.

(d) Comment on Herbert Marcuse's Paper. Pp. 192–201 in *Max Weber und die Soziologie heute.* (Proceedings of the Fifteenth Annual Meeting of the *German Sociological Society).* Edited by Otto Stammer. Tübingen: J. C. B. Mohr.

(e) "Dialogues Across the Centuries: Weber, Marx, Hegel, Luther." Pp. 149–65 in *The Origins of Modern Consciousness,* edited by John Weiss. Detroit: Wayne State University Press.

(f) "On Life's Way—Reflections on *Herzog*." *Soundings* (Spring): 148–54.
(g) "Max Weber and Talcott Parsons as Interpreters of Western Religious and Social Development." *Abstracts* of Papers presented to the Annual Conference of the *Society for the Scientific Study of Religion*. New York.
(h) "Storm Over Weber." *The New York Times Book Review*, January 3, p. 23 and February 28, pp. 34ff.
(i) "Mental Healers and Their Philosophies." [Reprint of 1962e]. *Literature, Religion, Psychiatry*. Special issue of *The Psychoanalytic Review* 52, 2:131–136.
(j) Review of *The Sociology of Religion*, by Max Weber (Boston: Beacon, 1963). *American Sociological Review* 30, 4:595–599.
(k) Editor. *Literature, Religion, Psychiatry*. Special issue of *The Psychoanalytic Review* 52, 2.

1967
(a) "Reflections on Michaelson's Scholarly Study of Religion." Pp. 26–32 in *The Study of Religion on the Campus Today*, edited by K. D. Hartzell and H. Sasscer. Washington, D. C.: Association of American Colleges.
(b) Reprint of [1954]. Pp. 563–76 in *Personality and Social Life*, edited by Robert Endleman. New York: Random House.
(c) *Usura e cristianesimo. Per una storia della genesi dell'etica* moderna. [Italian translation of *The Idea of Usury*]. Florence: Biblioteca Sansoni.
(d) *"The Early Modern Revolution in Science and Philosophy: Fictionalism, Probabilism, Fideism, and Catholic 'Prophetism'." Pp. 1–40 in *Boston Studies in the Philosophy of Science 3*, edited by R. S. Cohen and Marx Wartofsky. Dordrecht, Holland: Reidel.

1968
(a) "Scholastic *Rationales* of 'Conscience,' Early Modern Crises of Credibility, and the Scientific-Technocultural Revolutions of the 17th and 20th Centuries." *Journal for the Scientific Study of Religion* 7, 2:157–177.
(b) "The Avant-Garde Dramatist from Ibsen to Ionesco." *The Psychoanalytic Review* 55, 3:505–512.
(c) Introduction to *The Bourgeois: Catholicism versus Capitalism in Eighteenth-Century France*, by Bernard Groethuysen. New York: Holt, Rinehart and Winston, pp. vii–xiii.
(d) Editor. *Histories, Symbolic Logics, Cultural Maps*. Special Issue of *The Psychoanalytic Review* 55, 3.
(e) Co-editor with Marie Coleman Nelson. *Roles and Paradigms in Modern Psychotherapy*. New York: Grune and Stratton.

1969
(a) *The Idea of Usury: From Tribal Brotherhood to Universal Otherhood*. Second edition, enlarged. Chicago: University of Chicago Press.
(b) *"Conscience and the Making of Early Modern Cultures: *The Protestant Ethic* Beyond Max Weber." *Social Research* 36, 4:4–21.
(c) Introduction to *Madness and Society*, by George Rosen. New York: Harper Torchbooks, pp. vii-ix.
(d) "Metaphor in Sociology." Review of *Social Change and History*, by

Robert Nisbet (New York: Oxford). *Science* 116 (19 December): 1498–1500.

1970

(a) "Psychiatry and Its Histories: From Tradition to Take-Off." Pp. 229–59 in *Psychiatry and Its History: Methodological Problems in Research,* edited by George Mora and J. Brand. Springfield, Illinois: Charles C. Thomas.

(b) *"The Omnipresence of the Grotesque." *The Psychoanalytic Review* 57, 3:506–518. (Reprinted in *The Discontinuous Universe,* edited by Sallie Sears and Georgiana Lord. New York: Basic Books, 1972, pp. 172–185).

(c) "Is the Sociology of Religion Possible? A Reply to Robert Bellah." *Journal for the Scientific of Religion* 9, 2:107–111.

(d) Editor. *Philosophy, Technology and the Arts in the Early Modern Era,* by Paolo Rossi. New York: Harper Torchbook.

(e) Editor. *Cultural Revolutions and Generational Conflicts.* Special Issue of *The Psychoanalytic Review* 57, 3.

(f) Review of *The Sociology of Max Weber* by Julien Freund (New York: Pantheon). *American Sociological Review* 35, 3:549–50.

1971

(a) "The Medieval Canon Law of Contracts, Renaissance 'Spirit of Capitalism', and the Reformation 'Conscience': A Vote for Max Weber." Pp. 525–548 in *Philomanthes: Studies and Essays in the Humanities in Memory of Philip Merlan,* edited by Robert B. Palmer and Robert Hamerton-Kelly. The Hague: Martinus Nijhoff.

(b) Comment on Herbert Marcuse's Paper. Pp. 161–171 in *Max Weber and Sociology Today,* edited by Otto Stammer. New York: Harper Torchbooks. [English edition of 1965d].

(c) "Afterword: A Medium with a Message: R. D. Laing." Pp. 297–301 in *R. D. Laing and Anti-Psychiatry,* edited by Robert Boyers and Robert Orrill. New York: Harper and Row. [Originally published in *Salmagundi* 16 (Spring):199–201.]

(d) "Introduction" to "Max Weber on Race and Society." (Translated by Jerome Gittleman). *Social Research* 38, 1:30–32.

(e) "Introduction" to Emile Durkheim and Marcel Mauss, "Note on the Notion of Civilization" (translated by B. Nelson). *Social Research* 38, 4:808–813.

(f) Review of *The Study of Literate Civilization,* by F. L. K. Hsu. *American Anthropologist* 73, 2:319–20.

1972

(a) "Systems of Spiritual Direction." [An Autobiographical Essay]. *Criterion* (A Publication of The Divinity School of The University of Chicago) 11, 3:13–17.

(b) "Consciences, Sciences, Structures of Consciousness." *Main Currents in Modern Thought* 29, 2:50–53.

(c) "Droit Canon, Protestantisme et 'Esprit du Capitalisme.' A propos de Max Weber." [French translation of 1971a]. *Archives de sociologie des religions* 34:3–23.

(d) "Communities, Societies, Civilizations: Post-Millennial Views on the Masks and Faces of Change." Pp. 105–133 in *Social Development: Critical Perspectives,* edited by Manfred Stanley. New York: Basic Books.

(e) Review of *Science, Technology and Society in Seventeenth-Century England,* by Robert K. Merton (New York: Howard Fertig, 1970). *American Journal of Sociology* 78, 1:223–231. (Reprinted in *Varieties of Political Expression in Sociology* (An American Journal of Sociology Publication). Chicago: University of Chicago Press, pp. 202–210.)

1973

(a) *"Civilizational Complexes and Intercivilizational Encounters." *Sociological Analysis* 34, 2:79–105.

(b) "Weber's *Protestant Ethic:* Its Origins, Wanderings and Foreseeable Futures." Pp. 71–103 in *Beyond the Classics?* edited by Charles Y. Glock and P. Hammond. New York: Harper and Row.

(c) *"Priests, Prophets, Machines, Futures: 1202, 1848, 1984, 2001." Pp. 37–57 in *Religion and the Humanizing of Man,* second edition revised, edited by J. M. Robinson. Waterloo, Ontario: Council on the Study of Religion.

(d) "Introduction" to "Max Weber on Church, Sect, and Mysticism" (translated by Jerome L. Gittleman). *Sociological Analysis* 43, 2:140.

(e) With Vytautas Kavolis. "Comparative and Civilizational Perspectives in the Social Sciences and Humanities." *ISCSC Newsletter.* Genesco, New York. Excerpt in *Comparative Civilizations Bulletin* 5 (Spring):13–14.

(f) "An Overview." Communication to the authors of "Treatment of Psychosocial Masochism." *The Psychoanalytic Review* 60, 3:365–372.

(g) "The Games of Life and the Dances of Death." Pp. 113–131 in *The Phenomenon of Death: Faces of Mortality,* edited Edith Wyschogrod. New York: Harper and Row.

(h) With Dennis Wrong. "Perspectives on the Therapeutic in the Context of Contemporary Sociology: A Dialogue Between Benjamin Nelson and Dennis Wrong." *Salmagundi* 20 (Summer-Fall):160–195.

(i) With Donald Nielson. "Civilizational Patterns and Intercivilizational Encounters." (A selected and annotated bibliography.) *Bulletin of the International Society for the Comparative Study of Civilizations* 6 (Summer):3–15.

(j) With Jerome Gittleman. "Introduction" to "Max Weber, Dr. Alfred Ploetz, and W. E. B. DuBois (Max Weber on Race and Society II)." *Sociological Analysis* 34, 4:308ff.

1974

(a) *"*Eros, Logos, Nomos, Polis:* Their Shifting Balances and the Vicissitudes of Civilizations." Pp. 85–111 in *Changing Perspectives in the Scientific Study of Religion,* edited by Allan Eister. New York: Wiley-Interscience.

(b) *"Science and Civilizations, 'East' and 'West': Joseph Needham and Max Weber." Pp. 225–273 in *Boston Studies in the Philosophy of Science* 2, edited by R. S. Cohen and M. Wartofsky. Dordrecht-Holland: Reidel.

(c) "Max Weber's 'Author's Introduction' (1920): A Master Clue to His Main Aims." *Sociological Inquiry* 44, 4:269–278.

(d) *"De Profundis . . . Responses to Friends and Critics." *Sociological Analysis* 35, 2:129–141.

(e) *"On the Shoulders of the Giants of the *Comparative* Historical Sociology of 'Science'—In *Civilizational Perspective*." Pp. 13–20 in *Social Processes of Scientific Development*, edited by Richard D. Whitley. London: Routledge.

(f) With Harold Rosenberg. "Art and Technology: A Dialogue Between Harold Rosenberg and Benjamin Nelson." *Salmagundi* 27 (Summer-Fall):40–56.

(g) "Psychoanalysis and Psychohistory—An Analytic Perspective." *Book Forum: An International Transdisciplinary Journal* 1, 2:254–262.

1975

(a) *"The Quest for Certitude and the Books of Scripture, Nature, and Conscience." Pp. 355–371 in *The Nature of Scientific Discovery*, edited by Owen Gingerich. Washington, D. C.: Smithsonian Institution Press.

(b) "Copernicus and the Quest for Certitude: 'East' and 'West'." Pp. 39–46 in *Vistas in Astronomy 17*, edited by A. Beer and K. Aa. Strand. Oxford: Pergamon Press.

(c) "Max Weber, Ernst Troeltsch, Georg Jellinek as Comparative Historical Sociologists." *Sociological Analysis* 36, 3:229–240.

(d) "Quality of Life: Existence, Experience, Expression." Pp. 19–20 in *Systems Thinking and the Quality of Life*, edited by Clair K. Blong. Washington, D. C.: Society For General Systems Research.

1976

(a) "On Orient and Occident in Max Weber." *Social Research* 43, 1:114–129.

(b) "Max Weber as a Pioneer of Civilizational Analysis." *Comparative Civilizations Bulletin* 16 (Winter):4–6.

(c) "Vico and Comparative Historical Civilizational Sociology." *Social Research* 43, 4:874–881.

(d) Foreword to *The Narcissistic Condition: A Fact of Our Lives and Times*, edited by Marie Coleman Nelson. New York: Human Sciences Press, pp. 13–23.

(e) "Prêtres, prophètes, machines, futurs: 1202, 1848, 1984, 2001." [French translation of 1973c]. Pp. 227–246 in *Les Terreurs de l'an 2000*. Paris: Hachette.

(f) With Niklas Luhmann. "A Conversation on Selected Theoretical Questions: Systems Theory and Comparative Civilizational Sociology." *The Graduate Faculty Journal of Sociology* 1, 2:1–17.

(g) Editor. *Freud and Neo-Marxisms—Erikson and Psychohistories*. Special issue of *The Psychoanalytic Review* 63, 2.

1977

(a) *Der Ursprung der Moderne: Vergleichende Studien zum Zivilisationsprozess. [The Origins of Modernity: Comparative Studies of Civilizational Process]*. Frankfurt am Main: Suhrkamp.

(b) Review of *Business, Banking, and Economic Thought in Late Medieval and Early Modern Europe: Selected Studies,* by R. de Roover, edited by Julius Kirshner. (Chicago: The University of Chicago Press, 1974). *Journal of European Economic History* 6, 2:487–491.

(c) "Priester, Propheten, Maschinen, Zukünftiges: 1202, 1848, 1984, 2001." [German translation of 1973c]. Pp. 244–267 in *Die Schrechen des Jahres Zweitausend,* edited by Henry Cavanna. Stuttgart: Klett.

(d) "Tradition and Innovation in Law and Society: Comparative Historical and Civilizational Perspectives. A Paper in Progress." Prepared for the Conference at Freiburg [September 18–25, 1977], sponsored by the Institut für Historische Anthropologie, Unpublished.

(*) Reprinted in this volume.

B. GENERAL REFERENCES

Abelard, Peter (1079–1142). 1960. *Historia Calamitatum.* Paris.

——— 1962. *Dialectia.* Edited with an introduction by L. de Rijk. Assen.

——— 1971. *Ethics.* Edited and translated with an introduction and notes by D. E. Luscombe. Oxford.

Adelard of Bath (early 12th c.). 1903. *De Eodem et Diverso.* Edited by H. Willmer. *Beiträge zur Geschichte der Philosophie und Theologie des Mittelalters.* 4, 1. Münster.

——— 1934. *Quaestiones Naturales.* Edited by M. Müller. *Beiträge zur Geschichte der Philosophie und Theologie des Mittelalters* 31/2. Münster.

Adler, Alfred. 1939. *Social Interest: A challenge to mankind.* Translated by S. Linton and R. Vaughan. New York.

Agassi, Joseph. 1963. *Towards an Historiography of Science. History and Theory.* Beiheft 2. Middletown, Conn. Wesleyan University.

Alexander, Franz. 1960. *The Western Mind in Transition.* New York.

——— 1961a. *The Scope of Psychoanalysis: 1921–1961: Selected Papers.* New York.

——— 1961b. "Buddhist Training as an Artificial Catatonia." Pp. 74–89 in F. Alexander, *The Scope of Psychoanalysis.* New York.

Alexander, H. G. (editor). 1956. *The Leibniz-Clarke Correspondence. With Extracts from Newton's Principia and Optics.* Manchester

Al-Ghazali (1058–1111). 1955. *Tahafut al-Falasifah.* Translated by S. A. Kamali. Lahore.

——— 1962. *Book of Fear and Hope.* Translated by William McKane. Leiden.

Allport, Gordon. 1960. *Personality and Social Encounter.* Boston.

Ames, William. 1631. *De conscientia et eius, vel casibus, libri quinque.* Amsterdam.

Ancianx, Paul. 1949. *La théologie du Sacrament de Pénitence au XIIe siècle.* Louvain.

Anonymous. 1957. *The Cloud of Unknowing.* Edited with an introduction by Ira Progoff. New York.

——— (A Friend of God). *The Book of the Poor in Spirit.* Translated with an introduction by C. F. Kelley. New York (n.d.).

——— 1856. *Theologia germanica* (ca. 1350). Translated by S. Winkworth. Boston
Ansbacher, H., and R. Ansbacher. 1956. *The Individual Psychology of Alfred Alder*. New York.
Anselm of Canterberry, St. (c. 1033–1109). 1962. *Basic Writings: Proslogium; Monologium; Gaunilon's On Behalf of the Fool, Cur deus homo*. Translated by S. N. Dean. La Salle, Indiana.
——— 1967. *Truth, Freedom and Evil: Three Philosophical Dialogues*. Edited with a revised translation by J. Hopkins and H. Richardson. New York.
Arnold, E. V. 1911. *Roman Stoicism*. Cambridge.
Auden, W. H. 1950. *The Enchafed Flood*. New York.
Averroes (1126–1198). 1954. *Tahafut al-Tahafut (The Incoherence of the Incoherence)*. Translated by S. Van den Bergh. 2 vols. London.
Ayer, A. J. 1959. *Logical Positivism*. Glencoe, Illinois.
Babbitt, Irving. 1961 (1919). *Rousseau and Romanticism*. New York.
Bacon, Roger (ca. 1214–1292). 1859. *Opera Quaedam hactenus inedita*. (Contains *Opus tertiam* [vol. 1]; *Opus Minus* [vol. 2]; *Compendium philosophiae* [vol. 3].) Edited by J. S. Brewer. London
——— 1964. *The Opus Majus of Roger Bacon* (ca. 1267). 2 vols. Edited with an introduction by J. H. Bridges and supplement in Latin. Frankfort.
Bainton, Roland H. 1941. "The Struggle for Religious Liberty." *Church History* 10 (June): 95–124.
——— 1950. *Here I Stand: A Life of Martin Luther*. New York
——— 1958. *The Travail of Religious Liberty*. New York.
——— 1952. *The Reformation of the Sixteenth Century*. Boston.
——— 1956. *The Age of the Reformation*. Princeton.
——— 1962. *Collected Papers in Church History: Early and Medieval Christianity*. Series 1. Boston.
Baker, G. W. and D. W. Chapman (eds.). 1962. *Man and Society in Disaster*. New York.
Baker, Hershel. 1948. *The Dignity of Man*. Cambridge, Massachusetts.
Balakian, Anna. 1947. *The Literary Origins of Surrealism*. New York.
Baldwin, John W. 1970. *Masters, Princes and Merchants: The Social Views of Peter the Chanter and His Circle*. 2 vols. Princeton.
Barber, Bernard and Walter Hirsch (eds.) 1962. *The Sociology of Science*. Glencoe.
Barbu, Zevedi. 1960. *Problems of Historical Psychology*. New York.
Barker, Ernest. 1948. *Traditions of Civility*. Cambridge.
Barth, Karl. 1960. *Anselm: Fides quaerens intellectum (Anselm's Proof of the Existence of God in the Context of His Theological Scheme)*. Translated by I. W. Robertson. Richmond, Virginia.
Bates, F. L. 1962. "Some Observations Concerning the Structural Aspect of Role Conflict." *Pacific Sociological Review* 5, 2:75–82.
Bateson, Gregory and others. 1956. "Toward a Theory of Schizophrenia." *Behavioral Sciences* 1: 251–64.
Bauer, Raymond. 1952. *The New Man in Soviet Psychology*. Cambridge, Mass.
Baxter, Richard. 1925. *Chapters from a Christian Directory*. Edited by Jeanette Tawney. London.

Beaujouan, G. 1957a. *L'interdépendence entre la science scolastique et les techniques utilitaries (XIIe, XIIIe et XIVe siècles).* Paris.
——— 1957b. "La science dans l'occident médieval chétien." Pp. 468–532 in R. Taton (ed), *Ancient and Medieval Science.* New York.
——— 1963. "Motives and Opportunities for Science in Medieval Universities." Pp. 219–236 in A. Crombie (ed.), *Scientific Change.* New York.
Beck F., and W. Godin. 1951. *The Russian Purge and Extraction of Confession.* Translated by E. Mosbacher and D. Porter. London.
Becker, Ernest. 1962. *The Birth and Death of Meaning.* New York.
Bellah, Robert. 1964. *Modernization and Progress in East Asia.* New York.
——— 1968 (1963). "Reflections on the Protestant Ethic Analogy in Asia." Pp. 243–252 in *The Protestant Ethic and Modernization: A Comparative View,* edited by S. N. Eisenstadt. New York.
Ben-David, Joseph. 1960. "Scientific Growth: A Sociological View." *American Journal of Sociology* 65: 557–568.
——— 1971. *The Scientist's Role in Society: A Comparative Study.* Englewood Cliffs, New Jersey.
Benedict, Ruth. 1934. *Patterns of Culture.* Boston.
——— 1938. "Continuities and Discontinuities in Cultural Conditioning." *Psychiatry* 1: 161–167.
Bennett, Adrian. 1967. *John Fryer: The Introduction of Western Science and Technology into Nineteenth Century China.* Cambridge, Mass.
Berger, Peter. 1961. *The Precarious Vision.* New York.
Bernard, Henri. 1941. "Notes on the Introduction of the Natural Sciences into the Chinese Empire." *Yenching Journal of Social Studies* 3, 2:220–241
Berti, D. 1876. *Copernicano e le vicende del sistema Copernicano.* Rome.
Bertola, Ermenegildo. 1970. *Il Problema dell concienza nella teologia monastica del XII secolo.* Il Pensiero Medievale, Collana di storia della filosofia. Series 2, vol. 1. Padua.
Bettelheim, Bruno. 1943. "Individual and Mass Behavior in Extreme Situations." *Journal of Abnormal and Social Psychology* 38: 417–52.
Bierstedt, Robert. 1950. "An Analysis of Social Power." *American Sociological Review* 15, 6: 730–38.
Binswanger, Ludwig. 1956. *Drei Formen missglueckten Daseins.* Tübingen.
——— 1957. *Schizophrenie.* Pfullingen
Blake, R. N. 1960. See Madden (1960), chapters 1–3, 5.
Blakney, R. 1957. *Meister Eckhardt: A Modern Translation.* New York.
Bloch, Marc. 1964. *Feudal Society.* 2 vols. Translated by L. A. Manyon. Chicago.
Blunt, Anthony, 1938–39. "Blake's 'Ancient of Days': The Symbolism of the Compasses." *Journal of the Warburg Institute* 2: 53–63.
Boas, Marie, and A. R. Hall. 1959. "Tycho Brahe's System of the World." *Occasional Notes of the Royal Astronomical Society* 3, 21:253–263.
Boase, T. S. R. 1933. *Boniface VIII.* London
Bocheński, I. M. 1950. *Der Sowjetrussische Dialektische Materialismus (Diamat).* Bern.
——— 1956. *Contemporary Philosophy.* Translated from the 2nd revised German edition (1947) by D. Nicoll and K. Aschenbrenner. Berkeley.

Bochner, S. 1966. *The Role of Mathematics in the Rise of Science*. Princeton.
Boehmer, Heinrich. 1961 (1924). *Martin Luther: Road to Reformation*. Translated by J. W. Doberstein and T. G. Tappert. New York
Boehner, Philotheus. 1952. *Medieval Logic: An Outline of its Development from 1250 to ca. 1400*. Manchester.
Bohr, Niels. 1958. *Atomic Physics and Human Knowledge*. New York.
Boisen, Anton I. 1962. *The Exploration of the Inner World*. New York.
Bonaventura, Saint (d.1274). 1939. *On the Reduction of the Arts to Theology*. Translated by Sister E. T. Healy. New York.
——— 1956. *Itinerarium mentis in Deum*. Translated with an introduction by P. Boehner. New York.
Borkenau, Franz. 1934. *Der Übergang vom feudalen zum bürgerlichen Weltbild*. Paris.
Bornkamm, Heinrich. 1943. "Kopernicus in Urteil der Reformatoren." *Archiv für Reformationsgeschichte* 40: 171–183.
Boss, Medard. 1957. *Psychoanalyse und Daseinanalytik*. Bern.
Bozeman, Adda B. 1960. *Politics and Culture in International History*. Princeton.
Brémond, Henri. 1928. *A Literary History of Religious Thought in France from the Wars of Religion Down to Our Times*. Translated by K. L. Montgomery. New York.
Breuer, Josef, and S. Freud. 1957 (1895). *Studies on Hysteria*. Edited and translated by J. Strachey and A. Freud. New York.
Brodrick, Father James, SJ. 1950 (1928). *The Life and Work of Blessed Robert Francis, Cardinal Bellarmine, 1542–1621*. 2 vols. London.
——— 1961. *Robert Bellarmine: Saint and Scholar*. London.
——— 1964. *Galileo: The Man, His Work, His Misfortunes*. New York.
Bromberg, Walter, 1959. *The Mind of Man*. New York.
Brown, Norman O. 1959. *Life Against Death*. Middletown, Conn.
——— 1966. *Love's Body*. New York.
Buber, Martin. 1952. *Eclipse of God*. Translated by M. Friedman and others. New York.
——— 1955. *Between Man and Man*. Translated by R. G. Smith. Boston
Bunge, Mario (ed.). 1964. *The Critical Approach to Science and Philosophy*. New York.
Burckhardt, Jacob. 1958. *The Civilization of the Renaissance in Italy*. 2 vols. Translated by S. Middlemore, with a introduction by B. Nelson and C. Trinkaus. New York.
Burke, Kenneth. 1936. *Permanence and Change*. Los Altos, California.
——— 1946a. *A Grammar of Motives*. New York.
——— 1946b. *A Rhetoric of Motives*. New York.
——— 1961. *The Rhetoric of Religion: Studies in Logology*. Boston.
Burnaby, John. 1938. *Amor Dei*. London.
Burrage, Champlin. 1912. *The Early English Dissenters in the Light of Recent Research* . . . 2 volumes. Cambridge.
Burtt, E. A. 1925. *The Metaphysical Foundations of Modern Physical Science*. New York.
Butler, Cuthbert. 1966 (1922). *Western Mysticism*. New York.
Butler, H. E. (trans.). 1949. *The Chronicles of Jocelin of Brakeland (Concerning the Acts of Samson, Abbott of the Monastery of St. Edmund)*. New York.

Butterfield, Herbert, 1951. *The Origins of Modern Science, 1300–1800.* New York.

Caperan, Louis. 1934 (1912). *Le problème du salut des Infideles.* Toulouse.

Cary-Lundberg, Isabel. 1959. "On Durkheim, Suicide, and Anomie." *American Sociological Review* 24, 2: 250–1.

Cassirer, Ernst. 1953–57. *The Philosophy of Symbolic Form.* 3 vols. Translated by R. Manheim. New Haven.

——— 1966. *Individual and Cosmos in the Philosophy of the Renaissance.* Translated from the Italian by M. Domandi.

——— P. O. Kristeller, and J. H. Randall, (eds.). 1948. *The Renaissance Philosophy of Man.* Chicago.

Castellio, Sebastian. 1935. *Concerning Heretics.* Edited and translated by Roland Bainton. New York.

Chenu, Marie-Dominique. 1957. *La Theologie au douzième siècle.* Paris.

——— 1968. *Nature, Man and Society in the Twelfth Century.* Edited and translated by J. Taylor and L. K. Little. Chicago.

——— 1969. *L'éveil de la conscience dans la civilisation Médiévale.* Paris.

Christianson, J. R. 1973. "Copernicus and the Lutherans." *Sixteenth Century Journal* 4, 2:1–10.

Clagett, Marshall. 1955. *Greek Science in Late Antiquity.* New York.

——— (ed.). 1959a. *Critical Problems in the History of Science.* Madison.

——— 1959b. *The Science of Mechanics in the Middle Ages.* Madison.

——— G. Post, and R. Reynolds (eds.). 1961. *Twelfth Century Europe and the Foundations of Modern Society.* Madison.

——— 1967. "Some Novel Trends in the Science of the Fourteenth Century." Pp. 275–303 in *Art, Science and History in the Renaissance,* edited by C. Singleton. Baltimore.

——— (editor and translator). 1968. *Nichole Oresme and the Medieval Geometry of Qualities and Motions.* Madison.

Clark, G. N. 1937. *Science and Social Welfare in the Age of Newton.* London.

Clark, James M. 1949. *The Great German Mystics: Eckhart, Tauler and Suso.* Oxford.

Clark, Joseph, SJ. 1959. "The Philosophy of Science and the History of Science." Pp. 103–140 in *Critical Problems in the History of Science,* edited by M. Clagett. Madison.

Clavelin, Maurice. 1968. "Galilée et le refus de l'équivalence des hypothèses." Pp. 127–153 in *Galileé: Aspects de sa vie et son oeuvre.* Paris.

Climacus, Saint John (d. 649). *The Ladder of Divine Ascent.* Translated by Archimandrite L. Moore. New York (n.d.).

Cloward, Richard A. 1959. "Illegitimate Means, Anomie and Deviant Behavior." *American Sociological Review* 24, 2:164–176.

Coben, Stanley. 1971. "The Scientific Establishment and the Transmission of Quantum Mechanics to the United States, 1919–1932." *American Historical Review* 76, 2: 442–466.

Cohen, I. B., and René Taton (eds.). 1964. *L'aventure de la science: Mélanges Alexandre Koyré (Histoire de la Pensée* 12). 2 volumes. Paris.

Cohen, M. R. 1926. "The Myth about Francis Bacon and the Inductive Method." *Scientific Monthly* 23: 504–508. (Reprinted in Cohen 1949).

——— 1949. *Studies in Philosophy and Science.* New York.
Cohn, Norman. 1961. *The Pursuit of the Millennium.* New York.
Colby, Kenneth M. 1951. "On the Disagreement Between Freud and Adler." *American Image* 8: 229–38.
Cole, Stephen, and Jonathan Cole. 1967. "Scientific Output and Recognition: A Study in the Operation of the Reward System in Science." *American Sociological Review* 32: 377–390.
——— 1973. *Social Stratification in Science.* Chicago.
Coleman, Marie L. 1949. "An Integrative Approach to Individual and Group Behavior." *The Psychoanalytic Review* 36, 4: 389–401.
Colie, Rosalie L. 1957. *Light and Enlightenment.* Cambridge.
Collingwood, R. G. 1945. *The Idea of Nature.* New York.
Cornford, F. M. 1934. *Origins of Attic Comedy.* Cambridge.
——— 1968 (1912). *From Religion to Philosophy: A Study in the Origins of Western Speculation.* New York.
Conrad, Joseph. 1954. *The Portable Conrad.* Edited by M. D. Zabel. New York.
Cragg, Gerald R. 1950. *From Puritanism to the Age of Reason: A Study of Changes in Religious Thought Within The Church of England, 1660–1770.* Cambridge.
——— 1960. *The Church in The Age of Reason, 1648–1789.* Harmondsworth.
Cranefield, P. F. 1958. "Joseph Breuer's Evaluation of His Contribution to Psycho-Analysis." *International Journal of Psychoanalysis* (Sept-Oct): 319-322.
Crocker, Lester G. 1959. *An Age of Crisis: Man and World in Eighteen-Century French Thought.* Baltimore.
Crombie, A. C. 1950. "Galileo's Dialogues Concerning the Two Principal Systems." *Dominican Studies* 3:105–138
——— 1953. *Robert Grosseteste and the Origins of Experimental Science, 1100–1700.* Oxford.
——— 1959a. (1952). *Medieval and Early Modern Science.* 2 vols. Second edition. New York.
——— 1959b. "Commentary on Papers of Rupert Hall and Giorgio de Santillana." Pp. 66–78 in *Critical Problems in the History of Science,* edited by M. Clagett.
——— 1959c. "The Significance of Medieval Discussions of Scientific Method for the Scientific Revolution." Pp. 79–102 in *Critical Problems in the History of Science,* edited by M. Clagett. Madison.
——— 1961. *Augustine to Galileo.* Cambridge, Mass.
——— (ed.). 1963. *Scientific Change* (Symposium on the History of Science, Oxford 1961). New York.
——— (ed.). 1964. *Turning Points in Modern Science.* New York.
Crosby, H. Lamar. 1955. *Thomas of Bradwardine, his "Tractatus de proportionibus."* Madison.
Crozier, R. C. (ed.). 1970. *China's Cultural Legacy and Communism.* New York.
Curtius, E. 1953. *European Literature and the Latin Middle Ages.* Translated by W. R. Trask. New York.
Cushman, Robert E. 1958. *Therapeia: Plato's Conception of Philosophy.* Chapel Hill, North Carolina.
d'Abro, A. 1950. *The Evolution of Scientific Thought: From Newton to Einstein.* New York.

Davidson, Donald. 1968 (1963). "Actions, Reasons, and Causes." Pp. 44–58 in *Readings in the Philosophy of the Social Sciences,* edited by May Brodbeck. New York.

Davison, E. S. 1926. *Forerunners of St. Francis and Other Studies.* London.

de Blic, J., SJ. 1930. "Bartholomy de Medina et les origines du probabilisme." *Ephemerides theologicae Lovanienses.* 7:46–83, 264–291.

Dechanet, J. M. 1955. "La Christologie de Saint Bernard." In *Bernard von Clairvaux, Mönch und Mystikes.* Edited with an introduction by J. Lortz. Weisbaden.

D'Elia, P., SJ. 1960. *Galileo in China: Relations Through the Roman College Between Galileo and the Jesuit Scientist-Missionaries (1610–1640).* Cambridge, Mass.

Deman, Th. O.P. 1936. "Probabilisme." Pp. 417–619 in *Dictionnaire de théologie catholique* 13.

de Rijk, L. M. (ed.). 1962. *On the 12th Century Theories of Fallacy (Logica Modernorum 1).* Assen.

de Roover, Raymond. 1937. "Aux origines de'une technique intellectuale: la formation et l'expansion de la comptabilité à partie double." *Annales d'histoire économique et sociàle* 9: 171–193, 270–298.

de Rougemont, Denis. 1956. *Love in the Western World.* Translated by M. Belgion. New York.

Descartes, René. 1952. *Descartes' Philosophical Writings.* Selected and translated by Norman Kemp Smith. London.

Dewey, John. 1929. *The Quest for Certainty.* New York.

Dijksterhuis, E.J. 1961. *The Mechanization of the World-Picture.* Oxford.

Dill, Sir Samuel. 1957 (1898). *Roman Society in the Last Century of the Western Empire.* New York.

——— 1956 (1904). *Roman Society from Nero to Marcus Aurelius.* New York.

Dillenberger, John. 1960. *Protestant Thought and Nature Science.* New York.

Dilthey, Wilhelm. 1923. *Weltschauung und Analyse des Menschen seit Renaissance und Reformation.* Leipzig.

Dionysus the Areopagite (ca. 500). 1894. *On the Celestial Hierarchy.* Translated by Rev. John Parker . London.

Dodds, E.R. 1951 *The Greeks and the Irrational.* Berkeley.

Doellinger, I. von, and H. Reusch. 1889. *Geschichte der Moralstreitigkeiten in der roemisch-katholischen Kirche seit dem 16ten Jahrhundert.* 2 vols. Nördlingen.

Douie, Decima. 1932. *The Nature and Effect of the Fraticelli Heresy.* Manchester.

Dowey, Edward A., Jr. 1952. *The Knowledge of God in Calvin's Theology.* New York.

Drake, Stillman (editor and translator). 1957. *Discoveries and Opinions of Galileo.* New York.

——— 1967. "Mathematics, Astronomy, and Physics in the Work of Galileo." Pp. 305-330 in *Art, Science and History in the Renaissance,* edited by C. Singleton. Baltimore.

——— 1970. *Galileo Studies. Personality, Tradition, and Revolution.* Ann Arbor, Michigan.

Dubin, Robert. 1959. "Deviant Behavior and Social Structure: Continuities in Social Theory." *American Sociological Review,* 24, 2: 147–63.

Dubos, René. 1961. *The Dreams of Reason, Science and Utopia.* New York.
Dugas, René. 1958. *Mechanics in the Seventeenth Century. From the Scholastic Antecedents to Classical Thought.* Neuchâtel and New York.
Duhem, Pierre. 1905a. *L'evolution de la Mécanique. Paris.*
— *1905b (1954). The Aim and Structure of Physical Theory.* Translated by Philip P. Wiener. Princeton
——— 1906-13. *Etudes sur Leonard de Vinci.* 3 vols. Paris.
——— 1913-59. *Le Système du monde: Histoire des doctrines cosmologigues de Platon à Copernic.* 10 vols. Paris.
——— 1961 (1911) "History of Physics." Pp. 47–52 in Catholic Encyclopedia 12. Reprinted in *Toward Modern Science,* edited R. M. Palter, 2 pp. 115–13). New York.
——— 1969 (1908). *To Save the Phenomena: An Essay on the Idea of Physical Theory form Plato to Galileo.* Translated by E. Doland and C. Maschler with an introduction by S. L. Jaki. Chicago. (Originally published in *Annales de philosophie chrétienne* 79/156 (ser. 4, VI):113–38, 277–302, 352–77, 482–514, 576–92.)
Dumoulin, H. 1959. *Zen: Geschichte und Gestalt.* Bern.
Durkheim, Emile. 1933 (1893). *The Division of Labor in Society.* Translated by G. Simpson. New York.
——— 1915 (1912). *The Elementary Forms of the Religious Life.* Translated by Joseph Ward Swain. New York.
——— 1938 (1895). *The Rules of Sociological Method.* Translated by S. A. Solovay and J. H. Mueller. New York.
——— 1951. *Suicide. A Study in Sociology.* Translated by J.A. Spaulding and G. Simpson. New York.
——— 1955. *Pragmatisme et sociologie.* Paris.
——— and Marcel Mauss. 1963 (1901-02). *Primitive Classification.* Translated and edited with an introduction by Rodney Needham. Chicago.
——— and Marcel Mauss. 1971 (1913). "Note on the Notion of Civilization." Translated by Benjamin Nelson. *Social Research* 38, 4:809–913.
Duyvendak, J. J. L. 1948. "P. D'Elia's *Galileo in China.*" *T'oung Pao* 38: 321–329.
Dvorak, Max. 1928. *Kunstgeschichte als Geistesgeschichte.* Munich.
Eaton, J., and R. J. Weil. 1955. *Culture and Mental Disorder.* Glencoe, Illinois.
Eckhart, M. (d.1327). *Meister Eckhart: A Modern Translation.* Translated by R. Blakney. New York.
——— 1957. *An Introduction to the Study of His Works with an Anthology of his Sermons.* Translated and edited by James M. Clark. London.
Edgerton, Samuel, Jr. 1966. "*Alberti's Perspective:* A New Discovery and a New Evaluation." *Art Bulletin* 48 (Sept-Dec): 367–378.
Edie, J. M., and others (eds.). 1965. *Russian Philosophy.* 3 vols. Chicago.
Egenter, R. 1928. *Gottesfreundschaft: Die Lehre von du Gottesfreundschaft in der Scholastik and Mystik des 12 und 13 Jahrhunderts.* Augsburg.
Einstein, Albert. 1953. Foreword to Galileo, *Dialogue Concerning the Two Chief Systems.* Edited and translated by S. Drake. Berkeley.
——— and L. Infeld. 1950. *The Evolution of Physics.* New York.
Eisenstadt, S.N. 1965a. "Transformation of Social, Political, and Cultural Orders in Modernization." *American Sociological Review* 30: 659–673.
——— 1965b. *Essays on Comparative Institutions.* New York.

——— 1968a. "The Protestant Ethic Thesis in an Analytical and Comparative Framework." Pp. 3–45 in *The Protestant Ethic and Modernization*, edited by S. N. Eisenstadt. New York

——— (ed.) 1968b. *The Protestant Ethic and Modernization: A Comparative View*. New York.

Eissler, Kurt R. 1953. "The Effect of the Structure of the Ego in the Psychoanalytic Technique." *Journal of the American Psychoanalytic Association* 1, 1: 104–43.

Eliade, M. 1958. *Yoga: Immortality and Freedom*. New York.

——— 1959. *Cosmos and History: The Myth of the Eternal Return*. New York.

——— 1960. *Myths, Dreams, and Mysteries*. Translated by P. Mairet. New York.

Eliot, T. S.1950. *Selected Essays*. New York.

Erikson, Erik. 1956. "The Problem of Ego Identity." *Journal of the American Psychoanalytic Association* 4, 1:58–121.

——— 1957. "The First Psychoanalyst." Pp. 79–101 in *Freud and the Twentieth Century*, edited by B. Nelson. New York.

——— 1958. *Young Man Luther*. New York.

——— 1959. *Identity and the Life Cycle*. New York.

Esslin, Martin. 1961. *The Theatre of the Absurd*. New York.

Evans, Herbert (ed.). 1959. *Men and Moments in the History of Science*. Seattle.

Evans-Wentz, W. Y. 1949. *The Tibetan Book of the Dead*. London.

Fagnani, Prospero. 1661. *Commentaria super quinquos libros decretalium*. 5 books in 7 parts. Rome.

Fauconnet, Paul. 1928. *La Responsabilité: Etude de sociologie*. New York.

Favaro, Antonio. 1907. *Galileo e l'inquisizione*. Florence.

Febvre, Lucien. 1947. *Le Problème de l'incroyance au XVIe siècle: la religion de Rabelais*. Paris.

Feldman, Gene, and Max Gartenberg. 1959. *The Beat Generation and the Angry Young Men*. New York.

Feuer, Lewis, S. 1965. *The Scientific Intellectual*. New York.

Feuerbach, Ludwig. 1957 (1841). *The Essence of Christianty*. Translated by G. Eliot and edited by B. Nelson. New York.

——— 1967 (1844). *The Essence of Faith According to Luther*. Translated by M. Cherno. New York.

Feyerabend, Paul K. 1962. "Problems of Microphysics." Pp. 189–283 in *Frontiers of Science and Philosophy*, edited by Robert G. Colodny. Pittsburgh.

——— 1964. "Realism and Instrumentalism: Comments on the Logic of Factual Support." Pp. 280–308 in *The Critical Approach to Science and Philosophy*, edited by M. Bunge. New York.

Field, M. J. 1960. *Search for Security*. Evanston, Illinois.

Fifoot, C. H. S. 1949. *History and Sources of Common Law: Tort and Contract*. London.

Fijalkowski, Jürgen. 1968. "Carl Schmitt" Pp. 58–60 in D. L. Sills, (ed.), *International Encyclopedia of the Social Sciences*, 14. New York.

Fingarette, Herbert. 1963. *The Self in Transformation*. New York.

Firth, R. (ed.) 1957. *Man and Culture*. London.

Fiske, Adele, RCSJ. 1965. "Paradisus Homo Amicus." *Speculum* 40:436–459.

——— 1970 (1955). *The Survival and Development of the Ancient Concept of Friendship in the Early Middle Ages.* Ph.D. Diss. Now published as *Friends and Friendship in the Monastic Tradition.* CIDOC Cuaderno No. 51. Cuernavaca, Mexico.
Flew, R. N. 1934. *The Idea of Perfection in Christian Theology.* London.
Forest, Aimé, F. van Steenberghen, and M. de Gandillac. 1951. *Le mouvement doctrinale de IXe au XIVe siècle (Histoire de l'Eglise depuis les origines . . . jusqu'à nos jours.* Vol. 13). Paris.
Foscarini, P. 1615. *Lettera sopra l'opinione de' Pittagorica e Copernico della mobilita della terra.* Naples.
Foster, Michael, B. 1934. "The Christian Doctrine of Creation and the Rise of Modern Natural Science." *Mind* 43:446–468.
——— 1936. "Christian Theology and the Modern Science of Nature." *Mind* 45:1–28.
Fowlie, W. 1950. *Age of Surrealism.* Bloomington, Indiana.
Frank, Philipp. 1957. *The Philosophy of Science: The Link Between Science and Philosophy.* Englewood Cliffs, New Jersey.
Franke, W. 1967. *China and the West: The Cultural Encounter, 13th to 20th Centuries.* New York.
——— 1971. *A Century of Chinese Revolution, 1851–1949.* New York.
Freud, S. (with J. Breuer). 1895. *Studies on Hysteria.* [see S. E.: 2]
——— 1900. *The Interpretation of Dreams.* [S. E.: 4–6]
——— 1901. *Psychopathology of Everyday Life.* [S. E.: 6]
——— 1905. *Three Essays on the Theory of Sexuality.* [S. E.: 7]
——— 1910 [1909]. *Five Lectures on Psychoanalysis.* [S. E.: 11]
——— 1911–15. *Papers on Technique.* [S. E.: 12]
——— 1914. *On the History of the Psycho-Analytic Movement.* [S. E.: 14]
——— 1924–50. *Collected Papers.* 5 vols. Edited by J. Riviere and J. Strachey. London.
——— 1925. *An Autobiographical Study.* [S. E.: 20]
——— 1930. *Civilization and Its Discontents.* [S. E.: 21]
——— 1953–74. *The Standard Edition of the Complete Psychoanalytic Works of Sigmund Freud.* [S. E.]. 24 volumes. Edited by James Strachey and others. London
——— 1954 (1897–1902). *The Origins of Psychoanalysis: Letters to Wilhelm Fliess, Drafts and Notes.* Edited by M. Bonaparte, A. Freud and E. Kris. London.
Friess, Horace. 1929. "Wilhelm Dilthey: A Review of his Collected Works." *Journal of Philosophy* 3 (Jan.):5–25.
Fromm, E. 1941. *Escape from Freedom.* New York.
——— 1959. *Sigmund Freud's Mission.* New York.
——— D. T. Suzuki, and R. de Martino. 1960. *Zen Buddhism and Psychoanalysis.* New York.
Fueloep-Miller, R. 1956. *The Power and the Secret of the Jesuits.* Translated by F. S. Flint and D. F. Tait. New York.
Galilei, Galileo. 1890–1909. *Opere.* 20 volumes. Edited by A. Favaro. Florence.
——— (1615–1617). "Considerazioni circa l'opinione Copernicana." Pp. 349–370 in *Opere* 5.
——— 1953 (1632). *Dialogue Concerning the Two Chief World Systems—Ptolemaic*

and Copernican. Translated by Stillman Drake with a foreword by Albert Einstein. Berkeley, California.

——— 1965. *Sidereus Nuncius. Nachricht von neuen Sternen*. Edited with a introduction by Hans Blumenberg. Frankurt am Main.

Gandillac, Maurice de. 1933. "De l'usage et de la valeur des arguments probables dans les questions du cardinal Pierre d'Ailly sur les sentences." *Archives d'histoire littéraire et doctrinale du Moyen-âge* 1:43–91.

Garin, Eugenio. 1971. "A Proposition de Copernico . . . (Schede 3)." *Rivista di critica di storia di filosofia*. 26:81–96.

Garzend, Leon Abbé. 1912. *Galilée et l'inquisition*. Paris.

Gebler, Karl von. 1876. *Galileo Galilei und die römische Kurie*. Stuttgart.

Gebser, Jean. 1966. *Ursprung und Gegenwart*. 2nd edition. Stuttgart.

Gerber, Uwe. 1970. *Disputatio als Sprache des Glaubens*. Zurich.

Geyer, B. 1928. See Ueberweg.

Ghellinck, Joseph de. 1946. *L'essor de la littérature latine au XIIe siècle*. (Museum Lessianum, Section historique, iv–v.) Brussels.

——— 1948. *La mouvement théologégue du XIIe siècle. Nouvelle édition*. (Museum Lessianum, Section historique, X.) Brussels

Gieben, Servus. 1964. "Traces of God in Nature According to Robert Grosseteste." *Franciscan Studies* 24:144–158.

Gierke, O. von. 1900. *The Political Theory of the Middle Ages*. Translated with an introduction by F. C. Maitland. Cambridge.

Gilbert, N. 1963. "Galileo and the School of Padua." *Journal of the History of Philosophy* 1:223–231.

Gillispie, Charles C. 1960. *The Edge of Objectivity*. Princeton.

Gilson, Etienne. 1922. *La philosophie au moyen-age*. 2 vols. Paris.

——— 1930. *Etudes sur le rôle de la pensée medievale dans la formation du système cartésien*. Paris.

——— 1938. *The Philosophy of St. Bonaventure*. Translated by Dom Illtyd. New York.

——— 1940a. *The Mystical Theology of St. Bernard of Clairvaux*. Translated by A. H. C. Downes. New York.

——— 1940b. *The Spirit of Medieval Philosophy*. New York.

——— 1955. *History of Christian Philosophy in the Middle Ages*. New York.

——— 1960. *The Christian Philosophy of St. Augustine*. Translated by L. E. M. Lynch. New York.

——— 1963. *Dante and Philosophy*. Translated by David Moore. New York.

Ginsberg, Benjamin. 1936. "Duhem and Jordanus Nemorarius." *Isis* 25:341–362.

Gittleman, Jerome. 1974. "Structures of Consciousness, Universalization Processes, Civilizations." *Sociological Analysis* 35, 2:79–84.

Glacken, Clarence. 1967. *Traces on the Rhodian Shore*. Berkeley.

Glorieux, P. (ed.). 1963. "Walter of St. Victor's 'Contra Quattuor Labyrinthos Franciae'." *Archives d'histoire doctrinale et littéraire du Moyen Age* 19:187–355.

Glotz, Gustave. 1904. *La solidarité de la famille dans le droit criminel en Grèce*. Paris.

Goffman, Erving. 1959. *The Presentation of Self in Everyday Life*. New York.

Golino, Carlo (ed.). 1966. *Galileo Reappraised*. Berkeley.

Goode, William J. 1960. "A Theory of Role Strain." *American Sociological* Review 25, 4:483–96.
Goodman, Paul. 1960. *Growing Up Absurd.* New York.
Grant, Edward C. 1962a. "Hypotheses in Later Medieval and Early Modern Science." *Daedalus* 91, 3:612–616.
——— 1962b. "Late Medieval Thought, Copernicus and the Scientific Revolution." *Journal of the History of Ideas* 23:197–220.
——— 1971. *Physical Science in the Middle Ages.* New York.
Grant, R. M. (ed.). 1961. *Gnosticism.* New York.
Grinker, Roy (ed.). 1956. *Toward a Unified Theory of Human Behavior.* New York.
Groethuysen, B. 1927–30. *Die Entstehung der bürgerlichen welt-und Lebensanschauung* in Frankreich. 2 vols. Halle.
——— 1968. *The Bourgeois. Catholicism versus Capitalism in Eighteenth-Century France.* Translated from the French by Mary Ilford with an introduction by B. Nelson. New York.
Grosseteste, Robert. 1963 (ca. 1228–1233). *Commentarius in VIII Libros Physicorum Aristotelis.* Edited by Richard C. Dales. Boulder, Colorado.
Grossman, Henryk. 1935. "Die gesellschaftlichen Grundlagen der mechanistischen Philosophie und die Manufaktur." *Zeitschrift für Sozialforschung* 4:161–231.
Grünbaum, Adolf. 1960. "The Duhemian Argument." *Philosophy of Science* 27:75–87.
Grunebaum, G. E. von. 1953. *Medieval Islam: A Study in Cultural Orientation.* 2nd edition. Chicago.
——— 1961. *Islam. Essays in the Nature and Growth of a Cultural Tradition.* London.
——— 1962. *Modern Islam.* Berkeley.
——— (ed.). 1968. *The Islamic World.* London.
——— (ed.). 1970a. *Logic in Classical Islamic Culture.* Wiesbaden.
——— 1970b. *Classical Islam.* Chicago.
Habermas, Jürgen. 1976. "Geschichte und Evolution." Pp. 200–259 in *Zur Rekonstruktion des Historischen Materialismus.* Frankfurt am Main. [Now translated as "History and Evolution." *Telos* 12 (1979), 39:5–44.]
Hagstrom, W. O. 1965. *The Scientific Community.* New York.
Hall, A. Rupert. 1954. *The Scientific Revolution, 1500–1800.* London.
——— 1959. "The Scholar and the Craftsman in the Scientific Revolution." Pp. 3–21 in *Critical Problems in the History of Science,* edited by M. Clagett. Madison.
——— 1963a. *From Galileo to Newton: 1630–1720.* New York.
——— 1963b. "Merton Revisited or Science and Society in the Seventeenth Century." *History of Science* 2:1–16.
Hall, Marie Boas. 1962. *The Scientific Renaissance: 1450–1630.* New York.
——— (ed.). 1970. *Nature and Nature's Laws. Documents of the Scientific Revolution.* New York.
Hallowell, A. I. 1959. *Culture and Experience.* Philadelphia.
Hanson, N. R. 1958. *Patterns of Discovery.* Cambridge.
——— 1963. *The Concept of the Positron.* Cambridge.

——— 1965. "Galileo's Real Discoveries in Dynamics." Pp. 42–69 in M. Kaplan (ed.), *Homage to Galileo.* Cambridge, Massachusetts.

——— 1973. *Constellations and Conjectures.* Edited by W. D. Humphreys. Dordrecht-Boston.

Hardeland, A. 1897–98. *Geschichte der speciellen Seelsorge, in der vorreformatorischen Kirche und. . . .* Berlin.

Harnack, A. von. 1897. *History of Dogma.* 7 vols. Translated from the 3rd German edition. London.

Harrison, Jane. 1912. *Themis.* Cambridge.

——— 1962 (1921). *Epilegomena to the Study of Greek Religion and Themis: A Study of the Social Origins of Greek Religion.* New Hyde Park, New York.

Hart, H. L. A., and A. M. Honoré. 1959. *Causation in the Law.* Oxford.

Hartmann, H. 1958. *Ego Psychology and the Problem of Adaptation.* Translated by D. Rapaport. New York.

Haskins, Charles. 1957. *The Rise of the Universities.* Ithaca, New York.

——— 1960 (1924). *Studies in the History of Medieval Science.* New York.

——— 1961 (1927). *The Renaissance of the 12th Century.* New York.

Hatch, Edwin. 1957 (1888). *The Influence of Greek Ideas on Christianity.* New York.

Hazard, P. 1964 (1953). *The European Mind, The Critical Years, 1680–1715.* Translated by J. L. May. Cleveland.

Heer, F. 1961. *The Medieval World: Europe 1100–1350.* Translated by J. Sondheimer. Cleveland.

Hegel, G. W. F. 1955 (1807). *Phenomenology of the Mind.* Translated by J. B. Baille. New York.

Heidegger, Martin. 1962 (1927). *Being and Time.* Translated by J. Macquarrie and E. Robinson. New York.

Heiler, Friedrich. 1932. *Prayer: A Study in the History and Psychology of Religion.* Translated and edited by S. McComb. New York.

——— 1960. "Contemplation in Christian Mysticism." Pp. 186–238 in *Papers From the Eranos Yearbook* 4, edited by J. Campbell.

Heinemann, F. H. 1958. *Existentialism and the Modern Predicament.* 2nd edition. New York.

Heisenberg, Werner. 1958. *Physics and Philosophy. The Revolution in Modern Science.* New York.

Heller, E. 1957. *The Disinherited Mind: Essays in Modern German Literature and Thought.* New York.

Hellman, C. Doris. 1963. "Was Tycho Brahe as Influential as He Thought?" *British Journal for the History of Science* 1 pt. 4, 4:295–324.

——— 1970. "Tycho Brahe." Pp. 401–516 in *Dictionary of Scientific Biography* 2, edited by C. C. Gillispie. New York.

Helton, Tinsley (ed.). 1964. *The Renaissance: A Reconsideration of the Theories and Interpretation of the Age.* Madison.

Hessen, B. 1931. "The Social and Economic Roots of Newton's 'Principia'." Pp. 147–212 in *Science at the Crossroads.* London.

Hiltner, S. (ed.). 1958. "Rogers and Niebuhr." *Pastoral Psychology* 9 (Special number (85)).

Hilton, W. (d. 1396) 1948. *The Scale of Perfection.* Newly edited from ms. sources with an introduction by E. Underhill. London.
Hine, William. 1973. "Mersenne and Copernicanism." *Isis* 64, 221:18–32.
Hof, H. 1952. *Scintilla Animae: Eine Studie zu einem Grundbegriff in Meister Eckharts Philosophie.* Bonn.
Hofmann, Rudolf. 1941. *Die Gewissenslehre des Walter von Brügge O. F. M. und die Entwicklung der Gewissenslehre in der Hochscholastik. Beiträge zur Geschichte der Philosophie und Theologie des Mittelalters* 36. Münster.
Hofstadter, R. and W. Metzger. 1955. *The Development of Academic Freedom in the United States.* New York.
Holl, K. 1961. *The Cultural Significance of the Reformation.* New York.
Holton, Gerald. 1952. *Introduction to Concepts and Theories in the Physical Sciences.* Cambridge, Massachusetts.
——— 1965. *Science and Culture.* Boston.
——— 1973. *Thematic Origins of Scientific Thought: Kepler to Einstein.* Cambridge, Massachusetts.
——— and D. Roller. 1958. *Foundations of Modern Physical Science.* Reading, Massachusetts.
Holzner, Burkart. 1974. "Inter-Civilizational Encounters and the Construction of Social Actors. A Note." *Sociological Analysis* 35, 2:119–122.
Hopper, S. R. 1947. *The Crisis of Faith.* London.
——— (ed.). 1957. *Spiritual Problems in Contemporary Literature.* New York.
Horton, Robin. 1971. "African Traditional Thought and Western Science." Pp. 131–171 in *Rationality,* edited by B. Wilson. New York.
Huff, T. E. 1974. "Is A Theory of Sociocultural Process Possible Without Reference to 'Civilizational Complexes'?" *Sociological Analysis* 35, 2:85–94.
Huizinga, Jan. 1924 (1919). *The Waning of the Middle Ages.* New York.
Husserl, Edmund. 1970 (1954). *The Crisis of European Sciences and Transcendental Phenomenology.* Translated with an introduction by David Carr. Evanston, Illinois.
Huxley, A. 1944. *The Perennial Philosophy.* New York.
ICHSCDM (International Commission for a History of the Scientific and Cultural Development of Mankind). 1963–70. *History of Mankind: Cultural and Scientific Developments.* 6 vols. New York.
Inge, W. R. 1899. *Christian Mysticism.* New York.
Jabre, F. 1958. *La notion de certitude selon Ghazali dans ses origines psychologiques et historiques.* Paris.
Jackson, Don D. (ed.). 1960. *The Etiology of Schizophrenia: Genetics, Physiology, Psychology, Sociology.* New York.
James, W. 1958 (1901–02). *The Varieties of Religious Experience.* With a foreword by J. Barzun. New York.
Jaspers, K. 1949. *Von Ursprung und Ziel der Geschichte.* München.
——— 1962. *The Great Philosophers.* Edited by H. Arendt and translated by R. Manheim. New York.
Jhering, Rudolph von. 1915 (1879). *The Struggle For Law.* Translated by J. J. Lalor from the 2nd edition. Chicago.
Jocelin de Brakeland. (See Butler 1949).

John of the Cross, Saint (d. 1591). 1934–35. *The Complete Works of St. John of the Cross.* 3 volumes. Edited and translated by E. A. Peers. London.

Johnson, Francis R. 1959. "Commentary on the Paper of Derek J. de Solla Price." Pp. 219–221 in *Critical Problems in the History of Science,* edited by M. Clagett. Madison.

Jolivet, Jean. 1969. *Arts du language et theologie chez Abelard.* Paris.

Jonas, H. 1950. *The Gnostic Tradition.* Boston.

——— 1960 (1952). "Gnosticism and Modern Nihilism." *Social Research* 19, 4:430–452.

Jones, Christine (see Schofield).

Jones, Ernest. 1953–57. *The Life and Work of Sigmund Freud.* 3 volumes. New York.

Jones, Richard F. 1961. *Ancients and Moderns: A Study of the Rise of the Scientific Movement in Seventeenth Century England.* St. Louis.

Joravsky, David. 1961. *Soviet Marxism and Natural Science, 1917–1932.* New York.

Jung, C. (1931). *Modern Man in Search of a Soul.* New York. n.d.

——— 1938. *Psychology and Religion - West and East.* Translated by R. F. C. Hull. New York.

——— 1958. *The Undiscovered Self.* Translated by R. F. C. Hull. Boston.

——— 1962 (1912). *Symbols of Transformation.* 2 volumes. Translated by R. F. C. Hull. New York.

Kaplon, M. (ed.). 1965. *Homage to Galileo.* Cambridge, Massachusetts.

Kapp, Ernst. 1942. *Greek Foundations of Traditional Logic.* New York.

Kardiner, A. 1957. "Freud—The Man I Knew, The Scientist and His Influence." Pp. 46–58 in *Freud and the Twentieth Century,* edited by B. Nelson. New York.

Katzenellenbogen, Adolf. 1959. *The Sculptural Program of Chartres Cathedral: Christ, Mary, Ecclesia.* Baltimore.

Kaufmann, W. A. (ed.). 1956. *Existentialism from Dostoevsky to Sartre.* New York.

Kavolis, V. 1974. "On the Structures of Consciousness." *Sociological Analysis* 35, 2: 115–118.

Kempis, Thomas à (d. 1471). 1905–08. *The Imitation of Christ.* Volume 6 of *The Works of Thomas à Kempis.* London.

Kepler, Johannes. 1938. "Brief des Mathias Hafenreffer an Kepler vom 12 April 1598." Pp. 202–204 in *Gesammelte Werke* 13, edited by W. v. Dyck and M. Caspar. München.

——— 1967 (1600–01). "Apologia Tychonis contra Ursum." Pp. 216–287 in *Landmarks of Science.* New York (Readex Microprint from *Johannes Kepler's astronomia opera omnia* I [1858], pp. 216–287. Frankfurt.)

Kierkegaard, Soren. 1962 (1859). *The Point of View for my Work as an Author.* Translated by W. Lowrie, edited with a preface by B. Nelson. New York.

——— 1954. *Fear and Trembling, The Sickness Unto Death.* Translated by Robert Payne. New York.

Kirk, K. E. 1925. *Ignorance, Faith and Conformity.* London.

——— 1927. *Conscience and Its Problems. An Introduction to Casuistry.* London.

——— 1937. *The Vision of God: The Christian Doctrine of tne SUMMUM BONUM.* London.

Kline, Morris. 1972. *Mathematical Thought from Ancient to Modern Times.* New York.
Kluckhohn, Clyde, and A. Kroeber. 1952a. *The Nature of Culture.* Chicago.
——— 1952b. *Culture: A Critical Review of Concepts and Definitions.* Cambridge, Massachusetts.
Kluckhohn, Clyde, and Henry Murray. 1948. *Personality in Nature, Society and Culture.* New York.
Kluckhohn, Florence. 1956. "Value Orientations." Pp. 83–93 in *Toward a Unified Theory of Human Behavior,* edited by Roy Grinker. New York.
Kluckhohn, F. and F. L. Strodbeck. 1961. *Variations in Value Orientations.* Evanston, Illinois.
Kneale, William, and Martha Kneale. 1962. *The Development of Logic.* Oxford.
Knowles, David. 1963a. *The Historian and Character.* New York.
——— 1963b. *Saints and Scholars: Twenty-five Medieval Portraits.* Cambridge.
——— 1964. *The Evolution of Medieval Thought.* New York.
Knox, R. 1950. *Enthusiasm.* New York.
Koestler, A. 1945. *The Yogi and the Commissar.* New York.
——— 1959. *Sleepwalkers.* New York.
——— 1961. *The Lotus and the Robot.* New York.
Koester, Helmut. 1968. "ΝΟΜΟΣ ΦΥΣΕΩΣ: The Concept of Natural Law in Greek Thought." Pp. 521–541 in *Religions in Antiquity. Essays in Memory of Erwin Ramsdell Goodenough,* edited by Jacob Neusner. Leiden.
Kohlberg, L. 1971. "From Is to Ought: How to Commit the Naturalistic Fallacy and Get Away with it in the Study of Moral Development." Pp. 151–235 in *Cognitive Development and Epistemology,* edited by T. Mischel. New York.
Koyré, Alexandre. 1939. *Etudes galiléenes.* 3 volumes. Paris.
——— 1949. "Le vide et l'espace infini au XIVe siècle." *Archives d'histoire littéraire et doctrinale du moyen-âge* 24: 49–91.
——— 1957a. *From the Closed World to the Infinite Universe.* Baltimore.
——— 1957b. "Galileo and Plato." Pp. 147–175 in *Roots of Scientific Thought: A Cultural Perspective,* edited by P. P. Wiener and A. Noland. New York.
——— 1964. "The Exact Sciences." Pp. 11–104 in *The Beginning of Modern Science from 1450 to 1800,* edited by R. Taton. New York.
——— 1965. *Newtonian Studies.* London.
——— 1968. *Metaphysics and Measurement: Essays in the Scientific Revolution.* Cambridge, Massachusetts.
Kraft, Julius. 1957. *Von Husserl zu Heidegger.* Frankfurt.
Kris, E. 1952. *Psychoanalytic Explorations in Art.* New York.
Kristeller, Paul O. 1956. *Studies in Renaissance Thought and Letters.* Rome.
Kroeber, Alfred, and C. Kluckhohn. 1952. *Culture: A Critical Review of Concepts and Definitions.* Cambridge, Massachusetts.
Kroeber, Alfred and T. Parsons. 1958. "The Concepts of Culture and of Social System." *American Sociological Review* 23, 5: 582–83.
Kuhn, Thomas. 1957. *The Copernican Revolution.* Cambridge, Massachusetts.
——— 1962. *The Structure of Scientific Revolutions.* Chicago.
——— 1963. "The Function of Dogma in Scientific Research." Pp. 347–369 in *Scientific Change,* edited by A. C. Crombie. New York.
Kwok, D. W. Y. 1965. *Scientism in Chinese Thought: 1900–1950.* New Haven.

Laing, R. D. 1960. *The Divided Self*. Chicago.

Landgraf, Arthur. 1934. *Escrits théologiques de l'école d'Abelard*. Louvain.

Langer, Suzanne. 1942. *Philosophy in a New Key*. Cambridge, Massachusetts.

Langford, Jerome, O.P. 1960. "The Condemnation of Galileo." *Reality* 8: 65–78.

LaPiere, R. 1959. *The Freudian Ethic: An Analysis of the Subversion of American Character*. New York.

Laquer, W., and G. Lichtheim. 1958. *The Soviet Cultural Scene 1956–1957*. London.

Lasswell, H. 1932. "The Triple-Appeal Principle." *American Journal of Sociology* 37 (Jan.): 523–38.

Lea, Henry C. 1896. *A History of Auricular Confession and Indulgences*. 3 volumes. Philadelphia.

——— 1955 (1887). *A History of the Inquisition of the Middle Ages*. 3 volumes. New York.

Leclerq, Jean. 1945. "L'amitié dans les lettres au moyen-âge: Autour d'un manuscript de la bibliothèque de Pétrarque." *Revue du moyen-âge latin* 1: 391–410.

——— 1960. *The Love of Learning and the Desire for God: A Study of Monastic Culture*. Translated by C. Misrachi. New York.

Leeuwen, H. G. van. 1963. *The Problem of Certainty in English Thought: 1630–1690*. The Hague.

LeGoff, Jacques. 1957. *Les intellectuels au moyen-âge*. Paris.

——— 1964. *La civilisation de l'occident médiéval*. Paris.

Leibrecht, W. (ed.). 1959. *Religion and Culture: Essays in Honor of Paul Tillich*. New York.

Leites, Edmund. 1974. "Autonomy and the Rationalization of Moral Discourse." *Sociological Analysis* 35, 2: 95–101.

Lenin, W. I. 1927 (1908). *Materialism and Empirio-Criticism: Critical Notes Concerning a Reactionary Philosophy*. (Volume 13 of *Collected Works*). New York. (The second Russian edition of 1920 and the cited English translation carry an important epilogue by V. Nevsky, authorized by Lenin.)

Lenoble, Robert. 1943. *Mersenne et la naissance du méchanisme*. Paris.

Lerner, Ralph, and Mushin Mahdi (eds.). 1965. *Medieval Political Philosophy*. New York.

Levenson, Joseph. 1953. "The Abortiveness of Empiricism in Early Ch'ing Thought." *Far Eastern Quarterly* 13, 1: 155–165.

——— 1965. *Confucian China and Its Modern Fate*. 3 volumes. Berkeley.

Lévi-Strauss, C. 1960. "Four Winnebago Myths: A Structural Sketch." Pp. 351–362 in *Culture in History: Essays in Honor of Paul Radin*, edited by S. Diamond. New York.

——— 1963. *Structural Anthropology*. Translated by C. Jacobson and B. G. Schoepf. New York.

Lewin, B., and Helen Ross. 1960. *Psychoanalytic Education in the United States*. New York.

Lifton, R. J. 1961. *Thought Reform and the Psychology of Totalism: A Study of "Brainwashing" in China*. New York.

Lindberg, David. 1970. (see J. Pecham).

Linton, Ralph. 1936. *The Study of Man*. New York.

——— 1956. *Culture and Mental Disorders*. Edited by George Devereux. Springfield, Illinois.

Lipset, S. M. and L. Löwenthal (eds.). 1961. *Culture and Social Character: The Work of David Riesman in Review*. Glencoe, Illinois.

Lods, A. 1937. *The Prophets and the Rise of Judaism*. Translated by S. H. Hooke. London.

Loewith, K. 1941. *Von Hegel bis Nietzsche*. New York.

Longpré, E. (ed.). 1928. *Quaestiones disputatae du Gauthier de Bruges*. (*Les philosophes belges*, volume 10). Louvain.

Lopez, Robert S. 1971. *The Commercial Revolution of the Middle Ages, 950–1350*. Englewood Cliffs, New Jersey.

Lottin, Dom. O. 1942–60. *Psychologie et morale aux douzième et treizième siècles*. 6 volumes. Louvain.

Lovejoy, A. O. 1936. *The Great Chain of Being*. Cambridge, Massachusetts.

——— 1955. *Essays in the History of Ideas*. New York.

——— 1961. *The Reason, The Understanding and The Time*. Baltimore.

——— 1963. *The Thirteen Pragmatisms*. Baltimore.

Lovejoy, A. O.; Chinard, G.; Boas, G.; and R. S. Crane (eds.). 1935. *A Documentary History of Primitivism and Related Ideas*. Volume I: *Primitivism and Related Ideas in Antiquity*. Baltimore.

Lowinger, Armand. 1941. *The Methodology of Pierre Duhem*. New York.

Loyola, Ignatius (d. 1556). 1950. *Spiritual Exercises*. London.

Mach, Ernst. 1960 (1883). *The Science of Mechanics: A Critical and Historical Account of Its Development*. Translated by Thomas J. McCormack. La Salle, Illinois.

MacClintock, S. 1956. *Perversity and Error: Studies on the "Averroist" John of Jandun*. Bloomington.

Macquarrie, J. 1963. *Twentieth-Century Religious Thought*. New York.

Madden, Edward; Blake, R. N.; and C. J. Ducasse. 1960. *Theories of Scientific Method: The Renaissance Through the Nineteenth Century*. Seattle.

Maier, Annaliese. 1956–1968. *Studien zur Naturphilosophie der Spätscholastik* (2nd edition): Volume 1: *Die Vorläufer Galileis im 14. Jahrhundert;* Volume 2: *Zwei Grundprobleme der scholastischen Naturphilosophie, das Problem der intensiven Grösse und die Impetustheories;* Volume 3: *An der Grenze von Scholastik und Naturwissenschaft. Die Struktur der materiellen Substanz, das Problem der Gravitation, die Mathematik der Formlatituden;* Volume 4: *Metaphysische Hintergründe der spätscholastischen Naturphilosophie;* Volume 5: *Zwischen Philosophie und Mechanik. Studien zur Naturphilosophie der Spätscholastik*. Rome.

——— 1964–67. *Ausgehendes Mittelalter. Gesammelte Aufsätze zur Geistesgeschichte des 14. Jahrhunderts* (Volume 1, 1964; Volume 2, 1967). Rome.

Madge, John. 1962. *The Origins of Scientific Sociology*. New York.

Maine, Henry Sumner. 1963 (1861). *Ancient Law*. Boston.

——— 1875. *Village Communities: East and West*. London.

Maire, Albert. 1925. *Bibliographie des oeuvres de Blaise Pascal*. 5 volumes. 2nd edition. Paris. The first volume of the first edition (1911) has a preface by Duhem, which is reprinted here. Also preface by E. Picard.

Malinowski, B. 1962. *Sex, Culture, and Myth.* New York.
Mandonnet, Pierre, O.P. 1908–1911. *Siger de Brabant et l'averroisme latin au XIIe siècle: étude critique et documents inédits.* 2nd edition, 2 volumes. (*Les philosophes belges,* volume 6–7). Louvain.
Mannheim, Karl. 1936. *Ideology and Utopia.* Translated by L. Wirth and E. Shils. New York.
Marcuse, Herbert. 1966 (1955). *Eros and Civilization.* New York.
——— 1964. *One-Dimensional Man.* Boston.
Maritain, Jacques. 1928. *Three Reformers: Luther-Descartes-Rousseau.* London.
Marshall, S. L. A. 1953. *Men Against Fire.* New York.
Mathias, Peter (ed.). 1972. *Science and Society: 1600–1900.* Cambridge.
Mauss, Marcel. 1950. *Sociologie et Anthropologie.* With an introduction by C. Lévi-Strauss. Paris.
——— 1954. *The Gift.* Translated by Ian Cunnison. Glencoe, Illinois.
——— 1968a (1938). "A Category of the Human Spirit." Translated by Lawrence Krader. *The Psychoanalytic Review* 55, 3: 457–481.
——— 1968b. *Oeuvres.* 3 volumes. Paris.
May, Rollo; Angel, E.; and H. F. Ellenberger (eds.). 1958. *Existence.* New York.
Mayo, E. 1945. *The Social Problems of an Industrial Civilization.* Cambridge, Massachusetts.
——— 1952. *The Psychology of Pierre Janet.* London.
McKeon, Richard. 1935. "Renaissance and Method in Philosophy." *Studies in the History of Ideas* 3: 37–114.
McLaughlin, Mary M. 1954. *Intellectual Freedom and Its Limitations in the University of Paris in the Thirteenth and Fourteenth Centuries.* Ph.D. Thesis. Columbia University.
McMullin, Ernan. 1967. "Empiricism and the Scientific Revolution." Pp. 331–369 in *Art, Science and History in the Renaissance,* edited by Charles Singleton. Baltimore.
——— (ed.) 1968. *Galileo: Man of Science.* New York.
McNeill, J. 1951. *A History of the Cure of Souls.* New York.
Menniger, K. 1958. *Theory of Psychoanalytic Technique.* New York.
Merriam, Charles. 1934. *Political Power.* New York.
Merton, Robert K. 1938. "Social Structure and Anomie." *American Sociological Review* 3, 5: 672–82.
——— 1960. "Social Conformity, Deviation, and Opportunity-Structures. A Comment on the Contributions of Dubin and Cloward." *American Sociological Review* 24, 2: 177–89.
——— 1968 (1959). "Continuities in the Theory of Social Structure and Anomie." Pp. 215–248 in *Social Theory and Social Structure.* Enlarged edition. New York.
——— 1970 (1938). *Science, Technology and Society in Seventeenth Century England.* New York.
——— 1972. "Insiders and Outsiders: A Chapter in the Sociology of Knowledge." *American Journal of Sociology* 78, 1: 9–47.
Merton, Robert K., and Alice C. Kitt. 1950. "Contributions to the Theory of Reference Group Behavior." Pp. 40–105 in *Continuities in Social Research:*

Studies in the Scope and Method of "the American Soldier," edited by Merton and P. Lazarsfeld. New York.

Merton, Robert K.; Broom, L; and L. S. Cottrell (eds.). 1959. *Sociology Today.* New York.

Michalski, K. 1919–20. "Les courants philosophiques à Oxford et à Paris pendant le XIV siècle." *Bulletin internationale de l'Académie polonaise des Sciences et Lettres, Classe de philologie d'histoire et de philosophie.*

——— 1925. "Les courants critiques et sceptiques dans la philosophie du XIVe sicle." *Bulletin internationale de l'Académie polonaise des Sciences et Lettres . . .*

Michaud-Quantin, Pierre. 1962. *Sommes de casuistique et manuels de confession au moyen-âge.* Montreal.

——— 1970a. *Universitas: Expressions du mouvement communitaire dans le Moyen-Age latin.* Paris.

——— (with the collaboration of Michel Lemoine). 1970b. *Etudes sur le vocabulaire philosophique du moyen-âge.* Rome.

Miller, D. R., and Guy Swanson. 1960. *Inner Conflict and Defense.* New York.

Mills, Laurens J. 1937. *One Soul in Bodies Twain: Friendship in Tudor Literature and Stuart Drama.* Bloomington, Indiana.

Mizruchi, Ephraim. 1960. "Social Structure and Anomie in a Small City." *American Sociological Review* 25: 645–54.

——— 1961. "Reply to Lorna Mui." *American Sociological Review* 26, 2: 277–278.

Montaigne, Michel E. de (1533–1592). 1948. "Apology for Raymond Sebonde." Pp. 318–457 in *The Complete Works of Montaigne: Essays, Travel Journal, Letters.* Translated by D. M. Frame. Stanford, California.

Moody, Ernest A. 1935. *The Logic of William of Occam.* New York.

——— 1951. "Galileo and Avempace." *Journal of the History of Ideas* 12: 163–193, 375–422. Reprinted in *Roots of Scientific Thought,* edited by Wiener and Noland. New York, 1957.

——— 1966. "Galileo and His Precursors." Pp. 23–43 in *Galileo Reappraised,* edited by C. Golino. Berkeley.

Morrison, Karl F. 1969. *Tradition and Authority in the Western Church, 300–1140.* Princeton.

Mosse, George. 1957. *The Holy Pretence: A Study in Christianity and Reason of State from William Perkins to John Winthrop.* Oxford.

Mowrer, O. H. 1961. *The Crisis in Psychiatry and Religion.* Princeton.

Müller-Markus, S. 1965. "Soviet Discussion of General Relativity Theory." *Studies in Soviet Thought* 5: 204–222.

Munk, S. 1927 (1859). *Mélanges de philosophie juive et arabe.* Paris.

Munroe, Ruth. 1955. *Schools of Psychoanalytic Thought.* New York.

Murdoch, John. 1964. "Superposition, Congruence and Continuity in the Middle Ages." Pp. 416–441 in *L'aventure de la science: Mélanges Alexandre Koyré* 1, edited by Cohen and Taton. Paris.

——— 1969. "Mathesis in philosophiam scholasticam introducta." In *Arts libéraux et philosophie au moyen-âge.* Paris.

Murray, Albert V. 1967. *Abelard and St. Bernard: A Study in Twelfth Century "Modernism."* Manchester.

Murray, Gilbert. 1925. *Five Stages of Greek Religion*. Oxford.

Nagel, Ernest. 1961. *The Structure of Science*. New York.

Nakayama, S. 1973. "Science and Technology in China." Pp. 131–150 in *Half the World. The History and Culture of China and Japan*, edited by A. Toynbee. London.

Nardi, Bruno. 1942. *Dante e la cultura medievale*. Bari.

——— 1958. *Saggi sull'aristotelisma padavano dal secolo XIV al XVI*. Florence.

Needham, Joseph. 1954–71. *Science and Civilization in China*. 4 volumes in 7 parts, in progress. New York.

——— 1963. "Poverties and Triumphs of the Chinese Scientific Traditions." Pp. 117–153 in *Scientific Change*, edited by A. C. Crombie. New York.

——— 1969a. *The Grand Titration: Science and Society in East and West*. London.

——— 1969b. *Within the Four Seasons*. London.

——— 1970. *Clerks and Craftsmen in China and the West*. New York.

Nelson, Marie Coleman (ed.). 1963. *Paradigmatic Approaches to Psychoanalysis: Four Papers*. New York.

Niebuhr, H. R. 1957. *Social Sources of Denominationalism*. New York.

Nielsen, Donald. 1974. "Cities, Polities, and Ethnic Communities in Nelson's Civilization Analytic Perspective." *Sociological Analysis* 35, 2:102–114.

Nock, A. D. 1933. *Conversion: The Old and the New in Religion from Alexander the Great to Augustine of Hippo*. Oxford.

Nordström, Johann. 1933. *Moyen-âge et Renaissance, Essai Historique*. Translated by T. Hammer. Paris.

Nygren, A. 1953. *Agape and Eros*. Translated by P. S. Watson. Philadelphia.

Oakley, Francis P. 1961. "Christian Theology and the Newtonian Science: The Rise of the Concept of the Laws of Nature." *Church History* 30:433–457.

Oates, W. J. (ed.). 1940. *The Stoic and Epicurean Philosophers: The Complete Extant Writings of Epicurus, Epictetus, Lucretius, Marcus Aurelius*. New York.

Oberman, Heiko. 1963. *The Harvest of Medieval Theology: Gabriel Biel and Late Medieval Nominalism*. Cambridge.

——— 1975. "The Reformation and Revolution: Copernicus' Discovery in an Era of Change." Pp. 134–169 in *The Nature of Scientific Discovery*, edited by O. Gingerich. Washington, D.C.

Ockham, William of (d. 1349). 1957. *Philosophical Writings*. Edited and translated by Philotheus Boehner. London.

O'Donnell, J. R. 1931. "Nicholas of Autrecourt." *Medieval Studies* 1:179–280.

Olschki, L. 1942. "The Scientific Personality of Galileo." *Bulletin of the History of Medicine* 12:242ff.

——— 1949. *The Genius of Italy*. New York.

O'Meara, J. J. 1954. *The Young Augustine*. New York.

Opler, M. K. (ed.). 1959. *Culture and Mental Health*. New York.

——— 1961. "Cultural Definitions of Illness." Pp. 17–20 in *Conference on Medicine and Anthropology*. Harriman, New York.

Otto, Rudolph. 1950 (1917). *The Idea of the Holy*. Translated by J. W. Harvey. London.

——— 1957. *Mysticism, East and West: A Comparative Analysis of the Nature of Mysticism*. Translated by B. L. Bracey and R. C. Payne. New York.

Palter, Robert (ed.). 1961. *Toward Modern Science*. New York.

Panofsky, Erwin. 1967 (1957). *Gothic Architecture and Scholasticism*. New York.
Parsons, Talcott. 1937. *The Structure of Social Action*. New York.
——— and others. 1951. *Toward a General Theory of Action*. Cambridge, Massachusetts.
——— 1957. "Malinowski's Theory of the Social System." Pp. 53–70 in *Man and Culture: An Evaluation of the Work of Malinowski*, edited by R. Firth. London.
——— 1959. "A Rejoinder to Ogles and Levy." *American Sociological Review* 24, 2:248–249.
Parsons, Talcott; Shils, E.; Naegele, K. D.; and J. Pitts (eds). 1961. *Theories of Society*. 2 volumes. New York.
Pascal, Blaise. 1656 (1920). *Les lettres provinciales*. Edited with an introduction by H. F. Stewart. Manchester.
Passmore, John. 1957. *One Hundred Years of Philosophy*. New York.
Pastor, Ludwig. 1891–1940. *The History of the Popes, from the Close of the Middle Ages*. Translated by Dom E. Graf. London.
Pecham, John. 1970. *John Pecham and the Science of Optics. "Perspective Communis."* Edited, translated, and with an introduction and notes by D. C. Lindberg. Madison.
Pepper, Stephen. 1957. *World-Hypotheses*. Berkeley, California.
Petrus Hispani. 1947. *Summulae Logicales*. Edited by I. M. Bochenski. Turin.
Petrus Peregrinus (de Maricourt) (ca. 1269). 1898. *Epistola de Magnete*. Edited by G. Hellman. Berlin. (English translation by S. P. Thompson. London, 1902.)
Pettazzoni, R. 1929. *La confessione dei peccati*. Bologna.
Pfeiffer, Robert H. 1941. *Introduction to the Literature of the Old Testament*. New York.
Phillips, D. 1955. "The Way to Religious Perfection According to St. Bonaventura's 'De Triplice Via'." Pp. 31–58 in *Essays in Medieval Life and Thought*, edited by J. Mundy, B. Nelson and others. New York.
Picard, Emile. 1922. "La vie et l'oeuvre de Pierre Duhem." Pp. 1–39 in *Discours et Mélanges*. Paris.
Pirenne, Henri. 1956. *Economic and Social History of Medieval Europe*. New York.
Planck, M. 1949. *Scientific Autobiography and Other Papers*. New York.
Popkin, Richard H. 1960. *The History of Skepticism from Erasmus to Descartes*. Assen, Holland.
Popper, Karl. 1965 (1962). *Conjectures and Refutations. The Growth of Scientific Knowledge*. New York.
Pourrat, P. 1922. *Christian Spirituality*. New York.
Powell, Elwin H. 1959. "Rejoinder to Dr. Cary-Lundberg." *American Sociological Review* 24, 2:252–3.
Pratt, J. 1962. "Motivation and Learning in Medieval Writings." *American Psychologist* 17, 7:496–580.
Praz, M. 1956. *The Romantic Agony*. Translated by Angus Davidson. New York.
Previté-Orton, C. W. 1951 (1937). *A History of Europe, 1198–1378*. New York.
Price, Derek J. de Solla. 1959. "Contra-Copernicus: A Critical Re-estimate of

the Mathematical Planetary Theory of Ptolemy, Copernicus, and Kepler." Pp. 197–218 in *Critical Problems in the History of Science,* edited by M. Clagett.
——— 1962. *Little Science, Big Science.* New York.
Progoff, I. 1956. *The Death and Rebirth of Psychology.* New York.
Pumpian-Mindlin, E.; E. R. Hilgard; and L. S. Kubie. 1952. *Psychoanalysis as Science.* New York.
Ramon de Sabonde. 1966 (1436). *Theologia Naturalis seu Liber creaturarum.* Edited with an introduction by W. Stegmüller. Stuttgart.
Randall, John H. 1957a. "Scientific Method in the School of Padua." Pp. 139–146 in *Roots of Scientific Thought,* edited by Wiener and Noland.
——— 1957b. "The Place of Leonardo da Vinci in the Emergence of Modern Science." Pp. 207–218 in *Roots of Scientific Thought.*
——— 1962–64. *The Career of Philosophy: From the Middle Ages to the Enlightenment.* New York.
Rapaport, D. 1960. *The Structure of Psychoanalytic Theory.* New York.
Rashdall, Hastings. 1907."Nicolas of Autrecourt: A Medieval Hume." *Proceedings of the Aristotelian Society.* New series 8:1–27.
Raymond, M. 1950. *From Baudelaire to Surrealism.* New York.
Reichenbach, Hans. 1942. *From Copernicus to Einstein.* New York.
——— 1951. *The Rise of Scientific Philosophy.* Berkeley.
Renan, Ernest. 1852. *Averroes et Averroisme.* Paris.
Ricci, Matthew. 1953. *China in the Sixteenth Century: The Journals of Matthew Ricci: 1583–1610.* Translated from the Latin by L. J. Gallagher, SJ. New York.
Richard of Saint-Victor (d. 1173). [n.d.]. *Selected Writings.* Translated with an introduction by Claire Kirchberger. New York.
Rieff, P. 1959. *Freud: The Mind of the Moralist.* New York.
Riesman, David; R. Denney; and N. Glazer. 1950. *The Lonely Crowd.* New York.
Ritter, Gerhard. 1921–22. "Studien zür Spätscholastik." *Sitz. d. K. Akademia d. Wiss., phil-hist. Klasse* 4, 7.
Rosen, Edward. 1957. "The Ramus-Rheticus Correspondence." Pp. 287–292 in *Roots of Scientific Thought,* edited by Wiener and Noland.
——— 1959. *Three Copernican Treatises.* New York.
——— 1964. "Renaissance Science as Seen by Burckhardt and His Followers." Pp. 77–103 in *The Renaissance,* edited by T. Helton.
——— 1965. "The Debt of Classical Physics to Renaissance Astronomers, Particularly Kepler." *Proceedings of the Xth International Congress for the History of Science.* Paris.
Rossi, Paolo. 1970 (1962). *Philosophy, Technology and the Arts in the "Early Modern Era."* Translated by Salvator Attanasio. New York.
Rostow, W. W. 1962. *Stages of Economic Growth.* Cambridge.
Roszak, Theodore. 1969. *The Making of a Counter Culture: Reflections on the Technocratic Society and its Youthful Opposition.* New York.
Roussy, de Sales, R. de. 1938. "Love in America." *Atlantic Monthly* 161:645–51.
Rowbotham, A. 1942. *Missionaries and Mandarins.* New York.
Ruesch, Jurgen. 1961. *Therapeutic Communication.* New York.

Ruysbroeck, J. van (1293–1381). 1916. *Adornment of the Spiritual Marriage.* Translated by C. A. Wynschenck. London.
Sachs, W. 1947. *Black Hamlet.* Boston.
Sampson, R. V. 1956. *Progress in the Age of Reason.* London.
Santayana, G. 1905. *Reason in Science.* New York.
——— 1916a. *The Life of Reason.* 5 volume. New York.
——— 1916b. *Egotism in German Philosophy.* London.
——— 1922. *Soliloquies in England.* New York.
Santillana, Giorgio de. 1955. *The Crime of Galileo.* Chicago.
Sartre, J. P. 1953. *Existential Psychoanalysis.* Translated with an introduction by H. Barnes. New York.
Sarton, George. 1955. *Appreciation of Ancient and Medieval Science During the Renaissance (1450–1600).* Philadelphia.
——— 1968 (1927–48). *Introduction to the History of Science.* 3 volumes in 5 parts. Baltimore.
Schaar, J. H. 1961. *Escape from Authority: The Perspective of Erich Fromm.* New York.
Schapiro, Meyer. 1953. "Style." Pp. 287–312 in *Anthropology Today,* edited by A. Kroeber. Chicago.
Schein, E. H.; I. Schneier; and C. H. Barker. 1961. *Coercive Persuasion.* New York.
Scheler, Max, 1955 (1919). *Vom umsturz der Werte.* 2 volumes. Leipzig.
——— 1958. *Philosophical Perspectives.* Translated by Oscar A. Haac. Boston.
——— 1960. *On the Eternal in Man.* Translated by B. Noble. New York.
Schmitt, Carl. 1963 (1927). *Der Begriff der Politischen.* Berlin.
——— 1934. *Politische Theologie.* 2nd edition. Munich.
Schmitt, Charles. 1972. "The Faculty of Arts of Pisa at the Time of Galileo." *Physics* 14:243–272.
Schmitz, Mathias. 1965. *Die Freund-Feind–Theorie in der politischen Philosophie Carl Schmitts.* Cologne.
Schofield, Christine Jones. 1965. "The Geoheliocentric Mathematical Hypotheses in Sixteenth Century Planetary Theory." *British Journal for the History of Sciences* 2:291–296.
Schücking, Levin J. 1929. *The Puritan Family.* Translated by Brian Battershaw. New York.
Seeman, Melvin. 1959. "On the Meaning of Alienation." *American Sociological Review* 24, 2:783–91.
Seesholtz, Anna Groh. 1934. *Friends of God.* New York.
Selye, H. 1956. *The Stress of Life.* New York.
Senzaki, N., and R. S. McCandless (eds. and trans.). 1961. *The Iron Flute.* Rutland, Vermont.
Sextus, Empiricus. 1933. *Works.* Translated by R. B. Bury. 4 volumes. New York.
Shapiro, Hermann (ed.). 1965. *Medieval Philosophy: Selected Readings from Augustine to Buridan.* New York.
Shattuck, R. 1961. *The Banquet Years: The Origins of the Avant-Garde in France 1885 to World War I.* New York.
Sikes, J. G. 1932. *Peter Abailard.* Cambridge.

Singleton, Charles S. (ed.). 1967. *Art, Science and History in the Renaissance.* Baltimore.

Sivin, Nathan. 1968. Review of J. Needham, *Science and Civilization in China,* vol. 4, Part 2. *Journal of Asian Studies* 27, 4:859–864.

——— 1969. *Cosmos and Computation in Early Chinese Mathematical Astronomy.* Leiden.

——— 1970. Review of A. A. Bennett, *John Fryer: The Introduction of Western Science and Technology into Nineteenth Century China. Technology Review* (March):17–18.

——— 1971. Review of J. Needham, *The Grand Titration. Journal of Asian Studies* 30, 4:870–873.

——— 1972. Review of J. Needham. *Science and Civilization in China,* vol. 4, Part 3. *Scientific American* 226, 1:113–118.

——— 1973. "Copernicus in China." *Union internationale d'histoire et de philosophie des Sciences: Comité Nicolas Copernic: Colloquia II: Etudies sur l'audience de la théorie helicocentrique.* (Conference du Symposium de l'UIHPs.) Torun/Wroclam/Warsawa.

——— and H. Nakayama. 1973. *Traditional Science in China.* Cambridge, Mass.

Smith, T. V. 1934. *Beyond Conscience.* New York.

Smith, W. R. 1889. *The Religion of the Semites.* New York.

Solt, L. F. 1959. *Saints in Arms: Puritanism and Democracy in Cromwell's Army.* Stanford, California.

Sombart, Werner. 1915. *The Quintessence of Capitalism. A Study of the History and Psychology of the Modern Business Man.* Translated by M. Epstein. New York.

Sorokin, Pitirim A. 1928. *Contemporary Sociological Theory.* New York.

Southern, Richard W. 1953. *The Making of the Middle Ages.* New Haven.

——— 1970. *Medieval Humanism and Other Studies.* London.

Spektorsky, E. 1910–17. *The Problem of Social Physics in the 17th Century.* (Russian). 2 volumes. vol. 1: Warsaw 1910; vol. 2: Kiev 1917.

Spengler, Oswald. 1928. *The Decline of the West.* New York.

Spitzer, Leo. 1963. *Classical and Christian Ideas of World Harmony.* Edited by A. Granville Hatcher. Baltimore.

Stallo, J. B. 1960 (1881). *The Concepts and Theories of Modern Physics.* Edited by P. W. Bridgman. Cambridge, Mass.

Stein, M.; A. J. Vidich; and D. M. White. 1960. *Identity and Anxiety: Survival of the Person in Mass Society.* New York.

Sternagel, Peter. 1966. *Die artes mechanicae im Mittelalter. Begriffs– und Bedeutungsgeschichte bis zum Ende des 13. Jahrhunderts.* Kållmünz über Regensburg.

Stimson, Dorothy. 1917. *The Gradual Acceptance of the Copernican Hypothesis.* Diss. Hanover, New Hampshire.

Strong, Edward K. 1934. *Procedures and Metaphysics.* Berkeley, California.

——— 1957. "Newton's Mathematical Way." Pp. 412–432 in *Roots of Scientific Thought,* edited by Wiener and Noland.

——— 1959. "Hypothesis non fingo." Pp. 162–176 in *Men and Moments in History of Science,* edited by H. M. Evans.

Sullivan, J. 1959. "From Breuer to Freud." *Psychoanalysis and the Psychoanalytic Review* 46, 2:69–90.
Suso, H. (d. 1366). 1953. *Little Book of Eternal Wisdom and Little Book of Truth.* Translated with notes by James M. Clark. London.
Sutton, Robert. 1953. "The Phrase libertas philosophandi." *Journal of History of Ideas* 14:310–316.
Suzuki, D. T. 1956. *Zen Buddhism.* Edited by W. Barrett. New York.
——— 1957. *Mysticism: Christian and Buddhist.* New York.
Sypher, Wylie. 1962. *Loss of the Self in Modern Literature and Art.* New York.
Talbot, C. H. (editor and translator). 1959. *The Life of Christina of Markyate: A Twelfth Century Recluse.* Oxford.
Tanner, J. R., and others (eds.). 1957–59. *Cambridge Medieval History,* 8 volumes. Cambridge.
Taton, René (editor). 1963. *Ancient and Medieval Science.* New York.
——— (editor) 1964. *The Beginning of Modern Science from 1450 to 1800.* Translated by A. J. Pomerans. New York.
Tavard, G. H. 1954. *Transiency and Permanence: The Nature of Theology According to St. Bonaventure.* New York.
Tawney, R. H. 1926. *Religion and the Rise of Capitalism.* New York.
Teggart, Frederick, Jr. 1918. *The Processes of History.* New Haven.
Thackray, Arnold W. 1970. *Atoms and Power. An Essay on Newtonian Matter–Theory and the Development of Chemistry.* Cambridge, Mass.
Thode, Henry. 1904. *Franz von Assisi und die Anfänge der Kunst der Renaissance in Italien.* Berlin.
Thompson, A. H. 1957. "Medieval Doctrine to the Lateran Council of 1215." Pp. 634–698 in *The Cambridge Medieval History,* 6. Cambridge.
Thamin, R. 1884. *Un problème moral dans l'antiquité: étude sur la casuistique stoicienne.* Paris.
Tillich, Paul. 1945. "Conscience in Western Thought and the Idea of a Transmoral Conscience." *Crozer Quarterly* 22:289–300. Reprinted in *The Protestant Era.*
——— 1948. *The Protestant Era.* Edited by James L. Adams. Chicago.
——— 1959a. *The Religious Situation.* Translated by H. R. Niebuhr. New York.
——— 1959b. *Theology of Culture.* New York.
Tinbergen, Niko. 1951. *The Study of Instinct.* Oxford.
Tiryakian, Edward. 1974. "Reflection on the Sociology of Civilizations." *Sociological Analysis* 35, 2:122–128.
Tönnies, F. 1963. *Community and Society.* Translated by C. Loomis. New York.
Toynbee, Arnold (ed.). 1973. *Half the World. The History and Culture of China.* London.
Trémontant, C. 1955. *Etudes de métaphysique biblique.* Paris.
——— 1964. *La metaphysique du Christianisme et la crise du XIII siècle.* Paris.
Trinkaus, Charles. 1940. *Adversity's Noblemen: The Italian Humanists on Happiness.* New York.
——— 1955. "The Religious Foundations of Luther's Social Views." Pp. 71–90 in *Essays in Medieval Life and Thought,* edited by J. Mundy, B. Nelson and others.
Troeltsch, E. 1960(1911). *The Social Teachings of the Christian Churches.* 2 volumes. Translated by Olive Wyon. New York.

——— 1958 (1912). *Protestantism and Progress*. Translated by W. Montgomery. Boston.

Turner, R. Steven. 1971. "The Growth of Professional Research in Prussia, 1818 to 1848—Causes and Context." *Historical Studies in the Physical Sciences* 3:137–82.

Turner, Victor. 1968. *The Ritual Process: Structure and Anti-Structure*. Chicago.

Ueberweg, Friedrich. 1928. *Grundriss der Geschichter der Philosophie*. volume 2. *(Die patristische und scholastische Philosophie)*. Edited by Bernhard Geyer. Berlin.

Underhill, Evelyn. 1914. *Mysticism*. London.

——— 1960 (1933). *The Golden Sequence: A Fourfold Study of the Spiritual Life*. New York.

Vacandard, E. 1908. "Confession, ii. Du 1er au XIime siècle." *Dictionnaire de théologie catholique* 3. Paris.

Vaihinger, H. 1952 (1911). *The Philosophy of "As If."* Translated by C. K. Ogden. London.

Van der Loeuw, G. 1963 (1938). *Religion in Essence and Manifestation*. 2 volumes. Translated by J. E. Turner. New York.

Van Steenberghen, Fernand. 1942. *Siger de Brabant d'aprés ses oeuvres inédits. (Les philosophes belges, 12–13)*. 2 volumes. Louvain.

——— 1955. *Aristotle in the West*. Louvain.

Vignaux, Paul. 1938. *La pensée au moyen-âge*. Paris.

Vinogradoff, Paul. 1928. *Reason and Conscience in the 16th Century*. In *Collected Papers*, 2. Oxford.

——— 1967. *Roman Law in Medieval Europe*. New York.

Wach, J. 1944. *The Sociology of Religion*. Chicago.

——— 1951. *Types of Religious Experience, Christian and Non-Christian*. Chicago.

——— 1958. *The Comparative Study of Religions*. Edited by J. M. Kitagawa. New York.

Waelder, R. 1961. *Basic Theory of Psychoanalysis*. New York.

Wakefield, W. L., and Austin P. Evans (editors and translators). 1969. *Heresies of the High Middle Ages*. 2 volume. New York.

Wallace, A. F. C. 1959. "The Institutionalization of Cathartic and Control Strategies in Iroquois Religious Psychotherapy." Pp. 63–96 in *Culture and Mental Health*, edited by M. K. Opler.

Wallace, William, O. P. 1959. *The Scientific Methodology of Theodore of Freiburg*. Freibourg.

——— 1962. *The Role of Demonstration in Moral Theology*. Washington, D. C.

Walter of Bruges. (See Longpré).

Walter of St. Victor. (See Glorieux).

Ward, W. G. 1865. "Doctrinal Decrees of a Pontifical Congregation: The Case of Galileo." *Dublin Review* (N.S.) 5:376–425.

Weakland, John H. 1960. "The 'Double-Blind' Hypothesis as a Third Party Interaction." Pp. 373–88 in *The Etiology of Schizophrenia*, edited by Don Jackson, New York.

Weber, Max. 1920–21. *Gesammelte Aufsätze zur Religionssoziologie*. 3 volumes. Tübingen.

——— 1928. *General Economic History*. Translated by F. H. Knight. Glencoe, Illinois.
——— 1946a. *From Max Weber: Essays in Sociology*. Edited and translated by H. Gerth and C. W. Mills. New York.
——— 1946b (1913). "The Social Psychology of the World Religions." Pp. 267–301 in *From Max Weber: Essays in Sociology*.
——— 1946c (1915). "Religious Rejection of the World and their Directions." Pp. 323–359 in *From Max Weber: Essays in Sociology*.
——— 1949. *The Methodology of the Social Sciences*. Edited and translated by E. Shils and H. Finch. New York.
——— 1951 (1916–17). *The Religion of China: Confucianism and Taoism*. Translated by H. Gerth. Glencoe, Illinois.
——— 1952 (1917–19). *Ancient Judaism*. Edited and translated by H. Gerth and Don Martindale. Glencoe, Illinois.
——— 1958a (1904–05). *The Protestant Ethic and the Spirit of Capitalism*. Translated by T. Parsons. New York.
——— 1958b. *The Religion of India*. Edited and translated by H. Gerth and Don Martindale. New York.
——— 1958c. *The City*. Translated and edited by Don Martindale and Gertrud Neuwirth. New York.
——— 1958d. *The Rational and Social Foundations of Music*. Edited and translated by D. Martindale, J. Riedel, and G. Neuwirth. Carbondale, Illinois.
——— 1963. *The Sociology of Religion*. Translated by E. Fischoff. Boston.
——— 1968. *Economy and Society*. 3 volumes. Edited by G. Roth and C. Wittich. Totowa, New Jersey.
——— 1972 (1908). "Georg Simmel as Sociologist." (Translated with an introduction by Donald Levine). *Social Research* 39, 1:155–163.
——— 1975 (1903–06). *Roscher and Knies. The Logical Problems of Historical Economics*. Translated with an introduction by Guy Oakes. New York.
——— 1977 (1907). *Critique of Stammler*. Translated, edited and with an introduction by Guy Oakes. New York.
Weinberg, Julius. 1948. *Nicholas of Autrecourt*. Princeton.
Weingart, Richard E. 1970. *The Logic of Divine Love*. Oxford.
Weisheipl. J. A., O. P. 1955. *Nature and Gravitation*. River Forest, Illinois.
——— 1959. *The Development of Physical Theory in the Middle Ages*. London.
Weissberg, A. 1951. *The Accused*. Translated by E. Fitzgerald. New York.
Westman, Robert S. 1971. *Kepler's Adoption of the Copernican Hypothesis*. Ann Arbor, Michigan.
Wheelis, Allen. 1958. *The Quest for Identity*. New York.
White, Lynn Jr. 1947. "Natural Science and Naturalistic Art in the Middle Ages." *American Historical Review* 3:421–435.
——— 1962. *Medieval Technology and Social Change*. Oxford.
——— 1964. "Review of Needham's Work." *Isis* 58:248–251.
——— 1969. *Machina ex Deo: Essays in the Dynamism of Western Culture*. Cambridge, Mass.
——— 1971a. "Medieval Borrowings from Further Asia." *Medieval and Renaissance Studies* 5:3–26.
——— 1971b. "Cultural Climates and Technological Advance in the Middle Ages." *Viator* 2:171–250.

White, Morton. 1965. *Foundations of Historical Knowledge*. New York.
Whitehead, A. N. 1927. *Symbolism: Its Meaning and Effect*. New York.
——— 1956 (1933). *Adventure of Ideas*. New York.
——— 1968 (1925). *Science and the Modern World*. New York.
——— 1971. *The Concept of Nature (The Tarner Lectures 1919)*. New York.
Whorf, B. L. 1961. *Language, Thought and Reality*. New York.
Whyte, W. H. 1957. *The Organization Man*. Garden City, New York.
Wiener, P. P. and A. Noland (eds.). 1957. *The Roots of Scientific Thought: A Cultural Perspective*. New York.
Wightman, W. P. 1962. *Science and the Renaissance*. 2 volumes. Edinburgh.
Williams, L. P. 1962. "The Physical Sciences in the First Half of the Nineteenth Century, Problems and Sources." *History of Science* 1:1–15.
Williams, N. P. 1927. *The Idea of the Fall and Original Sin*. Oxford.
Williams, Watkin W. 1935. *Saint Bernard of Clairvaux*. Manchester.
Wilson, Curtis. 1956. *William Heytesbury: Medieval Logic and the Rise of Mathematical Physics*. Madison.
Wittfogel, Karl. 1931. *Wirtschaft und Gesellschaft Chinas*. Leipzig.
——— and Fêng Chia-Shêng. 1949. *History of Chinese Society: Liao, 907–1125*. Philadelphia.
——— and Fêng Chia-Shêng. 1957. *Oriental Despotism: A Comparative Study of Total Power*. New Haven.
Wolfson, Harry. 1964. "The Controversy Over Causality Within the Kalam." Pp. 602–618 in *L'aventure de la science: Mélanges Alexandre Koyré*, 2, edited by Cohen and Taton. Paris.
Wong, George, H. S. 1963a. "Some Aspects of Chinese Science before the Arrival of the Jesuits." *Chung Chi Journal* 2:169–180.
——— 1963b. "China's Opposition to Western Science during the late Ming and Early Ching." *Isis* 54:29–49.
Wood, T. 1952. *English Casuistical Divinity in the 17th Century*. London.
Woodhouse, A. S. O. (ed.). 1951. *Puritanism and Liberty: Being the Army Debates (1649–9) from the Clarke Manuscripts with Supplementary Documents*. Foreword by A. D. Lindsay. Chicago.
Wrightsman, A. B. 1970. *Andreas Osiander and Lutheran Contributions to the Copernican Revolution*. Diss. Ann Arbor, Michigan.
Wrong, Dennis. 1961. "The Oversocialized Conception of Man." *American Sociological Review* 26, 2:183–92.
Zaehner, R. C. 1961. *Mysticism—Sacred and Profane*. New York.
Zeuthen, H. B. 1966. *Geschichte der Mathematik im 16. und 17. Jahrhundert*. New York and Stuttgart.
Zilboorg, Gregory, and George W. Henry. 1941. *A History of Medical Psychology*. New York.
Zilsel, Edgar. 1942a. "The Sociological Roots of Science." *American Journal of Sociology* 47:544–562.
——— 1942b. "The Genesis of the Concept of Physical Law." *Philosophical Review* 51:245–279.
——— 1942c. "The Genesis of the Concept of Scientific Progress." *Journal of the History of Ideas* 6:325–349.

——— 1957a. "The Origins of Gilbert's Scientific Method." Pp. 219–250 in *Roots of Scientific Thought,* edited by Wiener and Noland.
——— 1957b. "Copernicus and Mechanics." Pp. 276—280 in *Roots of Scientific Thought.*
Zubov, V. P. 1968. *Leonardo da Vinci.* Translated by David H. Kraus, foreward by M. P. Gilmore. Cambridge, Mass.

Index